Newnes Guide to Television and Video Technology
D0626202
HULL COLLEGE

000185847

Newnes Guide to Television and Video Technology

Third edition

Eugene Trundle, TMIEEIE, MRTS, MISTC

Newnes

OXFORD AUCKLAND BOSTON JOHANNESBURG MELBOURNE NEW DELHI

Newnes
An imprint of Butterworth-Heinemann
Linacre House, Jordan Hill, Oxford OX2 8DP
225 Wildwood Avenue, Woburn, MA 01801-2041
A division of Reed Educational and Professional Publishing Ltd

A member of the Reed Elsevier plc group

First published 1988
Second edition 1996
Third edition 2001

British Library Cataloguing in Publication Data
A catalogue record for this book is available from the British Library.

ISBN 0 7506 48104

Typset by Keyword Typesetting Services Ltd, Wallington, Surrey
Printed and bound in Great Britain by MPG Books Ltd, Bodmin, Cornwall

PLANT A TREE
BTCV
British Trust for Conservation Volunteers
FOR EVERY TITLE THAT WE PUBLISH, BUTTERWORTH-HEINEMANN WILL PAY FOR BTCV TO PLANT AND CARE FOR A TREE.

Contents

Preface to third edition

The terms of reference for this book are very wide, and increase with each new edition. TV and video take an ever larger part in our leisure, educational and recreational activities; here the technology behind them is explored and explained.

Much use is made throughout the book of block diagrams. Since integrated circuits – silicon chips – have become so widespread, much service data are now presented in block diagram form. To explain principles and techniques I have sometimes used earlier discrete circuits and systems, in which the separate functions and circuit elements can be clearly discerned.

As this is being written the world of TV and video is in transition from analogue to digital operation in the realms of broadcast/reception, tape recording and disc systems, while a convergence of computer and TV/video technologies is under way. This is reflected in this new edition of the book, which is aimed at interested laypeople, students, technicians and those in allied fields seeking an insight into TV and VCR practice. I have assumed that the reader has a basic knowledge of electronics and mechanics. For further reading I can recommend my *Television and Video Engineer's Pocket Book*; and for those whose interest lies in fault-diagnosis and repair, my *Servicing TV, Satellite and Video Equipment*, both published by Newnes.

My thanks are due once more to my patient and loving wife Anne, whose moral support, coffee-brewing and keying-in services have kept me going through the three editions of this book.

Eugene Trundle

1 *Basic television*

For a reasonable understanding of colour television, it is essential that the basic principles of monochrome TV are known. As we shall see, all colour systems are firmly based on the original 'electronic-image dissection' idea which goes back to EMI in the 1930s, and is merely an extension (albeit an elaborate one) of that system.

Although there are few black and white TVs or systems now left in use, the compatible colour TV system used today by all terrestrial transmitters grew out of the earlier monochrome formats. In the early days it was essential that existing receivers showed a good black and white picture from the new colour transmissions, and the scanning standards, *luminance* signal, and modulation system are the same. What follows is a brief recap of *basic* television as a building block of the colour TV system to be described in later chapters.

Image analysis

Because a picture has two dimensions it is only possible to transmit all the information contained within it in *serial* form, if we are to use but one wire or RF channel to carry the signal. This implies a *dissection* process, and requires a timing element to define the rate of analysis; this timing element must be present at both sending and receiving ends so that the analysis of the image at the sending end, and the simultaneous build-up of the picture at the receiver, occur in synchronism. Thus a television picture may be dissected in any manner, provided that the receiver assembles its picture in precisely the same way; but the path link between sender and viewer must contain *two* distinct information streams: video signal, which is an electrical analogy of the light pattern being sent, and timing signals, or synchronisation pulses, to define the steps in the dissection process. The presence of a timing element suggests that each picture will take a certain period to be built up; how long will depend on how quickly we can serialise the picture *elements*, and this in turn depends on the bandwidth available in the transmission system – more of this later.

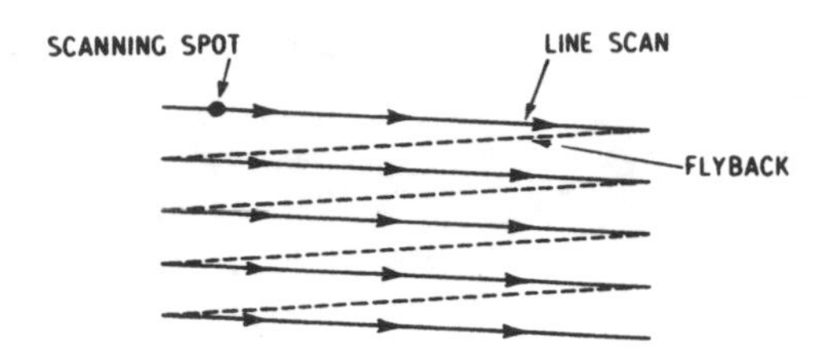

Figure 1.1 *The scanning process. Horizontal lines are drawn from left to right of the screen by horizontal direction, and 'stacked' vertically by the slower-moving vertical deflection field*

Scanning

If we focus the image to be televised on a light-sensitive surface we are ready for the next stage in the dissection process – the division of the pattern into picture elements or *pixels*. Each pixel is rather like the individual dots that go to make up a newspaper photograph in that each can only convey one level of shading. Thus the detail, or definition, in the reproduced picture is proportional to the number of pixels. In 625-line television we have approximately 450 000 pixels, adequate for a 67 cm-diagonal picture, but barely sufficient for much larger screens. These individual pixels are arranged in horizontal lines; there are 625 lines in the British TV system. Figure 1.1 shows how the image is scanned, line by line, to read out in serial form the pattern of light and shade which forms the picture. When half the lines have been traced out the scanning spot has reached the bottom of the picture and traced one *field*. It now flies back to the top of the screen to trace out the rest of the 625 lines in the spaces between those of its first descent. This is known as interlacing, and confers the advantages of a 50 Hz (Hz, Hertz, one cycle per second) flicker rate with the lower scanning speed and lesser bandwidth requirement of a 25 Hz *frame* rate. All TV systems use this 2:1 interlaced field technique; its success depends only on accurate triggering of the field scan.

Image sensor

In earlier designs of TV camera the image pick-up device was a thermionic tube whose light-sensitive faceplate was scanned by a sharply focused electron beam. Currently a solid-state device is used, as shown in Figure 1.2. Its faceplate is made up of an array of hundreds of thousands of silicon photodiodes mounted on a chip, typically 7 mm diagonal, arranged in lines and columns. Though a real sensor of this type may contain 750 diodes per line and 575 rows, our diagram shows a 12 × 9 matrix for simplicity. During the active field period each reversed-biased diode acts as a capacitor, and acquires an electrical charge proportional to the amount of light falling on it: the televised image is sharply focused on the sensor faceplate by an optical lens system. Each diode is addressed in turn by the sensor's drive circuit so that (as viewed from the front) the charges on the top line of photodiodes are read out first, from left to right. Each line is read out in turn,

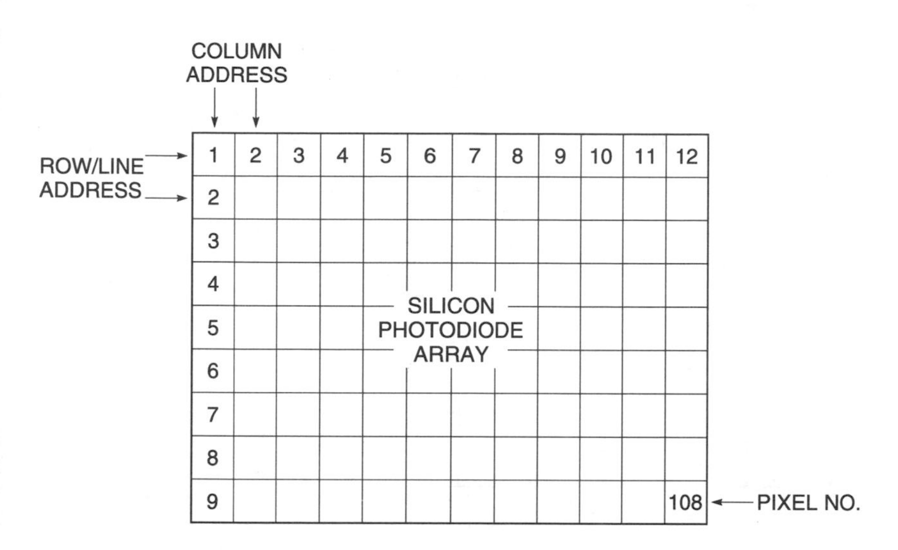

Figure 1.2 *Matrix of silicon photodiodes in an image pick-up chip*

progressing downwards, until the end of the bottom line is reached after 20 ms (ms, millisecond, 0.001 second). This first vertical scan involves every *other* line in the diode matrix, nos 1, 3, 5 and so on. Now another scan begins, this time addressing in turn all the photodiodes in lines 2, 4, 6 etc, and that is completed in a further 20 ms. At this point the entire picture has been scanned in two consecutive sweeps taking a total of 40 ms and simulating the pattern and sequence of Figure 1.1.

Charge coupling

The capacitors formed by the silicon photodiodes are represented by C1, C2, C3 and C4 in Figure 1.3, which portrays the first four pixels in one television line. C1 acquires a charge proportional to the light level on the first pixel; it appears as a voltage at the output of the first amplifier A1. If all the switches S1 to S3 are now momentarily closed the charge on C1 will be passed into C2, whose charge will be transferred into C3 and so on, all the way to the last in the line, at the right-hand side of the diagram. Thus by momentarily closing all the switches in synchronism the brightness level charge for each pixel can be made to 'march' like a column of soldiers to the right where an orderly sequential readout of the light pattern along that line can be picked up, at a rate dependent on the switching (clocking) frequency. This will form the video signal, with each TV line being read out in sequence, in similar fashion to a computer's *shift register*. This is the basis of the operation of a CCD. The amplifiers A1, A2 etc are fabricated onto the sensor chip so that the external connectors need only handle drive, clocking and transfer pulses. Emerging from the 'bottom right-hand corner' of the sensor chip, as it were, is an analogue video signal which – after filtering to remove remnants of the clocking frequency – is similar to that shown in Figure 1.4, an electrical

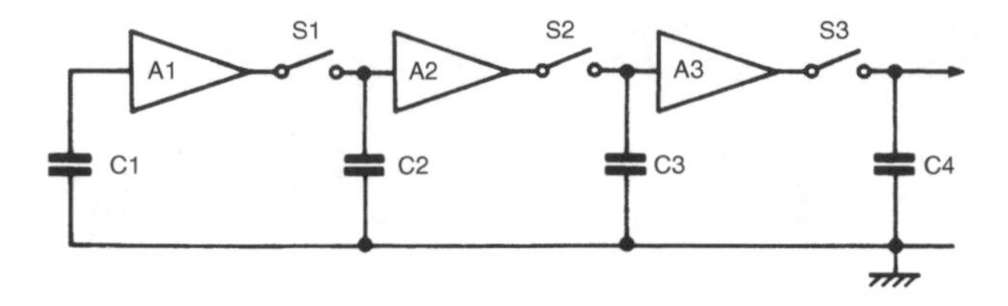

Figure 1.3 *The concept of a CCD image sensor. Light patterns are held as charges in the capacitors, one for each pixel*

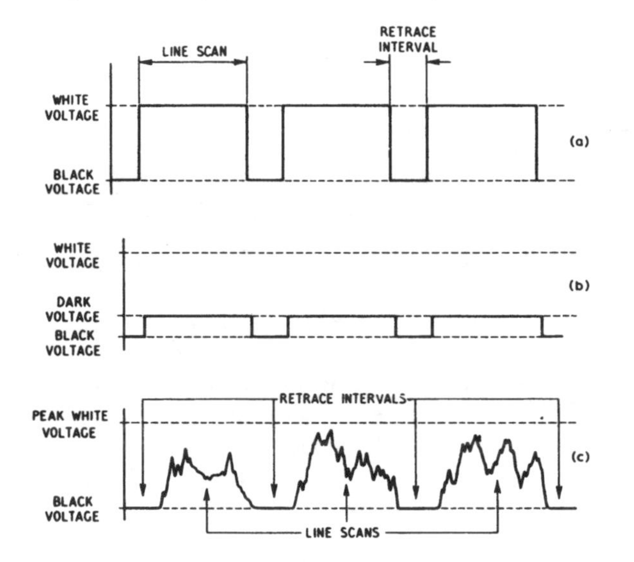

Figure 1.4 *Lines of video signal: (a) high brightness, (b) low brightness, (c) picture content*

analogy of the picture pattern, with 'blanks' at 64 μs and 20 ms intervals, during which flyback or *retrace* will take place in the display picture tube. In fact the image sensor scans a complete horizontal line in 52 μs (μs, microsecond, 0.000001 second) and rests for 12 μs before scanning the next line: the 12 μs interval contains no picture information. Similarly, vertical scan occupies 18.4 ms, with a retrace interval of 1.6 ms.

We now have one of our information streams (the video signal) and it's time to insert the second stream (timing pulses) to synchronise the TV receiver.

Synchronisation pulses

The receiver or monitor which we shall use to display the picture has scanning waveform generators which must run in perfect synchronism with the readout of pixel lines and columns in the image sensor. This ensures that each pixel picked up from the sensor is reproduced in the right place on the display. Plainly, if the

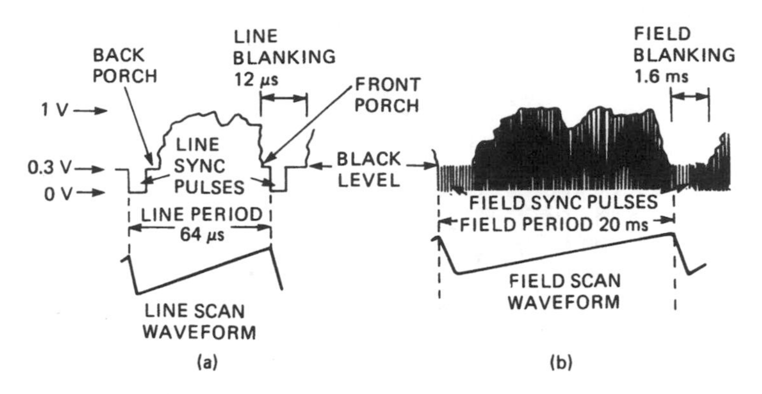

Figure 1.5 *The basic analogue TV signal – arrangement of video and sync information and the relationship between signal timing and scanning waveforms*

camera sees a spot of light in the top right-hand corner of the picture, and the monitor's scanning spot is in the middle of the screen when it reproduces the light, the picture is going to be jumbled up!

This is prevented by inserting synchronising pulses (sync pulses for short) into the video waveform at regular intervals, and with some distinguishing feature to enable the TV monitor to pick them out. To signal the beginning of a new line scan, we insert a 4.7 µs negative-going pulse into each line blanking period, and to initiate field flyback a *series* of similar, closely-spaced pulses is inserted into each field blanking period. These are shown in Figure 1.5, which represents what is called a VBS (video, black-level and sync) or *composite* video signal. Black level is established at 0.3 V (300 mV) from which the signal rises to 1 V for peak white with lesser brightness levels giving correspondingly lower voltage. Each time a sync pulse occurs, the signal voltage drops to zero for its duration. The timing of the field sync-pulse trains is very critical for good interlace in the displayed picture, and the sync pulse generator is carefully designed to achieve this. The short period preceding the line sync pulse is called the *front porch*, and the rather longer (5.8 µs) post sync-pulse period is termed the *back porch*. The time spent on porches and sync pulse is known as the *blanking period*, and it's 12 µs between lines, and 1.6 ms between fields, as shown in Figure 1.5. The lower section of the diagram indicates the relationship between sync pulses and scanning current for the picture tube.

Picture reproduction

We have now obtained a composite video signal, in a form which conveys both video and timing information. Let us now see how it is used to recreate an image on the screen of a picture-tube. For simplicity, we will assume we have a closed-circuit set-up, and that the camera and monitor are linked by a single coaxial cable. Figure 1.6 shows the arrangement. Here we have, at the sending end, an image sensor with the necessary lenses and drive electronics.

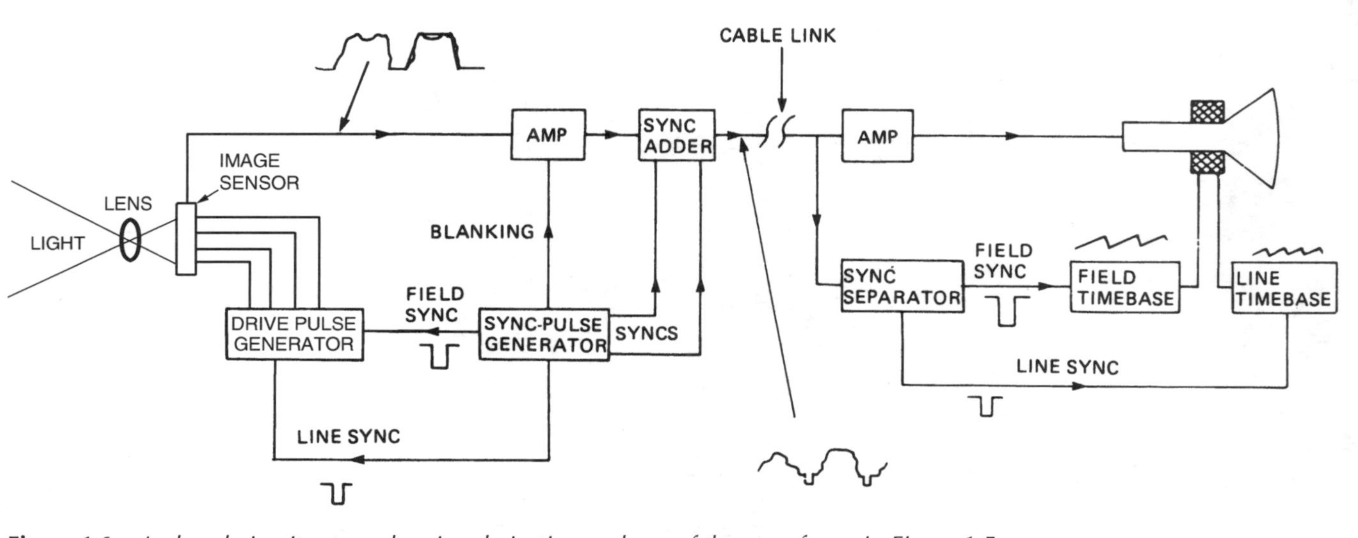

Figure 1.6 *A closed-circuit set-up showing derivation and use of the waveforms in Figure 1.5*

Attendant on it is a master sync pulse generator which triggers the drive-pulse generator in the camera and also provides a blanking signal for use in the video processing amplifier. A second pair of outputs from its sync-pulse section is taken to an *adder* stage for insertion into the video waveform, and the composite video signal is passed into the transmission cable.

On arrival at the monitor, the signal is first amplified, then passed to the cathode of the picture tube. A second path is to the *sync separator* stage which works on an amplitude-discriminating basis to strip off the sync pulses for application to the timebase generators. They work in just the same fashion as those in the camera to generate sawtooth currents in the scanning coils with which to deflect the scanning beam in the display tube. Thus we have the two scanning systems – one at the sender and one at the receiver – swinging to and fro and up and down in perfect synchronism; and a flow of constantly-changing voltage in the form of the video signal conveying the pattern of light and shade from the image sensor to the picture-tube screen.

The picture tube

In Chapter 7 we shall study the operation of colour tubes, and as an introduction to these, we need to examine the workings of monochrome tubes – they have much in common! Figure 1.7 shows the basics of a picture tube.

The process starts with a heated cathode, from which electrons are 'boiled off' by thermal agitation, to form a *space charge* around the cathode. Depending on the negative potential we choose for the 'grid' (in practice a cylinder surrounding the cathode, with a tiny hole in its otherwise-closed outer face), some of these electrons are attracted away down the tube neck towards the screen by the highly-positive anode cylinders, some of which are shaped and arranged to form an electron-lens whose focal point is the inner surface of the display screen. As the electron beam passes into the 'bowl' of the picture tube it comes under the influence of the scanning coils which deflect the scanning spot, at line and field rate, to trace out a rectangle of light, known as a *raster*, on the inner face of the tube. The screen is *aluminised*, and the thin aluminium layer is held at a potential

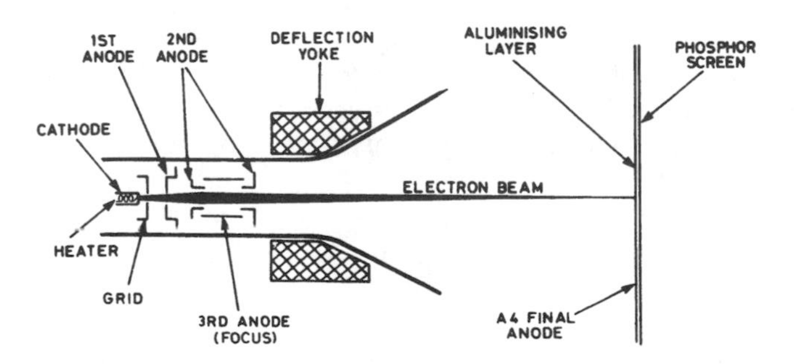

Figure 1.7 *The working principle of a monochrome picture tube*

of several thousand volts with respect to the cathode – this gives an accelerating 'boost' to the electrons in the beam, so that they collide with the phosphor screen surface at very high velocity indeed.

How is the light produced? The inner surface of the faceplate is coated with a continuous layer of *phosphor* which has the property of emitting white light when a high-velocity electron beam impinges on it. The aluminised screen backing reflects the light forward, and forms a barrier to prevent harmful ions burning the screen. The light output is regulated by varying the beam current, and as we have seen, this is the primary function of the 'grid' cylinder. We can arrange a fixed bias on this electrode to set the overall raster brightness, then feed a video signal to it (normally about 80 V peak-to-peak is required) to instantaneously vary the beam current to trace out, and faithfully copy, the brightness of each individual pixel as it is positioned on the screen by the scanning system. In practice the grid is held at a fixed potential and the video signal is applied to the cathode.

Bandwidth

The scheme outlined so far describes the stages in capturing, relaying and reproducing the picture. We have given little thought as yet to the requirements of the transmission medium, be it a cable, space or a glass fibre!

Mention has already been made of pixels, and the more we have, the faster the rate of change of the video facsimile signal. Much depends on the scanning rate, and this has to be faster than a certain minimum to avoid a disturbing flicker effect in the picture. By using the interlace technique we achieve 50 'flashes' per second, and this is just sufficient to accommodate the human eye's 'time-constant', known as *persistence of vision.* To capture reasonable detail we need all of 625 lines (in fact only 575 of them are used for the picture itself) and this means that if the 450 000 or so picture elements are each different from their neighbours, the rate of change of the video signal needs to be in the region of 5.5 MHz (MHz, megahertz, one million cycles per second).

Let us look more closely into the reasons for this. If we are televising a picture of a black cat on a white carpet, the rate of change of the video signal will be very low except at the point of the scanning spot's transitions from carpet to cat and back again, each of which will give rise to a sudden *transient.* At the other extreme, if the camera is looking at a pattern of fine vertical lines the video signal will be much 'busier' and contain a great deal of HF energy. In practice the frequencies in the video signal are mostly related to line and field scanning rates, and much of the energy in a video signal is concentrated into 'packets' centred on multiples of line and field frequency. This is an important point, and one to which we shall return.

Modulation

The word modulation, in our context, means the impressing of a signal waveform (usually video, sound or digital pulses) onto *a carrier* wave. The nature of the carrier wave is dependent on the medium to be traversed – space, cable, optical fibre; the distance required to be covered; and to a lesser extent, the nature of the information to be carried. Thus the medium-wave sound broadcast band (MF, medium frequency) is suitable for long-distance broadcasts of rather indifferent-quality sound signals, but quite useless for television; at the other extreme, the SHF (super-high frequency) band is well-suited to a 'beamed' television broadcast service from an orbiting satellite, but one wouldn't expect to find Radio Brighton up there as well! So far as the studio and viewer are concerned, the carrier is irrelevant, because it acts purely as a vehicle on which the wanted signal travels, being discarded at the receiving end once its usefulness has been realised.

Types of modulation

The four types of modulation used in communications to attach information to a carrier wave are (a) amplitude modulation, (b) frequency modulation, (c) phase modulation (which is a variant of *b*), and (d) pulse-code modulation. In ordinary sound broadcasting (a) or (b) is normally adopted, the former at LF, MF and HF, and the latter at VHF (very high frequency, around 100 MHz). In *terrestrial* TV broadcasts, (a) is used for vision and (b) for sound. In fact, television uses all four modulation modes, because a form of phase modulation is used for the colouring signal, as we shall see in later chapters, and modern TV links also make use of PCM.

Amplitude modulation

AM is amongst the earliest forms of modulation to be used, and is probably the easiest to understand. It is perhaps surprising that this was pre-dated, however, by PCM (*d* above) in that the very first crude spark transmitters used this mode, conveying the message by simply interrupting the transmitter's operation by means of a morse key! We use AM for the vision signal in terrestrial TV broadcasts; let's see how it is arranged.

For AM, we start with a stable oscillator to generate the carrier wave. For TV broadcasts, the crystal oscillator runs at a sub-multiple of the station frequency, and its output is multiplied up to the required frequency (corresponding to the specified channel number). The information to be sent is made to vary the *amplitude* of the carrier wave, so that in our example of a TV broadcast, the waveform of Figure 1.5 is fed to a *modulator* which may control the *gain* of an RF amplifier which is handling the carrier wave. Thus the video signal is impressed onto the carrier, and the polarity of the video signal is arranged to give *negative modulation*. This means that the tips of the sync pulses give rise to maximum carrier power, and whites in the vision modulating signal raise only about one-fifth of that level. This is illustrated in Figure 1.8 which represents the waveform fed to the UHF broadcasting aerial, and present in the tuner and IF sections of a conventional TV receiver.

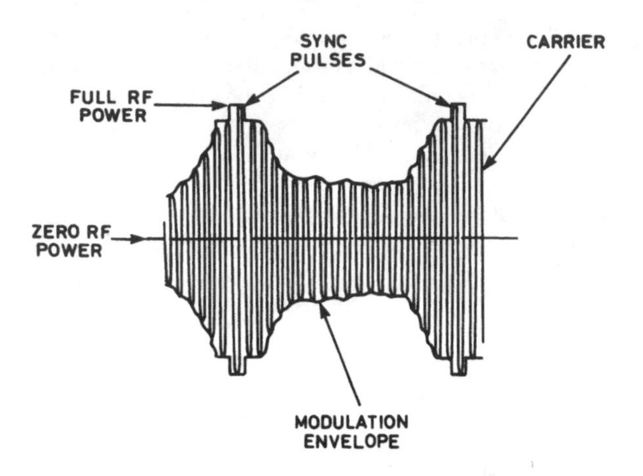

Figure 1.8 *AM modulation of the video signal onto an RF carrier for transmission*

One of the advantages of AM is the simplicity of the detection or *demodulation* process, which in its original form requires only a rectifier and filter to recover the modulating information.

Frequency modulation

This is the second most popular mode for broadcast use, and is used for high-fidelity (hi-fi) sound transmissions in VHF Band II. Most of these radio transmissions are multiplex-stereo-encoded. So far as we are concerned here, the most significant use of FM modulation is for the TV sound transmissions which accompany 625-line UHF picture broadcasts, and for vision broadcasts from satellites. In this type of modulation the carrier amplitude is kept constant and the frequency of the wave is varied at a *rate* dependent on the frequency of the modulating signal and by an *amount* (deviation) dependent on the strength of the modulating signal (see Figure 1.9). Thus the 'louder' the signal the greater the deviation, while the higher its frequency, the greater the rate at which the carrier frequency is varied either side of its nominal frequency.

Maximum deviation of the TV sound carrier is about ±50 kHz (kHz, kilohertz, one thousand cycles per second) corresponding to maximum modulation depth of this particular system, commonly described as 100 per cent modulation. 100 per cent amplitude modulation is when the troughs of the modulated waveform fall to the zero datum line. This is the absolute physical limit for AM.

Phase modulation

Here the phase of the carrier wave is altered according to the modulating signal. It has much in common with frequency modulation, but the deviation is very small

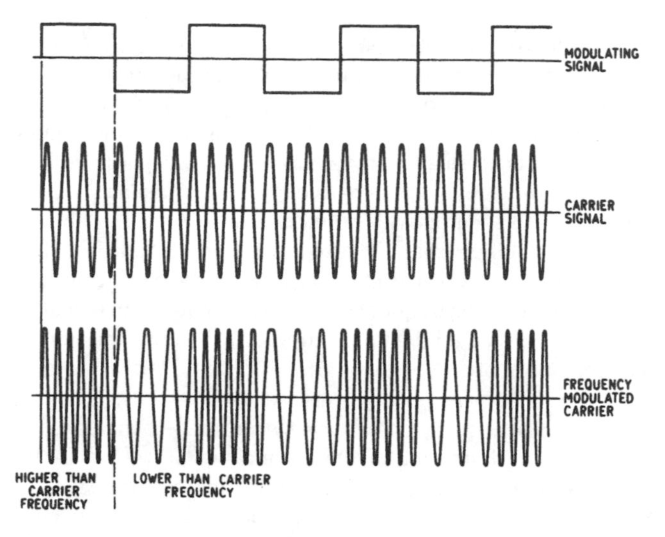

Figure 1.9 *Illustrating frequency modulation*

compared with the carrier frequency itself. In a television system, a common form of phase modulation is the transmission of a *reference* signal of constant period alongside a second carrier at the same frequency, but with phase, or timing, variations to convey the required information. This is the basis of the colour encoding system, fully dealt with in later chapters.

Pulse-code modulation (PCM)

An alternative method of modulation is a system of sampling, or quantising, an analogue signal to derive a transmission signal which has only two states, on and off. This is achieved by measuring the instantaneous amplitude of the sound or vision waveform at regular time intervals, short enough to ensure that the highest required frequency is adequately sampled. The amplitude of the signal at each 'sampling spot' is measured and assigned a number which represents the amplitude of the signal at that moment in time. This number can be represented in *binary* (two-state) form, as a series of pulses each of which has a closely defined time slot in a serial transmission. The presence or absence of a pulse in each time slot forms a pattern, unique to each number we may wish to send. The numbers are produced in a continuous string, as it were (known as *serial* transmission). In this form the signal is dispatched, a primary requirement of the transmission medium being sufficient bandwidth to enable it to resolve two recognisable levels (presence of pulse or absence of pulse) at high rates of change. Provided that this requirement is met, the sorts of distortion encountered by the signal on its path do not upset the final reproduction. The second requirement is that some form of *framing* or timing signal accompanies the digital transmission to synchronise the *decoder* at the receiving end.

On receipt, the PCM binary signal passes into a *decoder* which recreates 'numbers' from the pulse train. After processing in a D-A (digital to analogue) converter, the analogue signal is recreated, and its fidelity to the original waveform at the sending end depends on two factors – sampling rate and the number of quantising levels. The sampling rate needs to be at least twice the highest frequency of interest in the *baseband* signal (known as *Nyquist* rate). The number of quantising levels depends very much on the nature of the signal being sampled. For TV an impeccable picture is secured with no more than 256 levels of brightness. Surprisingly, perhaps, an audio signal, if it is to have an acceptable signal-to-noise (S/N) ratio and be free of spurious effects, requires in the region of 1024 quantising levels, though the sampling frequency will be much lower than that for TV pictures. As those who have studied digital electronics will know, this implies an eight-bit 'word' for each pixel of a TV signal ($256 = 2^8$) and a ten-bit word for sound encoding ($1024 = 2^{10}$). Some sound systems use 14-bit words.

PCM has many advantages over other modulation systems where the transmission path is long, and subject to distortion, reflections and similar hazards for the signal. It is finding increasing applications in television technology, especially in the areas of studio equipment, video tape recording, broadcasters' inter-location links, fibre-optic transmissions and domestic receiving equipment.

Sidebands

Whatever type of modulation is used, sidebands are produced. Treated mathematically, the modulated signal can be shown to become a 'group' consisting of a central frequency (the carrier wave) with a pair of symmetrical 'wings' consisting of a number of signals whose displacement from the carrier wave and amplitude depend on the excursions of the modulating signal. This results in a spread of signals across the frequency spectrum, shown for AM modulation in Figure 1.10.

These are the sidebands, and the maximum spread is determined by the highest frequency to be transmitted as modulation. Thus a high-frequency modulating signal will produce sidebands which are located at the limits of the transmission bandwidth. In a vision broadcast these high-frequency modulating signals correspond to the fine detail in the picture, and in a sound broadcast to the upper-treble notes.

Figure 1.10 shows the central carrier frequency with sidebands for four distinct modulating frequencies simultaneously present. The outer sidebands A and A′ would correspond to the high frequency (say around 5 MHz) components of the TV picture signal while D/D′ would arise from lower-definition parts of the baseband signal. C/C′ and B/B′ represent intermediate modulating frequencies. Also see page 30.

In FM modulation systems sidebands are also present; their nature is somewhat different, however. Even if only a single pure tone modulating frequency is present, a series of sidebands is produced, and in theory, the 'spread' of the sidebands is infinite. They are spaced on either side of the carrier by multiples of the modulating frequency, although only those relatively close to the carrier are significant in the reception process, depending on ultimate quality. In general an FM

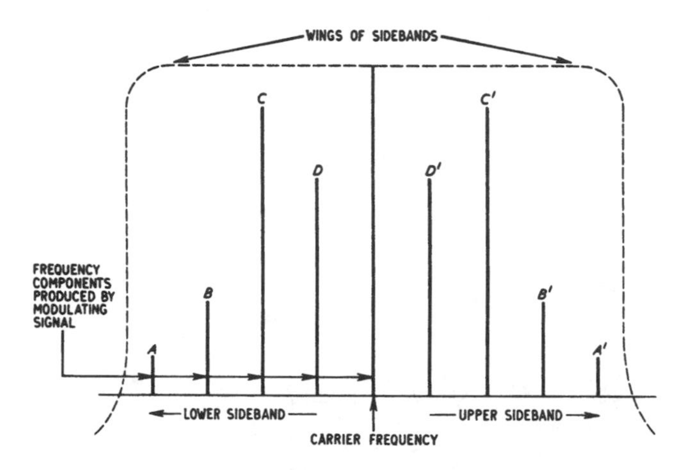

Figure 1.10 *Showing the sidebands of a modulated carrier wave*

transmission system needs more bandwidth than an AM one, as demonstrated by the radio broadcast bands, where a receiving 'window' 20 kHz wide (though it is often less than this in practice, to minimise adjacent channel interference) would afford most adequate reception of MF AM broadcasts, whereas VHF FM reception demands a 'window' of 200 kHz (wider for stereo). For picture transmissions, the comparison is similar – terrestrial AM TV transmitters operate within a bandwidth of 8 MHz, while FM picture signals from satellites demand 27 MHz of spectrum space, although in both cases cited the FM mode affords more 'detail' in the modulating signal, and the bands used for the FM examples are higher in frequency, giving more 'elbow room', so to speak, for the greater sideband spread. We shall return to the subject of sidebands several times in later chapters.

Transmission bandwidth

This band of frequencies has to pass through the various types of equipment, including amplifiers, aerial systems, detectors and so forth in its passage from the studio or programme source to the viewer's home. The bandwidth of such devices will determine just how much of the detail of the original signal gets through. If the bandwidth is too small it will eliminate some of the 'detail' information and also, perhaps, distort the signal in other ways. If, on the other hand, the bandwidth were too great it would allow undue 'noise' and other unwanted signals to enter the system and thus detract from the quality of the vision (and sound). In any case the bandwidth spread of any transmission must be limited to the minimum possible spectrum space (rigidly-kept channel widths are specified for all types of transmission) to conserve valuable band space and prevent mutual interference with adjacent channels. A descriptive 'model' illustrating these points is given in Figure 1.11.

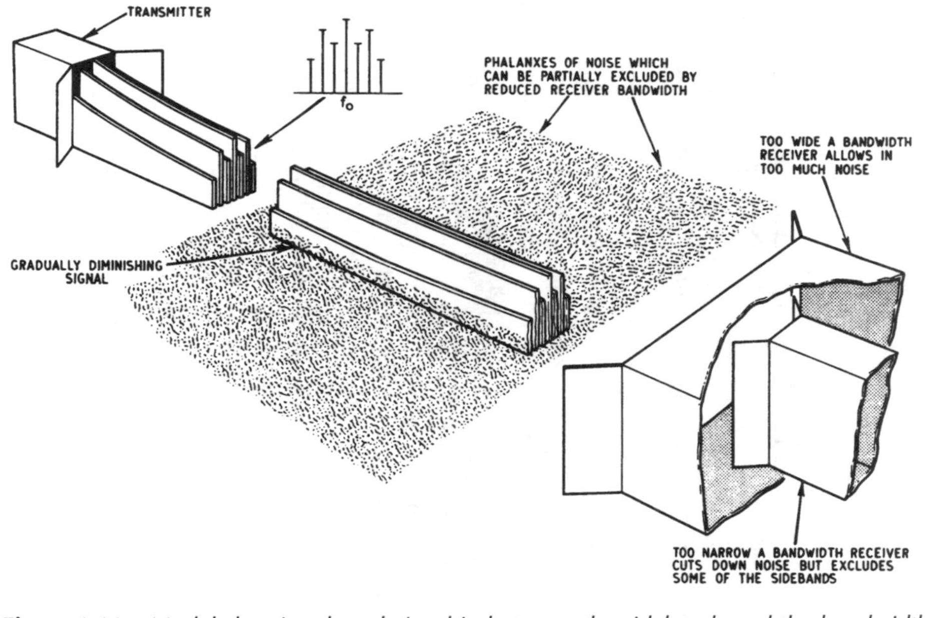
TRANSMITTER
f_o
PHALANXES OF NOISE WHICH CAN BE PARTIALLY EXCLUDED BY REDUCED RECEIVER BANDWIDTH
TOO WIDE A BANDWIDTH RECEIVER ALLOWS IN TOO MUCH NOISE
GRADUALLY DIMINISHING SIGNAL
TOO NARROW A BANDWIDTH RECEIVER CUTS DOWN NOISE BUT EXCLUDES SOME OF THE SIDEBANDS

Figure 1.11 *Model showing the relationship between the sidebands and the bandwidth*

2 *Light and colour*

We know radio and TV broadcasts as electromagnetic waves whose frequency determines their wavelength, and various broadcast bands have already been mentioned in Chapter 1. As we go up in frequency we pass through the bands allotted to radio transmissions, then terrestrial TV broadcast and space communications. Way beyond these we come into an area where electromagnetic radiation is manifest as heat, and continuing upwards we find infra-red radiation, and then a narrow band (between 380×10^6 and 790×10^6 MHz) which represents light energy. Beyond the 'light band' we pass into a region of ultra-violet radiation, X-rays and then cosmic rays.

Figure 2.1 gives an impression of the energy distribution curves for red, green and blue lights. It can be seen that they come in the same order as in a rainbow or from a prism. When we see white light it is in fact a mixture of coloured lights of all hues, and the 'splitting' effect of a prism demonstrates this by providing a different refractive index for each light wavelength. If the resulting colour spectrum is passed into a second prism it will be recombined into white light again!

This splitting and recombining process indicates that white light contains all the visible colours, and that by *adding* suitable proportions of coloured light we can create white light. Many combinations of colours may be added to render white, but in television three primary light colours are used: red, green and blue.

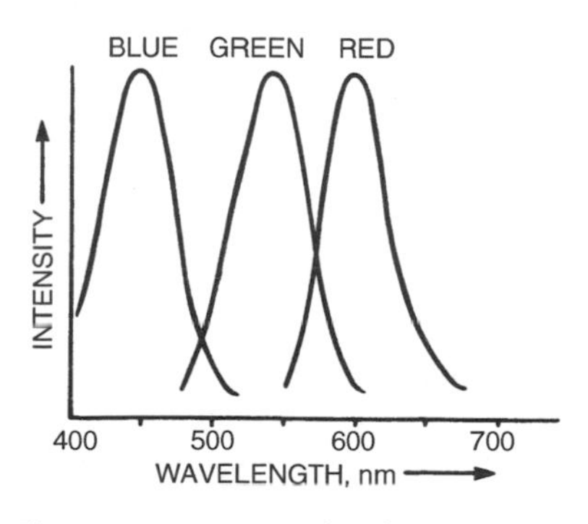

Figure 2.1 *Energy distribution curves for red, green and blue light*

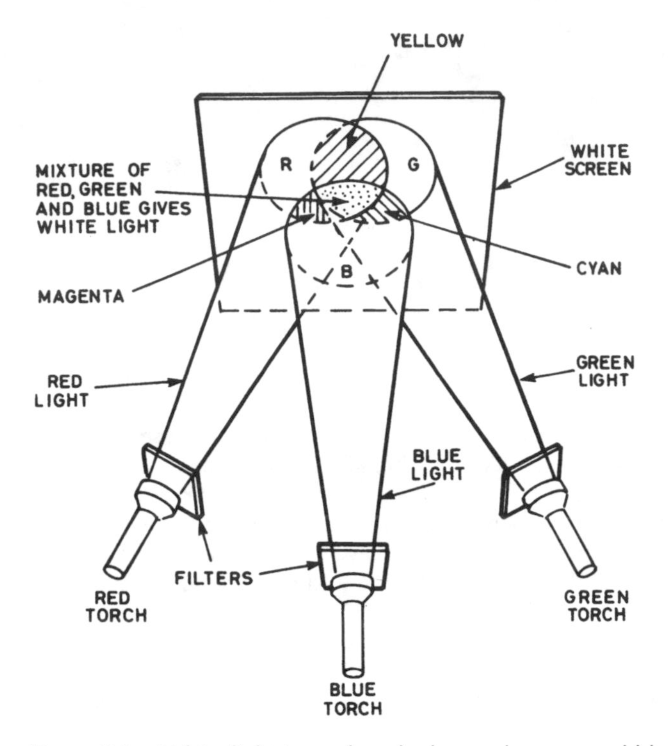

Figure 2.2 *White light is produced when red, green and blue lights are caused to overlap on a white screen. The nature of the white light depends on the intensities of the red, green and blue lights. Equal-energy white light is produced when the red, green and blue lights are of equal energy. This is a hypothetical 'white' used in colour studies. White light for colour television is called Illuminant D, which simulates direct sunlight and north-sky light mixture*

The principle of *additive* light mixing, as it is called, is shown in Figure 2.2, where light of these three colours is projected onto a screen; in the centre of the display all three colours are present and they add to give a white light.

Colour filtering

Figure 2.2 shows each torch as having a coloured filter over its faceplate. It's important to remember that a filter absorbs all colours except that required, so that the red filter, for instance, will offer little resistance to the low-frequency end of the visible light spectrum, but attenuate green, blue and all other colours. Similarly, a green filter absorbs red, blue etc., allowing only the green component of the white torch beam to pass.

Colour temperature

It is difficult to define just what 'white' light is! The appearance of the white depends entirely on the strength and wavelength of each of the component primaries. Many monochrome picture tubes glow with a bluish, rather cold white, the result of a predominance of high frequencies in the rendered light spectrum; this is because that type of phosphor is more efficient in terms of light output.

Fortunately, an exact definition of the 'whiteness' of a light is available in the form of a *colour-temperature* which relates the nature of the white light to an absolute thermal temperature, that to which a black body must be raised to render the same 'colour' of white. For TV applications this has for many years been standardised at 6500 K, known as Illuminant D, and simulating 'standard daylight.'

Complementary colours

White light may be regarded as a mixture of red, green and blue lights. With these three at our disposal, other colours can be generated by various combinations of two. By removing the blue light we would leave a mixture of red and green lights, which would render yellow, as in Figure 2.2. Yellow is a complementary colour. It is, in fact, complementary to blue since blue was the additive primary which had to be removed from white light to produce it. By similar tokens the complementaries of red and green are cyan and magenta, which means that cyan (akin to turquoise) is produced by the addition of green and blue, and magenta (violet/purple) by the addition of red and blue.

Thus we have the three primaries red, green and blue, and the complementaries cyan, magenta and yellow; all these colours are obtainable – with white light – from the three primary colour *lights*.

It is difficult to visualise the wide range of hues that can be obtained from the three television primaries by changing their relative intensities; but those who view a good-quality colour television picture under correct (low!) ambient lighting conditions will appreciate that almost the full range of natural colours can be reproduced. It is noteworthy that in all the discussions and proposals for TV system improvement and picture enhancement, no change to the primary-colour *additive* mixing scheme has been suggested.

The chromaticity diagram

The range of colours can be conveniently represented by a chromaticity diagram, pictured in Figure 2.3. This is in fact a more elaborate extension of Figure 2.2, showing an elliptical area of various 'whites' at the centre, with the wavelengths of the various colours shown around the periphery. The colours between red and blue have no wavelength references, being 'non-spectral' colours resulting from a mix of components from opposite ends of the visible light spectrum.

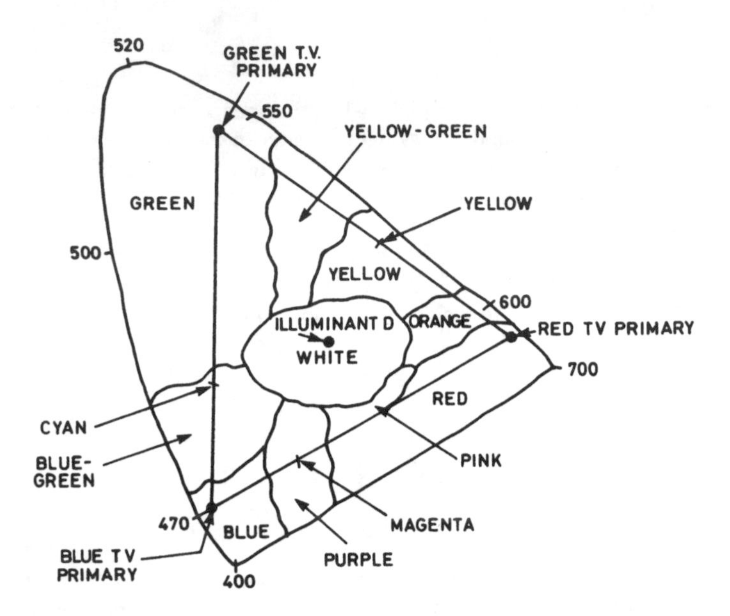

Figure 2.3 *The chromaticity diagram. The inner triangle rests on the TV primaries and the area enclosed by the triangle represents the range of colours in a TV display*

What we cannot show in a line drawing is the fact that the colours gradually merge into one another around the edges of the diagram. The three TV primaries are shown, and linked as a triangle – it can be seen that for any primary, its complement appears about half-way along the opposite side of the triangle.

Luminance, hue and saturation

Basic television, as we saw in Chapter 1, is concerned solely with brightness variations, and the video waveform conveys information which describes only the variations of light and shade in the picture, as does a black-and-white photograph. As such it works only at a point in the middle of our chromaticity diagram, and so far as the viewer is concerned, the precise point depends on the makeup of the phosphor used in his picture tube! The monochrome signal with which we have been dealing so far is called a *luminance* signal.

To describe a colour, two more characteristics must be defined, and these are hue (colour) and saturation ('strength' of colour). Let's look at each in turn. The hue is set by the dominant wavelength of the light radiation, be it red, blue, cyan or whatever, and corresponds to points around the edges of our chromaticity diagram. Figure 2.3, however, encloses an *area* and real colours can occupy any point within that area. Very few colours in nature are fully *saturated* (which would set them on the periphery of the diagram); most fall within it, and this is where the second parameter, saturation, becomes relevant.

Saturation describes the amount by which the colour is diluted by white light, which in terms of the colour triangle in Figure 2.3, tells us how far from the white centre is the colour in question. If one can imagine a pointer whose origin is Illuminant D in the centre, and whose head pointed to the colour of interest, its length would represent saturation, and its *angle* would describe hue. Comparing the healthy pink of a child's face and the bright red of a London bus, it's not hard to see that both have the same hue – red – but the saturation levels are very different, the colour coming from the child's face corresponding to a point near the centre of the chromaticity diagram. This can be graphically illustrated on some colour TV receivers by lowering brightness and contrast and grossly over-advancing the colour level – the face will be reproduced in as strong a red as the bus in a normal picture!

The human eye

The phenomenon of persistence of vision has already been touched upon. The eye tends to retain an image for around 80 milliseconds after it has disappeared, and advantage is taken of this in both television and cinematography, where a serious of still frames create the illusion of a continuous moving picture. Other important characteristics of the eye are its relative insensitivity to *coloured* detail in a scene, and its failure to respond equally to all colours. Its greatest sensitivity is in the region of yellow/green, with lesser response in the areas of red and particularly blue.

Contrast ratio

In nature the range of brightness levels is infinite, from the brilliance of the sun to the total darkness of an enclosed cave. When an image is reproduced in photographic or TV form, the difference between the brightest and darkest parts of the picture is greatly reduced, and so far as TV reproduction is concerned, much depends on the level of ambient light. The darkest parts of the picture can be no blacker than the unenergised phosphor screen, and even with no external lighting source to illuminate the screen, a degree of reflection of the highlight picture components is present from the viewing area, and indeed the viewers' faces! Because the maximum brightness from a picture tube is limited, a contrast ratio of about 50:1 is usual in normal conditions. With a large light-absorbing auditorium and much higher light energy available, the cinema does much better than this.

Gamma correction

In the previous chapter we briefly examined the camera tube and picture tube in their roles of pickup and display devices. In terms of light input to voltage output, the image sensor is reasonably linear so that equal increments of brightness in the televised scene will give rise to equal steps of voltage at the target output point. Unfortunately, the picture tube is not linear in its operation. If we apply an equal-increment staircase waveform to the electrical input of the picture tube, the light output will not go up in corresponding steps; at high brightness levels the graduations will be emphasised or stretched, whereas low-key steps will be compressed. This means that the video signal needs to pass through a compensating circuit, with a gain/level characteristic equal and opposite to that of the tube. It would be expensive to provide such a gamma-correcting amplifier in every receiver, so the process is carried out at the transmitting end, to 'pre-distort' the luminance signal and cancel out the display-tube non-linearity. Figure 23.2 shows how the gamma correction system works at the transmitting end.

3 *Reading and writing in three colours*

If we bring together the themes of the last two chapters we are well on the way to realising a form of colour television system. Because all the colours in the scene to be televised can be analysed in terms of the three television primaries, we can assemble a colour TV outfit by triplicating the 'basic' television system and assigning one primary colour to each of the three. Figure 3.1 shows the set-up. Three identical cameras are used, each with an appropriate filter in front of its pickup tube. The three transmission cables carry video signals which correspond to the three primaries: R, G, and B for short. The monitors also have colour filters fitted, and their pictures are superimposed either by projection onto a common screen, or by a series of dichroic mirrors.

The colour picture reproduced by this set-up would be very good indeed once perfect superimposition, or *register*, of the three images had been achieved at both ends. As a system, however, its shortcomings are immediately obvious: three cameras, three separate transmission channels using (taking normal 625/50 parameters) a total of 16 MHz bandwidth, three identical receivers and goodness

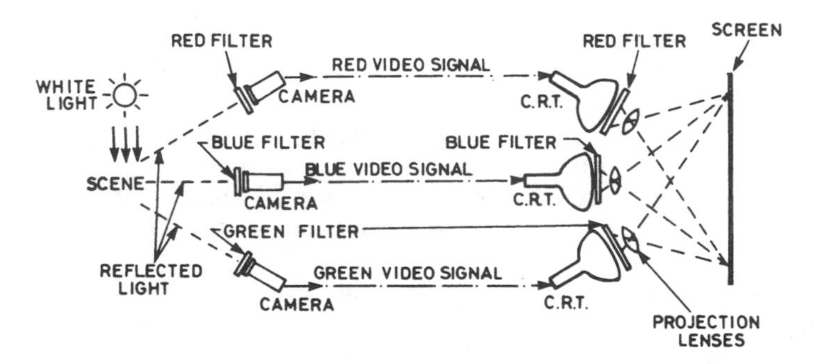

Figure 3.1 *Elementary colour TV system working on the 'simultaneous' principle*

knows what in the form of filters and optics at each end! Not only this, but the system is quite incompatible with any monochrome (black-and-white mode) apparatus at either end.

The sequential approach

We know that basic television is no more than an advancing series of still frames, following each other quickly enough to prevent visible flicker. An early idea was to extend this technique to colour reproduction by mechanically changing the colour filter over the pickup tube at field or frame rate in the sequence of RGBRGB . . . so that 'snapshots' of each of the primary-colour components of the televised scene are sent out in quick succession. The single transmission system would now require only one TV receiver feeding a single picture tube equipped with three sequentially-switched colour filters working in synchronism with those at the camera.

The problem here is one of flicker. Because the repetition rate of each primary colour is now only one-third of the (just adequate) TV field repetition frequency, coloured objects flicker alarmingly; even if a form of 'storage' were devised to overcome this, moving coloured objects would traverse the screen in a blur or a series of hops. To overcome the problem a threefold increase in field-scanning rate would be required, calling in turn for three times the signal bandwidth throughout the system. Plainly, a radically different method of sending colour pictures was required, and the fact that much of the information in the R, G and B signals is common to all three (oscilloscope examination of these waveforms on a normal picture defies anyone to tell them apart!) is a key factor in the solution devised.

Compatible colour television

Before we explore the practicalities of the colour TV system as it exists today, it is useful to provide a brief overview of how the problems described above are overcome in a modern system. In place of the three cameras we have a single camera which may contain between one and four pickup devices, with the necessary colour separation carried out by optical filters near or within the pickup tubes or sensors. The output from the camera is contained in one video transmission channel no more than 5.5 MHz wide by an encoding process, designed to render a signal which is compatible with monochrome receivers.

At the receiving end the encoded CVBS (colour, video, blanking and syncs) signal is handled in a single receiving channel identical to that of a monochrome set, so that only one tuner and IF amplifier are needed. The *luminance* and *chrominance* components of the signal are reassembled into RGB signals in a *decoder* and presented to a colour display device (usually a *shadowmask* picture tube) which is capable of reproducing all three primary-colour images on a single screen. This is summed up in Figure 3.2, and much of the first half of this book will be devoted to explaining the processes outlined above!

Three-tube camera

Probably the simplest type of camera to understand is the three-sensor type, so we shall adopt this as a basis for an account of the operation of the first link in the chain. A multi-sensor camera has only one lens, of course, so the incoming light has not only to be colour-filtered, but 'split' and distributed between the three image sensors. This is achieved by a system of dichroic filters and silvered mirrors as outlined in Figure 3.3. The dichroic filter is a precision-engineered optical device capable of reflecting one colour while appearing transparent to others. In this way each pickup tube is presented with an image on its active surface composed of one primary-colour component of the picture to be televised.

The output from each image sensor is thus the electrical fascimile of the amount of each primary colour present in the scene. These outputs are now ready for processing to produce the compatible CVBS signal referred to earlier.

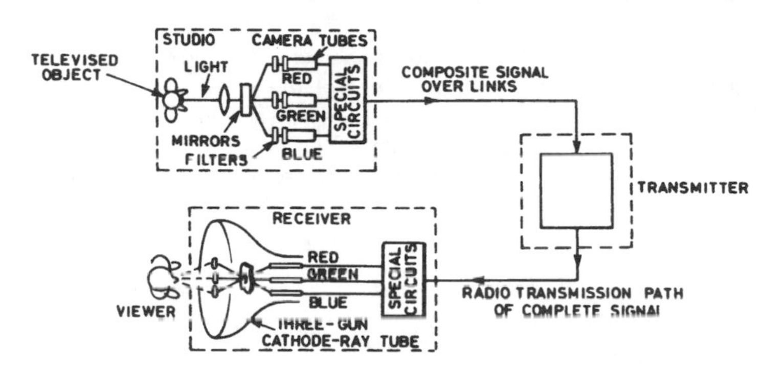

Figure 3.2 *Elementary concept of a complete colour television system*

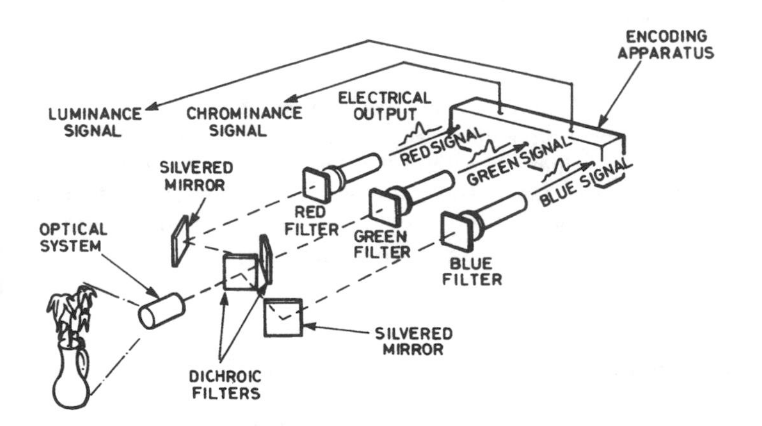

Figure 3.3 *Basic elements of the front end of a colour-transmitting system*

Deriving a luminance signal

It is essential that any colour system produces a signal which is recognisable to a monochrome receiver or monitor, so that the viewer has the choice between a colour or black-and-white set. This requires that the colour transmission conforms to all the specifications of the 'basic' TV system, so we must send primarily a luminance signal to convey brightness information, then add to it as best we can further information streams, used only in colour receivers, to describe the colour. This meets the compatibility requirement by permitting the use of monochrome picture sources at the sending end, and the choice of receiver at the receiving end. Our first task, then, is to derive a luminance signal from the RGB outputs of the three pickup sensors.

The luminance signal is produced by adding the red, green and blue signals from the image sensors in the respective proportions of 30, 59 and 11 per cent, making a total of 100 per cent. This is best seen in simple algebraic terms such that the luminance signal, denoted Y in colour television, is $Y = 0.3R + 0.59G + 0.11B$. Here R, G and B correspond to the primary colours of red, green and blue. If each primary-colour signal from the corresponding image sensor is initially adjusted on a pure peak white scene for 1 V, then by taking 30 per cent of the signal from the red sensor, 59 per cent of the signal from the green sensor and 11 per cent of the signal from the blue sensor and adding them all together we would obtain a total of 1 V Y signal. The image sensors, of course, are not themselves coloured, any more than the green channels, blue electron beams and red guns we shall discuss later in the book! The 'colour labels' so applied merely refer to the circuit or channel in which the component in question is operating.

It may be wondered why equal proportions of R, G and B are not combined to produce the luminance signal. In fact the proportions chosen correspond to the sensitivities to each colour of the human eye, discussed in Chapter 2, and this ensures that the luminance display (the only one available to a monochrome set) appears to the viewer to be *panchromatic*.

This luminance signal, then, corresponds closely with that termed the video signal in a monochrome system. In fact, from first principles a monochrome (single-sensor) camera adjusted to yield 1 V signal from a pure white 'object card' would produce (under the same lighting conditions) 0.3 V output from a saturated red card, 0.59 V from a saturated green card, and 0.11 V from a saturated blue card – the same proportions as given from the pickup sensors of a colour camera scanning a pure white card, the three adding to produce 'unity' white signal. In practice the idea works well, though the necessary gamma-correction process introduces mild brightness errors on monochrome receivers in areas of the picture where heavily-saturated colours are present.

CCD colour cameras

For use with a colour system the image-sensor may consist of three separate devices with an image-splitter/colour filter system like our Figures 3.2 and 3.3;

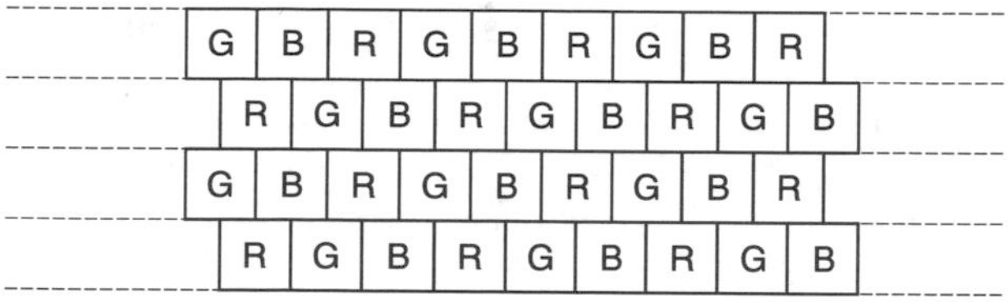

G B R G B R G B R
R G B R G B R G B
G B R G B R G B R
R G B R G B R G B

Figure 3.4 *Colour light filter for an image sensor. Each colour is positioned in front of a single photodiode on the surface of the sensor*

or a single sensor chip with a colour dot-matrix overlay as shown in Figure 3.4. In the latter case the resolution of the sensor chip must be much higher for colour applications since *three* silicon photodiodes are required for each pixel in the picture. The output of the colour-matrix image sensor contains information on the R, G and B components of the image for further processing, as we shall see next.

Colour-difference signals

We are now aware that a coloured scene possesses three important characteristics. One is the brightness of any part of it, already discussed as luminance; two is the hue of any part, which is the actual colour; and three is the saturation of the colour, that is the depth of colour. Any hue can vary between very pale and very deep, and the amount of the colour is basically its saturation. For example, a laurel leaf represents highly saturated green, while the pastel green of my blotting paper is the same hue but less saturated, which means that on a TV screen its image would have a greater contribution of white.

When an image sensor is scanning the red in a scene, as an example, it receives information on the luminance, hue and saturation because all three are obviously present in any colour. However, remember that the luminance signal is processed separately by proportioned addition of the primary-colour signals and transmitted effectively in 'isolation' so that it can be used by monochrome receivers. This, then, means that an additional signal has to be added to and transmitted with the luminance signal so that colour receivers will obtain the extra information they require to describe the hue and saturation of the individual pixels. This auxiliary signal may be regarded as the 'colouring' agent; it is called the *chroma* (short for chrominance) signal.

It is formed initially on a *subcarrier* which is then suppressed at the transmitter and recreated in the receiver, as we shall see later. The subcarrier is modulated in a special way by *colour-difference* signals, of which there are three but only two of them need to be transmitted.

The three colour-difference signals are red minus the luminance signal, green minus the luminance signal, and blue minus the luminance signal. By red, green and blue is meant the primary-colour signals delivered by the colour camera, while the luminance signal is the Y component as defined by the expression given earlier.

Thus, in simple algebraic terms the three colour-difference signals are: $R - Y$, $G - Y$ and $B - Y$.

It is not necessary to transmit all three colour-difference signals. If we send Y, $R - Y$ and $B - Y$, it is easy to recreate $G - Y$ at the receiver. Let's see how this comes about. At the receiver we have available the Y signal, and information on how *different* is the R signal (in $R - Y$) and how *different* is the B signal (in $B - Y$). When these *combined* differences are taken into account, any outstanding balance (or debit!) represents the G difference, so that a $G - Y$ signal can be derived by adding suitable proportions of *negative* $R - Y$ and $B - Y$ signals in the receiver's decoder.

The reason behind the choice of $R - Y$ and $B - Y$ signals for transmission is very simple. Although any two difference signals could be sent, we have seen that the Y signal draws more heavily on the green primary-colour component of the televised image than either of the other two (Y contains 0.59G), so it follows that the difference between G and Y is less than that for the others. This relatively small $G - Y$ signal would be more vulnerable to noise in the transmission system than the larger $R - Y$ and $B - Y$ signals.

It is interesting to observe that the colour-difference signals fall to zero amplitude when the televised scene is devoid of colour – that is, when greys and whites are being transmitted. This is not really surprising when we remember that the chroma signal is composed of colour-difference components – in a *monochrome* transmission there is no difference between the primary-colour signals and the luminance signal! Monochrome programmes are becoming rare now, consisting mainly of archive material and old movies.

We have seen that the red, green and blue signals from the image sensors of a colour television camera can be conveniently tailored to 'unity' (1 V) on a pure peak white input. Thus $R = 1$, $G = 1$, and $B = 1$. Y, we have seen, is equal to $0.3R + 0.59G + 0.11B$, which means that on a pure peak white input we have $0.3(1) + 0.59(1) - 0.11(1)$, which equals 1. Clearly, then, from this we get $R - Y = 1 - 1 = 0$ and $B - Y = 1 - 1 = 0$. The same conditions exist on greys when the red, green and blue signals from the sensors are less than 'unity' but still equal. On colour scenes, of course, the RGB signals are not equal and so colour-difference signals arise, and only when this happens is a chroma signal produced.

It is possible, of course, to calculate both the luminance signal and the colour-difference signals from the colour scanned by the colour camera, remembering that anything below full saturation means that white is added to the predominant hue in terms of the three primary colours in the proportions required for the luminance signal. Thus, while the R, G and B signals become unequal when the camera is scanning a coloured scene, the Y signal still retains the proportions of 0.3, 0.59 and 0.11 of the R, G and B signals respectively.

For example, purple of below full saturation comprises a mixture of red and blue with a little green signal too, so that the Y proportions of the red, green and blue provide the 'white' which reduces the saturation. Thus we may have $R = 0.6$, $G = 0.1$ and $B = 0.5$, meaning that the luminance signal Y is equal to $0.3(0.6) + 0.59(0.1) + 0.11(0.5)$, or $0.18 + 0.059 + 0.055 = 0.294$. Using this for Y, then, R – Y is 0.6 – 0.294 = 0.306 and B – Y 0.5 – 0.294 = 0.206. When the Y signal is of a greater voltage than the primary colour components of the colour-difference signal, the colour-difference signal as a whole then assumes a *negative* value as, of course, would be expected. The colour-difference signals, therefore, can swing from zero to a maximum in both the positive and negative directions. A little thought will confirm that if we have a white raster displayed, and wish to change it to blue or yellow, the means of doing so is to provide a positive or negative blue colour-difference signal. For saturated blue, the red and green colour-difference signals would also operate to turn off their respective colours.

Colour-difference signal bandwidth

The colour-difference signals are used, as we have seen, to *add* colour information to the basic black-and-white picture. Because the human eye cannot resolve fine detail in colour there is little point in transmitting high-definition colour-difference signals, which would be difficult to accommodate in the signal channel, and wasted on arrival! The luminance signal carries all the fine detail in the picture, then, and the colouring signal is 'overlaid' on the display in much coarser form. In practice the illusion is well-nigh perfect, and subjective viewing of the combination of high-definition luminance and rather 'woolly' chrominance is perfectly satisfactory. In the American colour system they go a step further, and transmit even lower definition in their 'Q' signal, corresponding to blue shades, where research shows that the eye is least able to discern detail. Reproduced colour pictures from domestic videocassette recorders offer even poorer chrominance definition, but still with (just) satisfactory results.

In the UK broadcast colour system, then, we limit the bandwidth of each colour-difference signal to about 1.2 MHz by means of electrical bandstop filters.

The encoder

The three primary-colour signals from the three image sensors are communicated to the input of the encoder, and this processes them ready for transmission. The process can be likened to the sending of an important message in code. If a code book is used at the dispatch end, the letter can be reduced to a shortened form and sent to the recipient who, using a similar code book, can decipher it and thereby recreate the original message.

The three primary-colour signals are first added together in the proportions required for the luminance or Y signal, as already explained. The resulting Y signal is then separately subtracted from the red and blue primary colour signals to give the two colour-difference signals R – Y and B – Y. These two signals are amplitude-modulated in quadrature (see anon) upon a subcarrier which is subsequently suppressed so that only the sidebands of the V and U signals remain.

In monochrome television (now largely confined to industrial surveillance and special-purpose applications) only the luminance signal and the sync and black level are modulated onto the broadcast carrier wave.

With the need to send a chroma signal in addition to the basic VBS (video, blanking and syncs) signal, coupled with the requirement to keep the combined signal within the channel bandwidth normally occupied by a monochrome signal, the method of fitting together all the components of a signal for compatible colour television is necessarily more complex. As already outlined, the scheme utilises a colour subcarrier of a much lower frequency than the main carrier wave. The latter, in fact, may be in the hundreds of MHz or the GHz (GHz, gigahertz, one thousand million cycles per second) range, whereas the former is a few MHz only. It is within the luminance bandwidth range in fact, the actual frequency being geared to the line and field timebase repetition frequencies.

We shall be seeing later that to get the chroma signal to integrate neatly with the luminance signal, the frequency of the subcarrier must be related to line- and field-scanning frequences. The extra 'colouring' information is then squeezed into discrete intervals of low energy between the luminance sidebands in the overall frequency spectrum. It is by this technique that the extra chroma information can be carried in an ordinary 625-line television channel with the least mutual interference, especially to black-and-white pictures produced by monochrome receivers working from a colour signal.

This clever technique, of course, calls for a comprehensive type of sync-pulse generator, for its master oscillator must be correctly related to the frequency of the subcarrier generator. However, in practice the master oscillator generates subcarrier frequency, which is then divided down by counter circuits to derive line and field synchronisation pulses. This ensures that the relationship between the three is correct. Figure 3.5 outlines the processes described thus far.

Composition of the chroma signal

Having discussed the derivation of luminance and colour-difference signals, and seen how all the characteristics of a full colour picture can be carried in them, it's time now to explore the way in which they are combined to make up a CVBS (chroma, video, blanking, syncs) signal. We have already referred to the process of quadrature modulation onto a subcarrier – just what does this mean?

The basics of amplitude modulation have been discussed in Chapter 1. The modulation signal of lowish frequency modulates the carrier wave of higher frequency by causing its amplitude to vary in sympathy with the modulating signal. Figure 3.6 illustrates the effect. Here a single sinewave signal is shown modulating a higher-frequency carrier wave. This modulation is quite easy to understand from an elementary viewpoint, and it can be achieved by the use of simple circuitry. For example, the modulating signal can be applied to a transistor so that it effectively alters the supply voltage on the collector. Thus when the carrier wave is applied to the base, the amplitude of the output signal alters to the pattern of the modulation waveform. When a pure sine wave is the modulation signal, therefore, the carrier-wave amplitude varies in a like pattern and the carrier

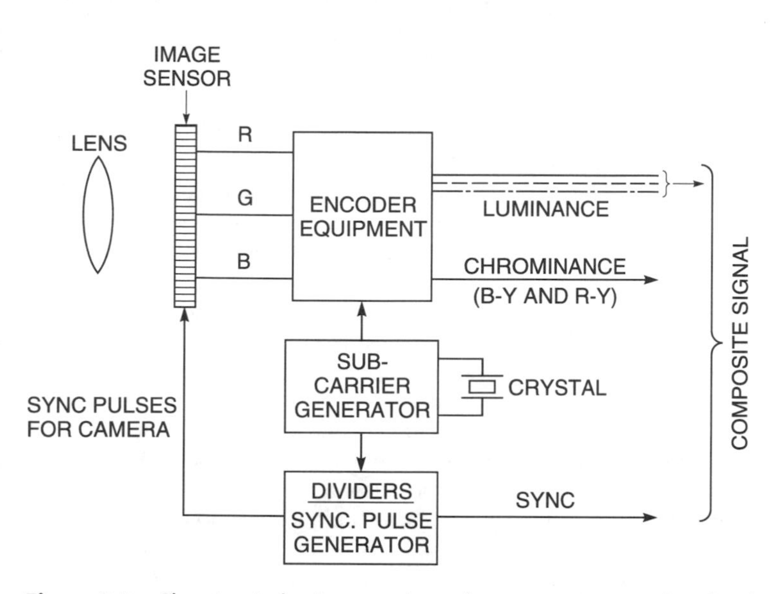

Figure 3.5 *Showing in basic terms how the composite signal is developed. The subcarrier generator provides a reference for both sync and colour signals, and the composite signal developed contains all information necessary for the recreation of a colour picture*

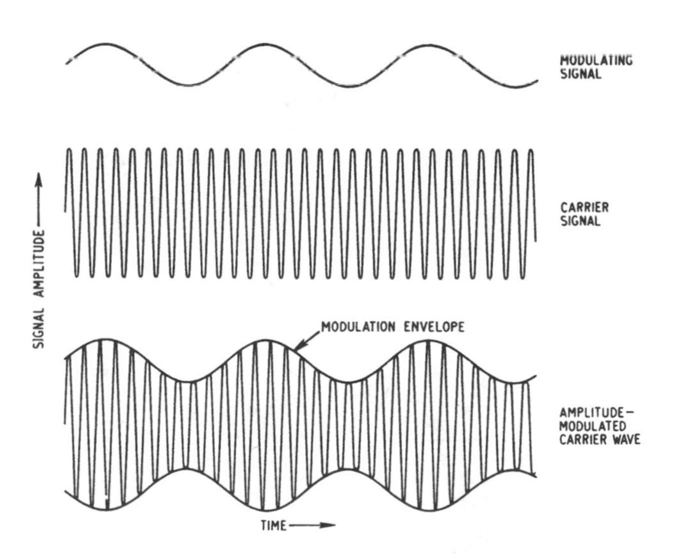

Figure 3.6 *The amplitude modulated waveform is produced by the carrier signal being modulated with the modulating signal. Note that the waveform inside the modulation envelope results from the addition of the carrier wave (fc), the upper sideband (fc + fm) and the lower sideband (fc – fm), where fm is the modulation frequency*

wave is then said to possess a modulation *envelope* as shown at the bottom of Figure 3.6.

We have already seen that any type of modulation gives rise to sideband signals, and for amplitude modulation there is an upper and a lower sideband for each modulation frequency. If the carrier wave is, say, 10 kHz and the modulating signal a pure 1 kHz sine wave (e.g. a single modulation frequency) then the upper sideband will be 10 + 1 = 11 kHz, while the lower sideband will be 10 – 1 = 9 kHz. This simple arithmetic follows for all single modulation frequencies. Thus the modulator delivers three signal components, the carrier wave, the upper sideband and the lower sideband. The information due to modulation is effectively present in the sidebands, so it is possible to suppress the carrier after modulation for transmission, though for demodulation at the receiver the carrier wave will need to be somehow reintroduced very accurately.

So much, then, for the amplitude modulation of one set of information, but what about the modulation of two sets of information which convey the R –Y and B – Y colour-difference signals? This is where the quadrature amplitude modulation of the subcarrier comes in. In the UK system of colour television the subcarrier frequency is accurately controlled at 4.43361875 MHz. This is usually referred to roughly as 4.43 MHz. The numerous decimal places (the last of which defines the frequency to one-hundredth of one cycle per second!) are necessary for various reasons, one of which is to minimise interference of 'dot-pattern' type which can mar received pictures due to a beat effect arising between subcarrier and line timebase frequencies. The rigid frequency relationship set up between these two gives optimum performance not only in minimising the dot pattern, but in preventing a disturbing 'crawl' effect of such pattern elements as remain. It is also necessary to have the subcarrier within the video bandwidth, and at as high a frequency as possible for minimum interference. In the PAL system both the upper and lower chrominance sidebands are exploited equally, and this means that subcarrier frequency must be chosen so that both sidebands, each extending to about 1.2 MHz, can be fully accommodated within the video spectrum, as shown in Figure 3.7.

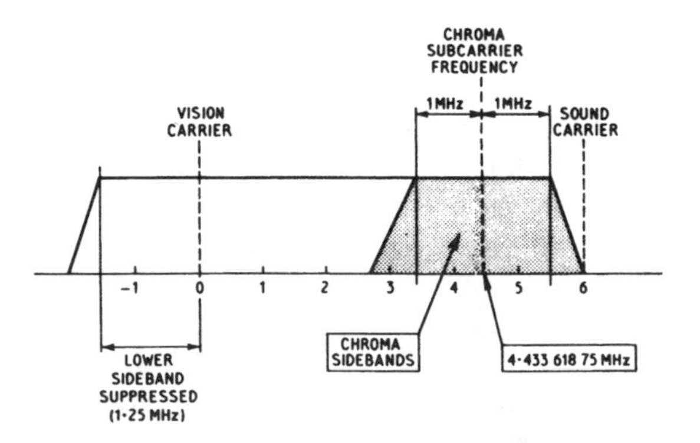

Figure 3.7 *Spectrum of 625-line PAL signal in UK television channel*

The master sync-pulse generator, then, produces an output at subcarrier frequency, and this is fed into *two* amplifiers which produce two subcarrier outputs, but with one having a 90 degrees phase difference with respect to the other.

'Quadrature' and degrees of timing

A complete cycle of alternating current or voltage can be regarded as occupying 360 degrees – a circle! Although this statement is somewhat arbitrary it is supported by sound reasoning. Consider a generator, for example; this will, in the simplest case, rotate through 360 degrees and during that period produce a complete sine wave. If a second generator is set going but a quarter turn ahead of the first, then the sine wave produced by this will be 90 degrees out of phase with that yielded by the first one.

Since in the case of the colour television the one oscillator drives two amplifiers, the frequency of their two outputs will be absolutely identical and, moreover, the synchronism will be maintained, but always with one output 90° ahead of the other owing to the deliberately contrived 90° phase shift introduced by the circuit elements.

Figure 3.8 highlights the situation, where the two full-line sine waves are of exactly the same frequency but 90° apart in phase. The X axis is calibrated from 0 to 360° corresponding to a complete cycle of the first full-line sine wave. The second one starts a little later in time; in fact, when the first has arrived at the 90° mark. Thus we have a direct illustration of the 90° difference between the two subcarriers which, remember, are derived at the transmitter from a common oscillator or generator. Hence phase can be seen to be a function of *timing* between two signals. The term quadrature comes from the fact that there are four 90° segments in a circle; two signals with a 90° phase relationship are said to be in quadrature.

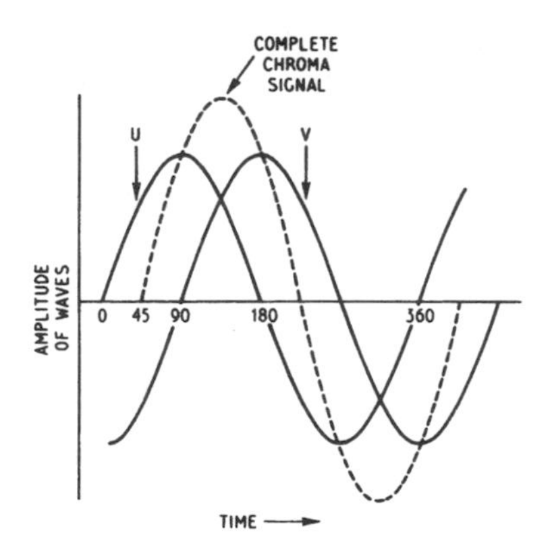

Figure 3.8 *The V and U waveforms have the same frequency and amplitude but differ in phase by 90°. The broken-line waveform represents the chroma signal complete which is the quadrature addition of the V and U signals*

Colour modulation

One of our subcarrier signals is amplitude-modulated with the R –Y signal, then, and its quadrature companion is amplitude-modulated with the B – Y signal. After suitable *weighting* they become *V* and *U* chroma signals of the PAL colour system. Weighting is a process of reducing the amplitudes of the R – Y and B – Y subcarriers. At a later stage they will be added to the luminance signal, and where a large subcarrier signal (corresponding to a highly-saturated colour) coincides with an extreme excursion of the luminance signal (corresponding to a very light or very dark part of the picture) the combination of the two could lead to overmodulation of the transmitter. The weighting values are: V = 0.877 (R – Y) and U = 0.493 (B – Y). After recovery in the receiver's decoder, the subcarrier signals are restored to normal proportions by simple adjustment of amplifier gain in each colour-difference signal path.

The next move in the game is to add the V and U signals to form the chroma signal proper, and this signal is shown by the broken-line sine wave in Figure 3.8. The fundamental feature of this mode of signal addition is that by special detection at the receiving end it becomes possible to isolate the V and U signals again and thus extract the original R – Y and B – Y modulation signals. This is facilitated by the fixed 90° phase difference between the two original carriers. Signals of the same frequency and phase lose their individual identity for all time when added.

In colour TV engineering, the various colouring signals are regarded in terms of vectors. These are mathematical devices for displaying the specific features of amplitude and phase of a signal at one instant in time. The basic vector 'background' is given in Figure 3.9 where our complete timing circle is represented by the four quadrants. Motion or time is regarded as anti-clockwise, so starting from the zero-degree datum we follow the angles of phase as shown by the arrowed circle. Colour signals are generally more complex than implied by simple vectors

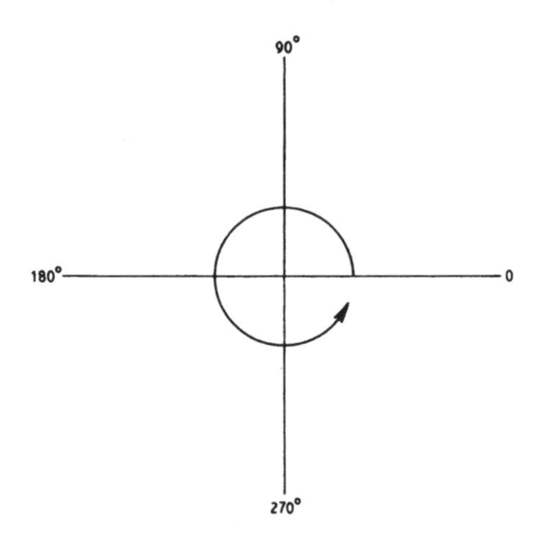

Figure 3.9 *Vector 'background' diagram in four quadrants. Time or motion is regarded by convention as being anticlockwise from the 0-degree datum, as shown by the arrowheaded circle*

and phasors, so this method of presentation may be regarded as a simplification, though an adequate one for our purposes here.

Developing the vector theme in terms of the V and U signals in Figure 3.8 we get the vector diagram of Figure 3.10. The arrowed lines here correspond to the three signals – the V and U signals and the resulting chroma signal. The amplitude of this result is obtained by completing the square (or rectangle, when V and U signals are not equal) as shown. The diagram clearly shows that the V and U signals have a 90° phase difference and that the amplitudes of these two signals are in this case equal. The angle changes (but within the same quadrant) when V and U amplitudes differ. Should the amplitudes alter together and in sympathy, then the original angle is maintained, though the vector line will now be shorter to describe lesser saturation in the transmitted colour. That the complete chroma signal is 45 degrees relative to either the U or V signal is proved in Figure 3.8. However, it is more convenient to work with vectors than with complex waveforms and the remainder of our account will be on a vector basis.

We can easily discover the amplitude of the complete chroma signal which, in colour television at least, is often referred to as the phasor, from the expression: phasor amplitude $= \sqrt{(V^2 + U^2)}$.

Thus when the V and U signals have equal amplitude, the phasor length is 1.4 times that of V or U. Readers with some knowledge of trigonometry will also see that the angle ϕ of the phasor in the quadrant of Figure 3.10 is equal to $\tan^{-1}V/U$.

We next need to get clear what effect negative colour-difference signals have on the vector diagram. The quadrant of Figure 3.10 can only accommodate reds, purples and blues, and we need to describe all the colours within the triangle in Figure 2.3. To enlarge on our earlier discussion of colour-difference signals, we saw that they can move from zero in a positive or negative direction. In terms of fully-saturated colours, let's take one or two examples to illustrate this. Refer to Figure 3.11, which shows a standard colour-bar display, consisting of a white bar, followed by saturated bars of yellow, cyan, green, magenta, red and blue, then a black (zero luminance and zero colour) bar, reading from left to right. On the first bar, all three lights are on to give a full-brightness equal mix of R, G and B. There is no colour present, so the colour-difference signals are at zero as shown in the time-related luminance and colour-difference waveforms below. Moving on to

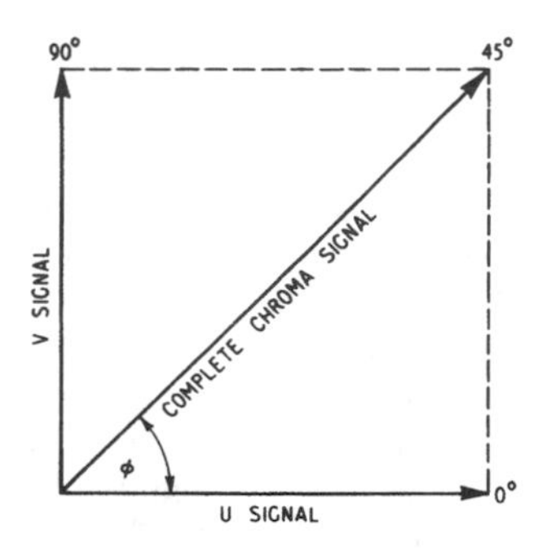

Figure 3.10 *Vector representation of the V, U and complete chroma signals of Figure 3.8 (see text)*

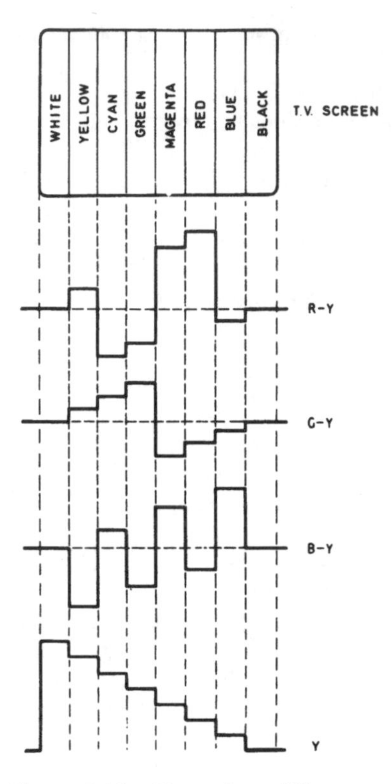

Figure 3.11 *The colour-difference signals and their relationship to the TV screen display on standard colour bars*

the yellow bar, we know that this consists of a mixture of red and green light, so the colour-difference signals for these two will have to act to turn them up, giving rise to positive colour-difference signals, while the blue light needs to be turned off, calling for a *negative* B – Y signal. The result is the complement of blue–yellow. For the next (cyan) bar, we are looking to turn off red, and turn up blue and green (to compensate for the loss of *brightness* output from the now-absent red light), so here we have positive B – Y and G – Y signals, along with a negative R – Y signal. The other bars call for different combinations of colour-difference signals, but their relationship is plain from the waveforms below. Relating these to the vector diagram, let us take the case of the yellow bar. Here the B – Y signal necessarily moves in a negative direction, into the second quadrant of the vector diagram (Figure 3.12). This indicates how the chroma subcarrier reflects the polarity of the colour-difference signals in terms of its phase. In the case of yellow, the R – Y signal is still positive, so the resultant vector angle reflects the presence of a small positive value of R – Y and a large negative value of B – Y. Figure 3.12 shows the vectors for all three complementary colours, along with that for green, which (as should now be plain) calls for negative R – Y and negative B – Y signals.

To summarise, however, Table 3.1 gives all the parameters so far discussed relative to the hues of the standard colour bars. All these values are based on 100 per cent saturation and amplitude.

This brings us to the concept of a 'colour clock' whose face is the vector diagram, and which has a single 'hand' able to rotate to any angle to describe the hue of the picture and having a 'telescopic' feature, whereby the longer the hand the greater the saturation. With the hand pointing north, as it were, fully saturated red would imply a long hand, whereas pale pink would give rise to a short one. Similarly, a bright orange colour would bring the hand to a 'long north-easterly' aspect and a green colour would direct it to the south-west. It will be

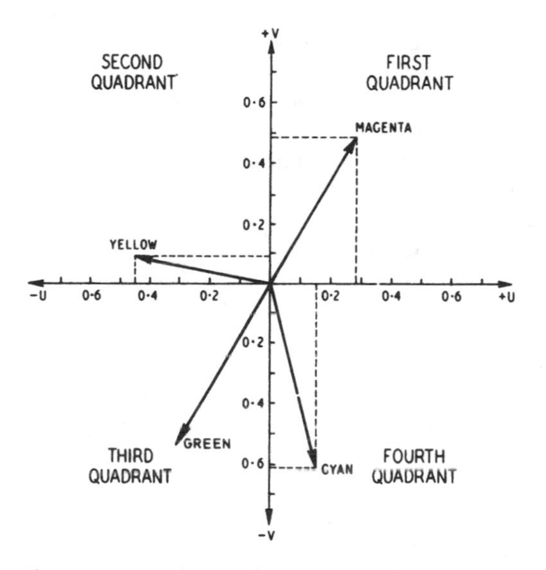

Figure 3.12 *Phasors for various colours: the vector angles and lengths for green and complementary colours megenta, yellow and cyan*

Table 3.1

Colour	Y	B – Y	R – Y	U	V	*Phasor Amplitude*	*Angle (deg)*
Yellow	0.89	–0.89	+0.11	–0.4388	+0.0965	0.44	167
Cyan	0.7	+0.3	–0.7	+0.1479	–0.6139	0.63	283
Green	0.59	–0.59	–0.59	–0.2909	–0.5174	0.59	241
Magenta	0.41	+0.59	+0.59	+0.2909	+0.5174	0.59	61
Red	0.3	–0.3	+0.7	–0.1479	–0.6139	0.63	103
Blue	0.11	+0.89	–0.11	+0.4388	–0.0965	0.44	347
White	1.0	0	0	0	0	0	–
Black	0	0	0	0	0	0	–

remembered that the same 'telescopic pointer' effect was used in our description of the chromaticity diagram, Figure 2.3. Thus as each line of picture is scanned, so can be visualised the phasor changing in both amplitude and angle to describe the saturation and the hue of each individual pixel in the line in turn. When there is no colour, the phasor shrinks to zero and the picture is then under the sole control of the luminance part of the signal.

Integrating luminance and chrominance

We have now seen how all the colouring information is contained in a single chroma signal, by means of (a) leaving out the G – Y component and (b) quadrature-modulating the remaining two colour-difference signals onto a single carefully-chosen subcarrier. The next step is to combine this with the luminance signal in a way which causes the minimum mutual interference – a daunting task, since the luminance signal appears to occupy all the available channel bandwidth! Close examination of the frequency spectrum of the luminance signal, however, reveals that most of the picture information is centred on 'energy packets' at multiples of line- and field-scanning frequencies, as intimated in Chapter 1. The spaces between these packets are relatively quiet. Let us now give some thought to the chrominance subcarrier signal. It is describing the same picture, and many 'detail' features of that picture will be common to chrominance and luminance signals. Thus the chroma sidebands will have a packet energy-distribution characteristic similar to that of the main signal, and this is the key to the interleaving process which is used. If we can *offset* the chroma subcarrier frequency to place it exactly between two of the luminance packets, as it were, the chroma sidebands will fall between the similarly-spaced luminance energy packets, rather like the teeth of two interlocking combs. Figure 3.13 illustrates the principle; to achieve it

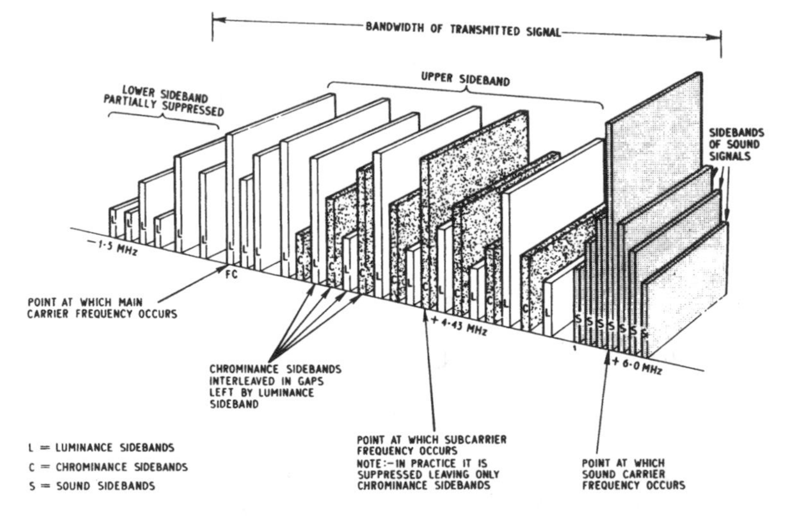

Figure 3.13 *The sidebands can be looked upon as packets of energy at various frequencies. This diagram shows the interleaving of chrominance and luminance, and the situation of the sound channel in the transmitted signal*

the subcarrier frequency is related to the line-scanning frequency thus:

$$\text{Line frequency} \quad (f_1) = \frac{f_{sc} - \frac{1}{2}f_f}{284 - \frac{1}{4}}$$

where f_{sc} is the subcarrier frequency (4.43361875 MHz for PAL), f_1 is the line frequency (15.625 kHz) and f_f the field-scanning frequency (50 Hz).

Suppression of subcarrier

Because the modulating information is wholly present in the sidebands of the subcarrier, it is possible to suppress the subcarrier itself at the transmitter, and by so doing we can significantly reduce the severity of the dot-pattern introduced at the receiver by the presence of the colour signal. It will be recalled that the waveform inside the modulation envelope such as that shown in Figure 3.6 results from the addition of the carrier wave and the upper and lower sidebands. As would be expected, a modulation waveform devoid of its carrier wave differs significantly from that with the carrier wave intact since the former is composed of only the upper and lower sidebands. Figure 3.14 attempts in one way to reveal the salient features.

Compared with the bottom diagram in Figure 3.6 it will be seen that the envelope has effectively 'collapsed', so that the top and bottom parts intertwine, the sine wave in heavy line representing the top of the original and that in light line representing the bottom of the original. Further, the high frequency signal inside the collapsed envelope has also changed character. The frequency, however, is just the same as the original carrier wave because it is composed of the original upper and lower sidebands, the average of which is the carrier frequency; but it can be seen that phase reversals occur each time the sine waves representing the top and bottom parts of the original envelope pass through the datum line. It is difficult to show these diagramatically; but the high-frequency signal changes phase by 180 degrees at each 'envelope crossover' point, and this has a vital significance in colour encoding, as we shall see.

The subcarrier modulation constitutes the B – Y and R – Y signals, which change continuously during a programme. However, in a colour-bar signal the

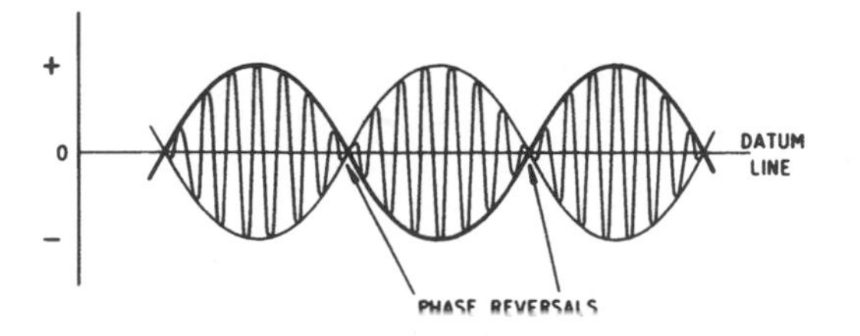

Figure 3.14 *Modulation waveform when the carrier wave is suppressed. Compare this with the waveform at the bottom of Figure 3.6. An important aspect is the phase reversal of the high-frequency wave each time the modulation sine waves cross the datum line. This waveform is the addition of the upper and lower sidebands only*

signal is less 'busy' because the colour signals remain constant over each bar per line scan, changing only from bar to bar.

Figure 3.15 shows the weighted B – Y signal produced by the yellow and cyan bars at (a), the U chroma signal modulation waveform due to the bars at (b) and the reference subcarrier signal at (c). The U chroma modulation signal at (b) is with the subcarrier suppressed, and since the modulation signal is a stepped waveform going from – 0.33 (corresponding to the yellow bar) to + 0.1 (corresponding to the cyan bar) through the zero datum, it follows that the polarity change between the two bars will cause a 180° phase reversal of the high-frequency signal just the same as when the modulation is a sine-wave signal, shown in Figure 3.14.

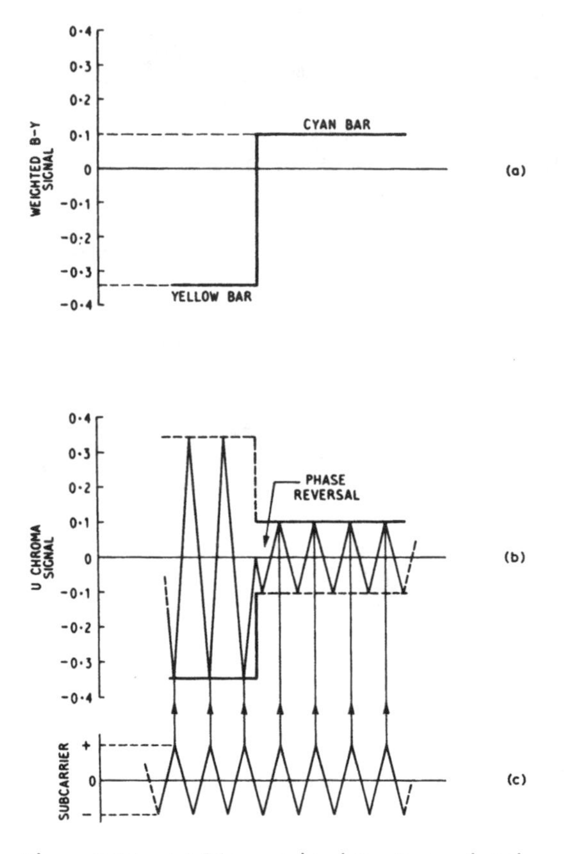

Figure 3.15 *(a) PAL-weighted B – Y signal at the yellow/cyan bar transition. (b) The waveform resulting from suppressed carrier modulation. (c) Subcarrier signal in correct phase. At the receiver the reintroduced sub-carrier samples the peaks of the modulation sideband components to give an output corresponding to the amplitude and polarity of the original colour-difference signal. The polarity is given because the phase of the sideband components reverse each time the colour-difference modulation signal crosses the zero datum line (also see Figure 3.14)*

The phase reversal is indicated, and more clearly shown in Figure 3.15 than in Figure 3.14. For example, it will be seen that the positive tips of the subcarrier correspond to the negative tips of the U chroma signal during the yellow bar, and to the positive tips (owing to the phase reversal) during the cyan bar. Thus if we can recreate the original subcarrier at the colour receiver, and make it accurately sample the peaks of the chroma signal in this way, information is recovered on both the amplitude *and* the polarity of the colour-difference signal. This, in fact, is how the chroma signals are demodulated at the receiver. The operation of its subcarrier generator will be described in the next chapter.

Composite colour signal

When we add our carefully-contrived suppressed subcarrier signal to the luminance waveform we form the composite colour signal (CVBS signal) referred to near the beginning of this chapter. We shall have the basic television picture signal of our Figure 1.4 with the addition of the chroma signal, which latter will tend to ride on the luminance level, whatever that may be. Figure 3.16 indicates the form of the CVBS signal as it leaves the studio en route to the transmitting site. This is for a colour-bar signal, showing its luminance step waveform and the large superimposed subcarrier signal characteristic of fully-saturated bars. In a typical 'real' picture, saturation will not be nearly so high, and as a result the subcarrier amplitude will be correspondingly lower. The luminance signal, too, will be of a random nature, so that the equivalent composite waveform would appear fuzzy and gauze-like.

There is one other feature of Figure 3.16 which we have not yet considered, and that is the burst of what appears to be a chrominance signal on the back porch. This is in fact a reference subcarrier timing signal for the receiving decoder, and it will be considered in detail in the next chapter.

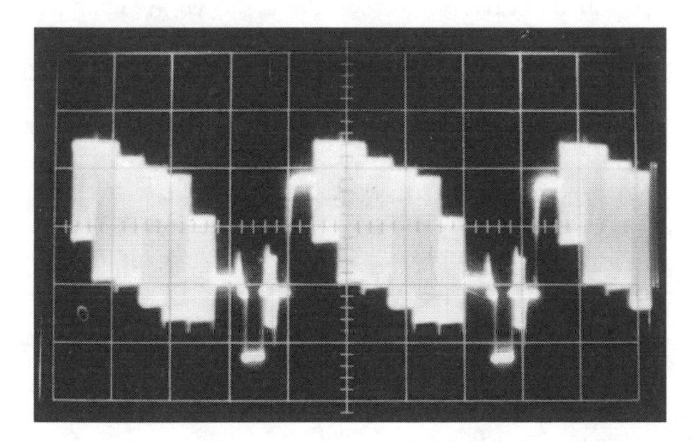

Figure 3.16 *Oscillogram showing the composite signal over two lines of a colour-bar transmission*

The deciphering process at the receiver

Assuming that our CVBS signal passes unscathed through the transmission system, whatever form that may take, it will arrive at the receiver's detector output (or the monitor's vision input) in the same form, and now requires sorting out to derive the original RGB signals as were present at the output from the camera. The processing of the Y signal, as in a monochrome set, consists purely of amplification as it is already in a form recognisable to a picture tube. Our concern here is to examine the overall process of colour demodulation, and see how the chrominance signal is unscrambled and recombined with luminance to recreate R, G and B signals; and finally how they are displayed on the viewing screen.

The colour subcarrier is selected from the composite signal by means of a bandpass filter, on the left of Figure 3.17. This has a response about 2.4 MHz wide, centred on 4.43 MHz to embrace the entire sideband spread of the chrominance signal. We now need to demodulate the signal, and to achieve this the subcarrier is routed to two separate switches, each capable of operating at subcarrier rate. Imagine that the switches are normally off, and that each has a storage capacitor connected to its output. In this state the stores will be empty and nothing will happen until the switches close. In fact the switches are closed for a brief instant once every subcarrier cycle. Consider the upper switch S1 which is driven by a local source of subcarrier reference. If the phasing of the reference is correct, the switch will close momentarily at the instant when the V (carrying R – Y) carrier is at its peak, and the storage capacitor C1 will acquire a charge corresponding to the instantaneous level of the V signal during the *sampling* phase. At this time the U signal (carrying B – Y information) will be passing through zero because of the quadrature relationship between V and U subcarriers, so that the U signal cannot affect the charge acquired by C1. If this storage capacitor is paralleled by a resistor to give an appropriate time constant, the signal appearing across the resistor will accurately follow the V signal modulated onto the subcarrier at the sending end.

The U detector works in like fashion, but here the sampling phase must match that of the U signal. This is easily achieved by introducing a matching phase shift in the reference subcarrier path to the switch, as shown in Figure 3.17. S2, then, will switch on briefly at the instant that the U subcarrier is passing through its zenith and this of course represents the time of passage of the V subcarrier through

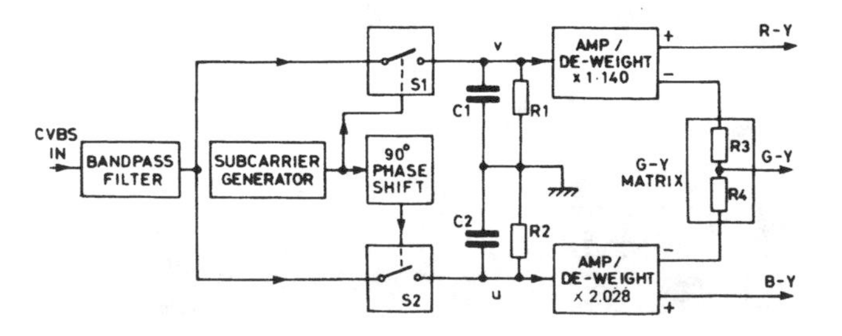

Figure 3.17 *A basic decoder, showing the synchronous detector-switch system and the derivation of a G – Y signal*

the zero datum line. The signal appearing across R2 will be a replica of the U signal at the studio.

C1 and C2 can charge to negative levels as well as positive ones, and will do so whenever the narrow and precisely-timed 'sampling window' catches the subcarrier below the zero datum line. It will be recalled that we discussed the ability of a subcarrier signal to reflect polarity changes in the colour-difference signals in terms of phase (Figure 3.15), and here we can see how the demodulators recover this information to present negative or positive signals at their outputs, dependent on the phase of the incoming subcarrier.

In practice we shall only see maximum positive outputs from the V and U demodulators on a saturated magenta picture; on the green bar, both will be at a maximum negative level, giving rise to a highly positive G – Y signal which, remember, is derived by adding *inverted* R – Y and B – Y signals. We shall come shortly to the G – Y recovery matrix. It is important that the triggering pulses for the switches, i.e. the local subcarrier signal, are precisely in phase with the broadcaster's subcarrier generator. If any timing error is present here, the sampling phases will take place at the wrong times relative to the incoming chroma signal, and this will cause incorrect or random colour-difference signal levels and polarities, making nonsense of the hues in the reproduced picture!

After detection and filtering, the V and U signals are de-weighted by passing them through amplifiers whose gain is the reciprocal of the weighting factor. Thus the V signal will undergo amplification by 1/0.877 = 1.140 to render R – Y, and the U signal will be amplified by 1/0.493 = 2.028 to render B – Y.

G – Y and RGB matrixing

Now that we have the signals back in R – Y and B – Y form, we can recover the missing G – Y signal. It will be recalled that 100 per cent Y signal is equal to 0.3R + 0.59G + 0.11B which, of course, is equal to 0.3Y + 0.11Y. Now if we subtract the second expression from the first we get 0 = 0.30(R – Y) + 0.59(G – Y) + 0.11(B – Y), which gives –0.59(G – Y) = 0.30(R – Y) + 0.11(B – Y) and from which is obtained – G – Y = 0.30/0.59(R – Y) + 0.11/0.59(B – Y). Inverting both sides of this final equation renders: G – Y = 0.3/0.59 – (R – Y) + 0.11/0.59 – (B – Y). Thus to arrive at a correct G – Y signal we need to add 30 fifty-ninths of an *inverted* R – Y signal to 11 fifty-ninths of an *inverted* B – Y signal. This is shown in Figure 3.17, where R3 and R4 select the correct proportions.

Now we are back to the three separate colour-difference signals plus a Y signal and all is plain sailing! We merely add Y separately to each of the colour-difference signals to render RGB signals ready for the display device.

Display

In our earlier example three separate display devices were used, one for each primary colour. Except in some forms of projection display this is not common, and as is well known a colour picture tube is generally used, in which all three primary colour signals are handled simultaneously. We shall cover colour picture tubes in detail in Chapter 7; suffice it here to say that the shadowmask picture tube manages to simulate the effect of three superimposed tubes, each working in one primary colour. It has a single electron-gun assembly and is set up with accelerating voltage and deflection fields in just the same way as a monochrome tube, but its three cathodes accept the RGB signals with which we started this chapter. Thus we are able to read and write in three colours to capture, transmit and display a full colour picture, with but a single link between programme source and viewer.

There is more to the mechanics of encoding and decoding the colour signal, and indeed other ways of going about it. We shall explore these in the next chapter.

4 *The PAL system*

In the last chapter, the principle of chroma encoding was explained with particular reference to the chroma signal itself, and the method of 'dovetailing' the chroma and Y signals. Reference was made to the necessity of accurately reproducing the original subcarrier reference signal at the receiver so that chroma signal phase detection can be carried out with precise timing to faithfully reproduce the hue and saturation of the colour picture. To recreate the subcarrier at the receiving end, we must transmit some reference signal to regulate the 'clock' which generates *local* subcarrier to time the sampling phases of the V and U detectors.

The colour burst

Since the subcarrier is of constant and unvarying phase, we need not send a continuous reference signal. Provided we can arrange a very stable oscillator at the receiving end (invariably a high-Q crystal oscillator) we need only send a sample for comparison at regular intervals. A place has to be found for this sample in the CVBS waveform, and the only space left which can be utilised is the 5.8 μs back porch between the line sync pulse and the start of picture information. Onto this back porch is inserted ten cycles of subcarrier signal, with a peak-to-peak amplitude equal to the height of the sync pulse itself (300 mV in the standard-level CVBS signal). This *colour-burst* can be seen in Figure 3.16 and is the key, so far as the receiver's decoder is concerned, to the accurate regeneration of the subcarrier signal.

At the decoder, a *burst-gate* is present, triggered from the transmitted line sync pulse to open during the back porch. This isolates the burst signal and directs it into a *phase-lock-loop* (PLL) circuit which compares the frequency and phase of the local crystal oscillator with that of the bursts coming at 64 μs intervals: see Figure 4.1. Any discrepancy gives rise to an error signal from the phase detector; applied to a reactance stage associated with the crystal oscillator it can pull the oscillator into lock with the transmitted colour burst so that its frequency and phase are identical with that at the transmitting end. Figure 4.1, then, gives the bones of a complete 'basic' decoder system as an enlargement of Figure 3.17, and this basic method of encoding and decoding colour in a compatible system is that

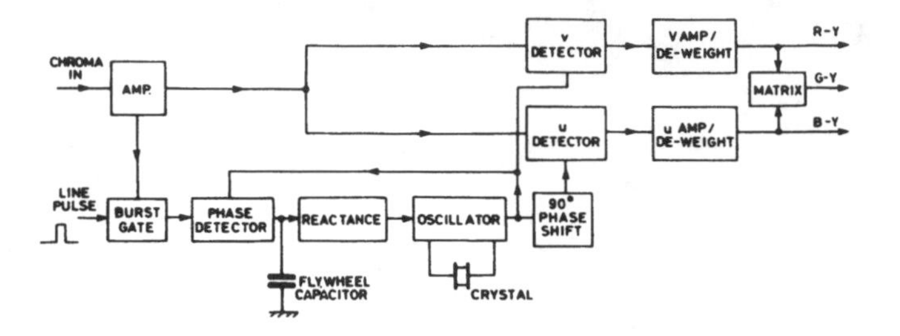

Figure 4.1 *Driving the synchronous demodulators. The two transmitted components of chroma and burst are processed to render locked and correctly-proportioned R, G and B signals*

adopted in the USA, where it is known as the *NTSC* system (short for National Television Systems Committee, who recommended it for America in the early 1950s).

NTSC and PAL

The basic colour system outlined above works well, and has been in successful use in the USA for well over 40 years. Accurate decoding is facilitated by the fact that any distortions or timing errors encountered and picked up by the signal in the transmission system are imposed equally on the burst and chroma signals, so that generally their phase relationship remains constant and all is well with reproduced hues.

If we have any non-linearity in the signal's path, however, a problem can arise. We have seen that the chroma signal proper rides on the luminance signal which can be anywhere between black and peak white levels. The colour-burst, however, is always sitting on the black-level and thus passes through the system at a low level. Any source of timing error which is *level-dependent* will deal differently with the burst and chroma signals, leading to wrong-axis detection and consequential hue errors. This effect, known as *differential phase distortion*, can be quite severe, and to counter it, NTSC receivers are fitted with a hue control with which the phase of the subcarrier regenerator can be adjusted to somewhere near the correct axis, as subjectively judged on flesh-tones. This is somewhat haphazard, and certainly inconvenient, as hue adjustment is often necessary on channel-changing.

To overcome the differential phase problem, a modification of the NTSC system was suggested by Dr Walter Bruch and developed by his team in the Hanover, Germany, laboratories of the Telefunken company. This, the PAL system, is used in the UK and some European countries for terrestrial broadcasting with great success. PAL counteracts any phase error which may be present on one line by introducing an equal and opposite error on the next line. The errors are cancelled electrically by an 'averaging' process in a *delay-line* matrix, to be described later, before being demodulated and used to write colours into the display with great accuracy.

Phase Alternation, Line

The scheme is achieved by the reversal in phase (effective inversion) of the V chroma signal for the duration of alternate scanning lines. What happens is that on one line of a field the phase of the V chroma subcarrier is normal, then on the next line of the same field the phase reverses. These can be regarded as 'normal' and 'inverted' lines. While this is happening the phase of the U signal remains normal. *This is not phase-inverted on alternate lines.*

The diagrams in Figure 4.2 show how this corrects the effect of phase distortion. Diagram (a) shows a 'normal' line and phase distortion on a green element causing the phasor angle to lag 45° from the correct 241°. The full-line phasor represents the correct phase and the broken-line phasor (in all drawings) the error phase as received. The green element on this line, therefore, is displayed towards yellow.

Now, on the next line of the field we have the 'inverted' line which is shown in diagram (b). Notice here that the effect of the phase inversion is to invert the diagram and to reverse the direction of the phasor, so that the error now leads the correct 241° by 45°. The green element on this 'inverted' line, therefore, is displayed towards cyan.

Diagram (c) clearly shows how the average phase of the two errors (196° on the 'normal' line and 286° on the 'inverted' line) works out to the correct phasor for green, which as we have seen is 241°. In Chapter 6 we shall see how this averaging process is achieved by means of a glass delay line and an adder network.

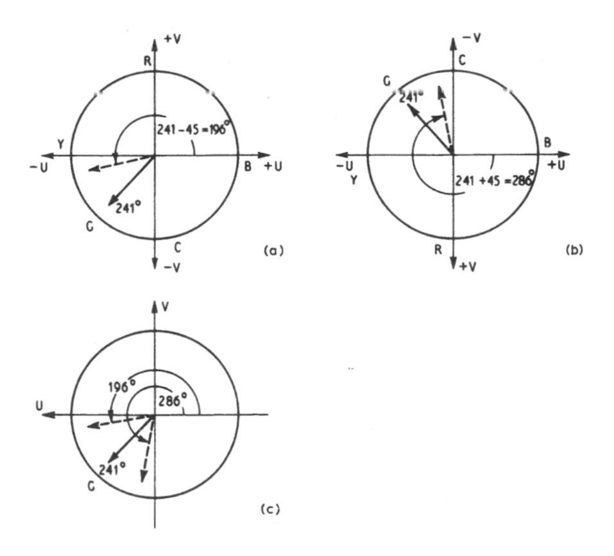

Figure 4.2 *Diagrams revealing in simplified form the phase-error combating artifice of the PAL system. (a) Phasor due to green element occurring at 196° instead of 241° on a 'normal' line. (b) The same phase error on a 'reversed' line. The error is now reversed, and diagram (c) shows that the average of the two errors is 241°, corresponding to the correct phasor angle for green*

V chroma detector switching and synchronising

Perhaps the reader has now become aware that, owing to the V chroma subcarrier phase reversals, the R – Y output from the V chroma detector would also reverse in phase unless some form of compensatory switching were introduced at that particular detector. PAL receivers therefore incorporate an electronic switch to cancel out the PAL characteristic once its usefulness has been realised; it takes the form of an inverter operating on either the V chroma signal or the subcarrier reference to the V detector – either method will achieve the desired result. The switch can be operated by pulses from the set's line timebase which will 'set' and 'reset' the switch on a line-by-line basis. More difficult are the circuit and control required to ensure that the detector switching is synchronised to the 'normal' and 'inverted' lines as transmitted, for clearly if the detector was switched to work from a 'normal' line when the input was an 'inverted' line the displayed colours would be totally wrong! Thus we need to send an identification (*ident* for short) signal to synchronise the receiver's PAL switch.

Swinging burst

This synchronising signal is conveyed by the colour bursts of the PAL signal. It will be recalled that the fundamental purpose of the bursts of the NTSC system is to frequency- and phase-lock the subcarrier regenerator at the receiver (other functions of the bursts include colour-killer switching and automatic chroma control, both dealt with in Chapter 6). The PAL bursts, however, are made to swing in phase 45° either side of the – U chroma axis in synchronism with the phase alternations of the V chroma subcarrier. This is revealed in Figure 4.3, where the burst phase on a 'normal' line is indicated at (a) and that on an 'inverted' line at (b).

Clearly, the *average* phase of the bursts is coincident with the – U chroma axis (180°) and it is this onto which the reference oscillator at the receiver locks, which is the requirement for correct colour reproduction on a PAL receiver.

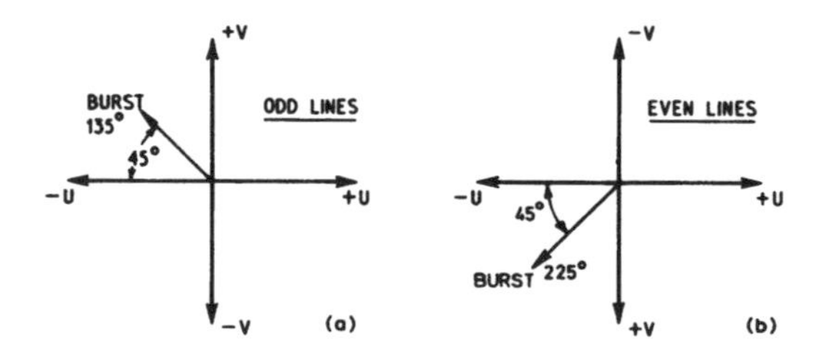

Figure 4.3 *Showing how the bursts swing ±45° relative to the – U axis on successive lines. That is, from 135° on odd lines to 225° on even lines. This means that the average phase of the bursts is 180°, coincident with the phase of the – U chroma axis. Notice how the burst swings are geared to the V chroma phase alternations*

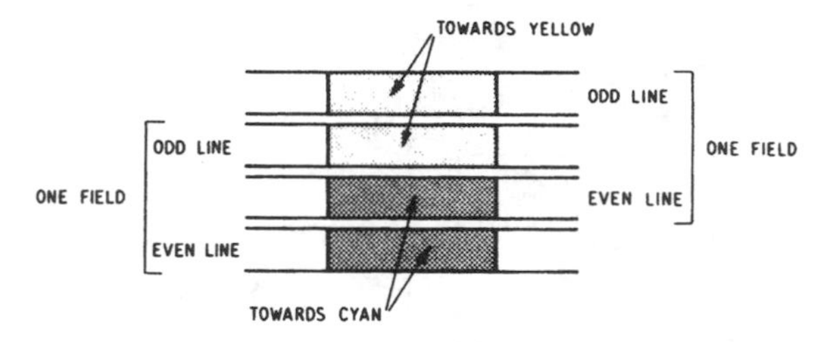

Figure 4.4 *Owing to interlaced scanning, pairs of adjacent lines carry the real and opposing hue errors, and when the error is large this results in the display of Hanover bars*

The NTSC bursts do not swing in phase like this; their phase is fixed to that of the B – Y axis and is constant on all lines.

Figure 4.3 shows the 'normal' lines corresponding to odd-numbered TV lines and the 'inverted' lines to even-numbered TV lines, allowing us to refer the V chroma subcarrier and burst phases to odd and even lines. Thus from *first principles* on, say, line 7 of a field we get the conditions at (a), on line 8 the conditions at (b), on line 9 the conditions at (a) and so on.

Hanover bars

In cases of malfunction in the receiver's decoder the PAL makeup of the chroma signal becomes visible. If alternately-transmitted colour lines are displayed in different hues due to a phase error, the fact that the TV lines are interlaced, as described in Chapter 1, means that two adjacent lines of the picture display carry a hue error, then the next two carry a complementary hue error, and so on, as shown in Figure 4.4. This gives a coarse structure to the Hanover bar pattern, and makes it more noticeable to the eye.

PAL phase identification

Because one PAL switching cycle occupies two lines, the switching frequency is at half line rate, i.e. about 7.8 kHz. This ident frequency is generated in correct phase at the receiver by a circuit which responds to the swinging bursts, and it is this which identifies 'normal' and 'inverted' V chroma signals at the receiver and maintains the PAL switch in correct synchronism. The 'PAL' switch proper is usually a pair of diodes or transistors, internal to an IC, which are alternately switched on by the action of a *bistable*, toggled by line-rate pulses and 'steered' by the ident signal.

PAL encoding system

The block diagram of Figure 4.5 illustrates the basic PAL encoding system. From the camera the RGB signals pass through a gamma correction system into a

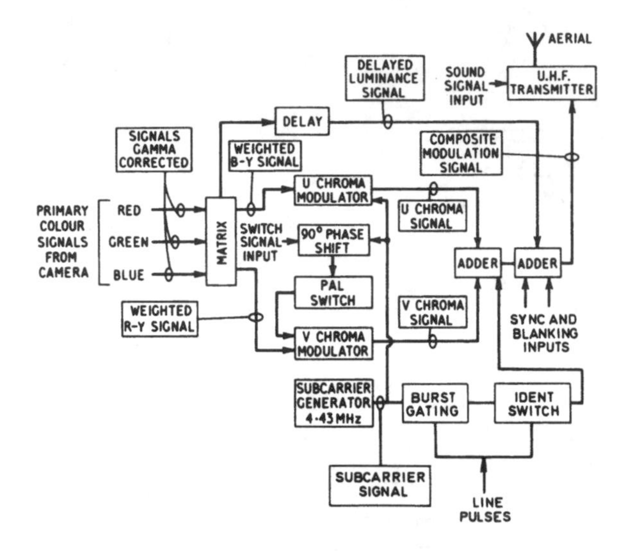

Figure 4.5 *Block diagram of basic PAL transmission system. Note that U and V modulators are balanced types, which means that output from them is only present when both inputs are present*

matrix, wherein the Y signal is derived according to the formula given earlier. This signal passes out of the top of the matrix block in our diagram, then through a delay line on its way to the adder.

This ensures that the luminance signal emerges from the adder in time-coincidence with the chroma signals. Without this delay the luminance signal would arrive early, because the signal is delayed less as the bandwidth of a circuit is increased, and the Y signal occupies a greater bandwidth than the low-definition chroma signals. A similar delay is present in the luminance chain at the receiver for the same reasons, and must not be confused with the chroma delay line!

Returning now to the colour-difference signals, these emerge from the matrix and are weighted before passing into the chroma modulators, each of which is supplied with subcarrier signal; the U modulator direct from the subcarrier generator, and the V modulator via a 90° phase shift (to set up the quadrature condition) and the PAL switch (to impart the alternate line inversions). Modulator outputs are combined in an adder where the burst signal is also gated in, coming from the subcarrier generator via a gate (opened during the appropriate section of the back porch) and a switched phase-shift network to swing the burst phase through a total of 90° line by line, as per Figure 4.3.

We can see now that in essence the PAL variant differs from the NTSC signal in that alternate line phase inversions of the R – Y subcarrier, and synchronised swings of the burst phase, are incorporated as a method of facilitating the cancellation of signal phase errors and hence hue error at the receiver. Let's sum up the PAL system as we have discussed it so far. The luminance signal is transmitted within the parameters of a conventional monochrome transmission. The chrominance signal is carried in the sidebands only of a subcarrier which is suppressed at the encoder; this suppressed subcarrier carries hue information by virtue of its phase and saturation information by virtue of its amplitude. To facilitate demodulation the subcarrier needs to be regenerated at the receiver, and a reference burst is transmitted once per line on the luminance back porch to synchronise the subcarrier regenerator. The V signal is phase-inverted on alternate lines to give a phase-correction-by-averaging feature, and these 'inverted' lines are identified by the swinging characteristic of the burst phase. The whole shooting-match is then sent to the transmission point, where it is reunited with the studio sound signal for transmission to the viewers' homes.

From PAL to MAC

The original NTSC colour-coding system was devised around 1950 and its PAL variant adopted about 15 years later, as was another form of colour encoding, SECAM. The latter was developed in France and is now in use in that country, much of Eastern Europe, and others. All these systems have the primary objective of compatibility, and all achieve it by some form of spectrum-interleaving of the Y and chrominance signals. We have seen how, in the 625/PAL mode, the colour subcarrier frequency is carefully chosen to ensure that luminance and chrominance sidebands interleave.

What problems are introduced by this spectrum-sharing of the Y and C components of the colour TV signal? So far as the viewer is concerned, the main one is the introduction of *cross colour*, in which luminance signals at frequencies near the subcarrier are accepted by the decoder and interpreted as chroma signals. They are demodulated to produce random colour patterns superimposed on the fine-detail areas of the picture, and in subjective viewing are most noticeable on televised objects like an actor's striped shirt or a distant picket-fence, and particularly the finer definition gratings of a test card. Cross-colour manifests itself as random herringbone patterning in red/green and blue/yellow. The effect is graphically demonstrated in the two photographs of grating-patterns in Figure 4.6. On the left is an off-screen photograph of a green test pattern embracing virtually the entire luminance bandwidth. The photograph on the right shows the display after passing this pattern through a conventional PAL *codec* (coder/decoder)

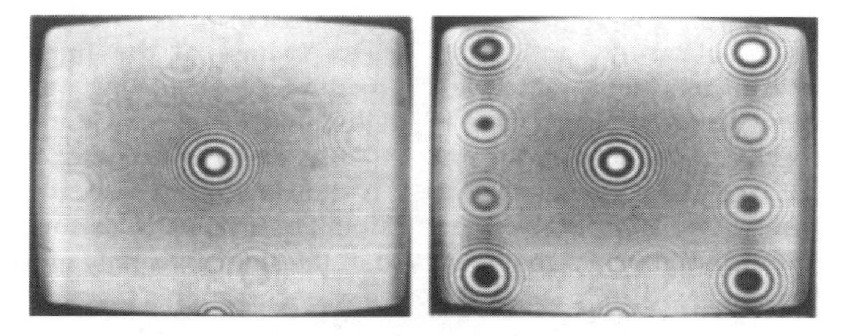

Figure 4.6 *Cross-colour: the interference patterns resulting from band-sharing by luminance and chrominance signals. On the left, the signal in green as generated; on the right the display from a PAL codec system*

system. Interference patterns are very obtrusive and the 'beat' effects, centred on 4.43 MHz, can be clearly seen.

There are several other defects inherent in the spectrum-sharing scheme such as cross-luminance (a luminance dot-pattern adjacent to chrominance changes) and the relatively poor chrominance definition we have already discussed – most noticeable on small captions, where horizontal bars of letters are coloured while vertical bars are colourless. The latter defects are not too objectionable on small screens, but become so on larger displays. The new digital TV encoding and transmission systems discussed in Chapter 12 overcome these problems by totally different techniques, and give images free from these problems.

5 *Transmission and reception*

At present most of our programmes come to us via the very large network of terrestrial UHF colour transmitters which has been built up since 1967 when the first UHF/colour broadcasts started on BBC2. In 1969, UHF colour broadcasting on BBC1 and ITV began, to be joined by Channel Four in 1982. In the early days, there were only a few *main* transmitters, each centrally sited in its service area, and radiating at high power. Secondary and relay stations followed, providing service in 'shadow' and poor reception areas. Today there are over 1000 UHF TV transmitting sites in the UK, each radiating four channels at ERP (effective radiated power) levels ranging from 1000 kW at Crystal Palace, London, to 0.7 W at Glyncorrwg in Wales. The effective coverage of these ground stations is well over 99 per cent of the population, a remarkable achievement when it is considered that UHF waves get little further than line-of-sight!

UHF bands and channels

The terrestrial TV-transmitting spectrum is divided into two sections: Band IV containing channels 21 to 34 from 471.25 MHz to 581.25 MHz, and Band V containing channels 39 to 68, 615.25 MHz to 853.25 MHz. The sound carrier in each channel is spaced 6 MHz above the vision carrier. Each local group of channels works within a given UHF spectrum, usually 88 MHz wide. Some groups have a wider spectrum, and each group is known by a letter and colour, mainly to facilitate aerial selection, since aerials are designed to have a bandwidth appropriate to the group for which they shall be used. Thus colour TV viewers need only one UHF aerial to receive all four programmes from the *co-sited* transmitters.

Transmitter links

The sound and vision signals from the studio centre are sent over a cable or radio link to main transmitters, and to many secondary transmitters; most transmitters are in high and relatively inaccessible places. The cable link may be a co-axial or glass-fibre wideband system, often provided and maintained by British Telecom for the transmitting authorities. Where a radio link is utilised it consists of a microwave system, with dish aerials at each end, very accurately aligned with each other. The receiving dish can often be seen about halfway up the mast at the transmitting site. Studio signals are directed through a control centre where *network* distribution is carried out; this acts as a 'switchboard' for incoming and outgoing transmissions.

Transmitter range

At UHF the distance over which the signals can be transmitted is limited to a little beyond line-of-sight, due to the curvature of the earth's surface. Thus reliable reception can only be assured within 30–40 km of a high-power transmitting aerial even when no obstructions are present. On a high receiving site 60 km or more may be achieved, though weather and propagation variations would probably affect reception.

To 'illuminate' those areas out of range of the main transmitters or shaded by local hills or mountains, *relay* transmitters are used, typically with an ERP of 1 kW or less per channel. Most work on the 'transposer' principle, in which the transmissions from the nearest main station are picked up by a suitable aerial, and each carrier is beat against a stable local oscillator to produce a third frequency which is modulated in the same way as the incoming transmission, and will form (after filtering and amplification) the 'relay output signal' for local use. This avoids the necessity to demodulate and subsequently remodulate the baseband signals, and makes for good economy and a high standard of rebroadcast signal.

Polarisation and ERP

When UHF signals are launched from a transmitting aerial, the geometry of the latter can be designed to impart either a vertical or horizontal polarisation to the electric field created in the air. By suitably polarising transmitting and receiving aerials, a great reduction in the level of interference between transmitters working on the same channel can be achieved, and this property is fully exploited in arranging the 4000 or so (1000 x4) transmitters in the UK to share the 44 UHF channels available with a minimum of interference, even in the face of tropospheric abnormalities which tend to increase the transmission range. A precision *offset* of carrier frequency is also employed to minimise the subjectivity of any co- or adjacent-channel interference which may occur.

We have referred several times to ERP in discussing transmitters, and the effective radiated power is that which is actually *radiated* by the aerial. Because transmitting aerials are designed to have a *directional* field, the suppression of radiation at some angles allows a greater concentration of energy in the wanted direction, so that a more concentrated beam is achieved, with a useful gain in transmitter range. Because all reception takes place at ground level, upward radiation into space is suppressed; many transmitters have horizontal directional properties, too, so that wasteful transmission over the sea or into a mountainside is avoided. By this means the ERP takes account of the directional power gain of the aerial system. This 'gain' characteristic is a feature of receiving aerials, too, as we shall see later.

Vision transmitter

Each station usually has a standby transmitter which comes into operation should the main one fail. In main transmitters the sound and vision signals are dealt with separately until they are in high-power RF form, when they are routed into a combining unit, from which a single co-axial cable takes the UHF signals to the common radiating aerial.

Some large transmitting stations include basic studio equipment such as slide scanners and caption generators for producing test signals or 'breakdown apology' captions. The broadcasters maintain a number of ROCs (Regional Operations Centres) at main transmitter sites, from which all the main and secondary transmitters in a region are monitored and controlled.

The video signals from the studio centres are received, monitored and then 'clamped'. This latter procedure is very important in television. The various components of the composite signal have to be maintained in specific proportions. The black level intervals, for example, have to hold steady at all times, as does the amplitude of the sync pulses and colour bursts. Variations in proportions and levels can occur in the signal after leaving the studio centre, and so it is arranged that the video signal is clamped electrically to a stable and fixed level before it is applied to modulate the carrier wave. Teletext signals in the field blanking interval are extracted, 'cleaned' and reinserted at this stage, then the baseband signal is ready to pass into the modulator.

As we have seen in Chapter 1, amplitude modulation is used for the vision signal as shown in Figure 1.11. The high-power AM UHF signal is launched into space by a stack of dipoles or slot elements at the top of the mast, each radiating element being spaced by half the wavelength ($\lambda/2$). The aerial's directional properties are imparted by carefully adjusting the phase of the carrier signal fed to each tier of elements. To achieve a downward tilting of the beam, for instance, a progressive phase lag is introduced into the carrier as we move downwards from the topmost tier, and in practice this is done by suitable trimming of the feeder cables to each tier. The entire radiating panel is then encased in a closed fibreglass cylinder for protection from the weather – this gives the characteristic 'candle' appearance that we see crowning the transmitting mast.

Sound transmitter

The sound signal is also brought into the transmitting station and monitored before being fed to the preamplifiers which supply the modulation input to the sound transmitter. In the UK TV system, terrestrial transmitters use FM for sound, which has several advantages over AM. The dynamic range of the audio information (e.g. the range between the softest and the loudest sounds) can be more readily retained, and because most of the interference experienced on sound reception (and indeed vision) results from amplitude variations of the modulated carrier wave, this can be eliminated by amplitude *limiting*, that is shaving off the top and bottom of the waveform at the receiver, when the sound is carried by FM. Moreover, high-quality sound reproduction is facilitated by an FM system, and compatible stereo is possible by several different techniques. The UK has settled upon the *Nicam 728* system for stereo TV sound. Here an additional carrier at + 6.552 MHz carries pulse-code modulation signals containing information on 'L' and 'R' sound channels, quite separate from the established FM carrier (see Chapter 11).

Other desirable features stem from the use of FM sound in the system as a whole. For example, it makes possible at the receiver a common IF (intermediate-frequency) channel for the sound and vision signals, owing to the lack of interaction between the two types of modulation. It also facilitates the system of *intercarrier sound*, where a 6 MHz FM sound signal is produced from the beat between the sound and vision carriers, spaced as they are by exactly 6 MHz. Since the accuracy of this 6 MHz signal is ensured by the crystal-controlled references at the transmitter, it follows that the intercarrier sound scheme eliminates the effect of any tuning-frequency drift at the receiver on the sound signal – it would be far otherwise if the sound IF at 33.5 MHz were directly demodulated.

Because of its lesser bandwidth and the FM modulation system there is no need for the sound transmitter to have so great a power as its vision counterpart. In system engineering, the ideal is that in circumstances of signal attentuation, interference etc., all components of the system should have about the same threshold of failure; there is little point in maintaining a perfect sound signal when the picture is completely obliterated by 'snow' or synchronisation is lost in the presence of heavy interference. This was a guiding principle in the design of the colour-encoding system and sync/vision ratio. For sound, it has been found that the 'common threshold' is achieved when the ratio of vision to sound carrier power is about 10:1 (sound carrier 10 dB below vision carrier power), and this is commonly adopted for UHF TV transmitters throughout the UK.

Vestigial sidebands

We have seen in Chapters 1 and 3 that the information transmitted by a modulated carrier wave lies in both the sidebands (see Figure 5.1a). Each sideband carries the same information (they are mirror-images of each other) and so a duplication is present, and more bandwidth than necessary is used up. It is possible to send a complete signal using one sideband only; and this is necessary in the UHF bands to conserve precious spectrum space. However, it is not possible (by filtering) to sever one sideband as cleanly as shown at (b) in Figure 5.1, so suppression usually takes the form of a 'tail' or a gradual roll-off of the unwanted portion. The choice is to have this occurring either in the wanted sideband or in the unwanted one, as shown by (c) and (d) in Figure 5.1. Neither of these alternatives is really satisfactory. In the first case low-frequency energy is lost if the roll-off occurs in the wanted sideband, while in the second case incomplete suppression takes place and the bandwidth of the signal as a whole is still fairly large. In television a compromise is sought, where the lower sideband is partially

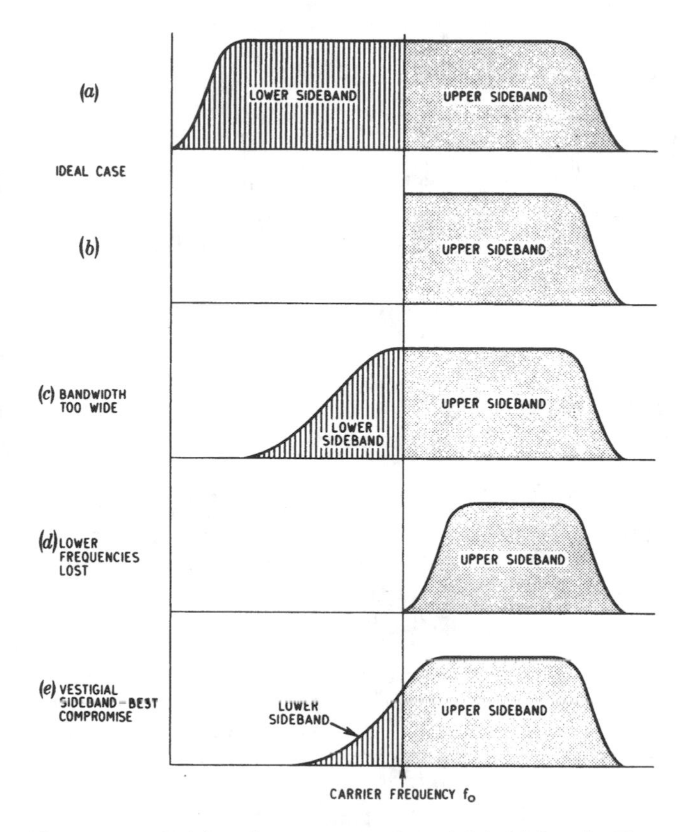

Figure 5.1 *Sideband structure and vestigial sidebands. These are explained in the text*

suppressed, and because there is a vestige of it remaining, the scheme is known as *vestigial sideband transmission.*

The net effect of this type of transmission is that the TV transmission uses the practical minimum of band-space, but when picked up and processed by a receiver suitably designed to accept it the asymmetrical sideband energy is compensated for and equalised to give reception as good as that from a double-sideband system. The spectrum of energy for a single transmission channel is given in Figure 5.2, where the sound and vision carriers can be seen with sideband energy distribution and limits. This diagram also shows the interleaving effect of the luminance and chroma components of the vision signal within the upper sideband of the vision carrier.

It should be borne in mind that this is but a single 8 MHz-wide channel in the broadcast spectrum. One has to imagine the diagram of Figure 5.2 reproduced many times, with each laid side by side, stretching to right and left of our diagram, to get a picture of the UHF band in action. Obviously, receiver bandwidth must be not only carefully shaped in its response curve, but also provision must be made to reject (by means of notch filters) out-of-band signals in adjacent channels.

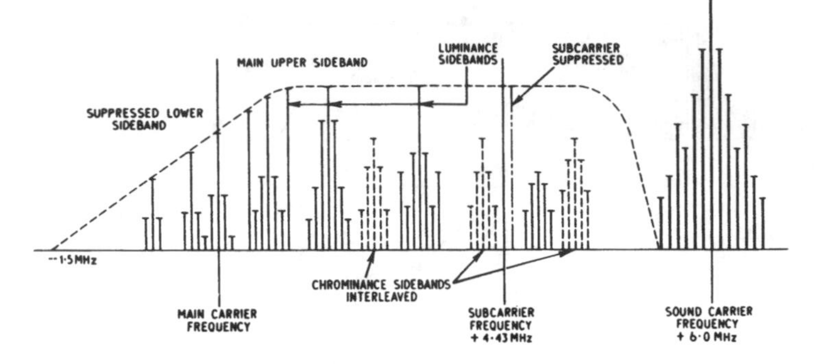

Figure 5.2 *Detailed diagram showing how the sidebands of the chroma signal are interleaved with those of the luminance signal. The diagram also shows the sidebands of the sound signal and the suppressed lower sideband*

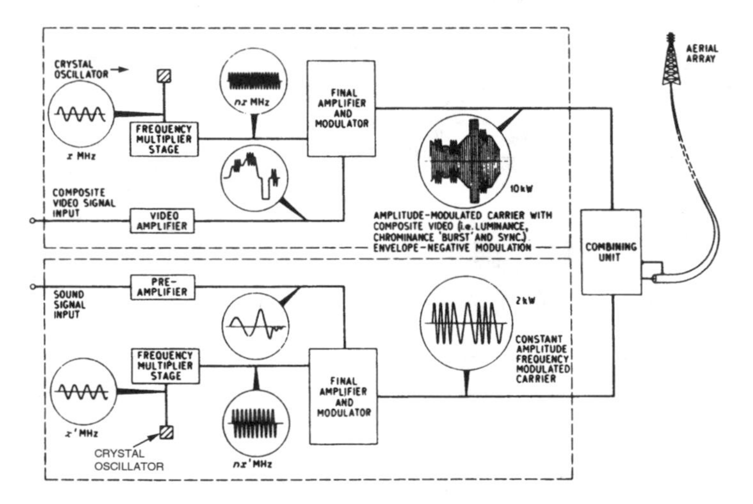

Figure 5.3 *The prime elements of a terrestrial colour television transmitting station*

To round off our description of terrestrial transmitters, the diagram of Figure 5.3 shows in simplified form the basics of a broadcasting set-up.

Terrestrial digital TV transmissions have a more even spread of energy in the 8 MHz-wide transmission slot, and carry a *multiplex* of several programmes. More information on this is given in Chapter 12.

UHF reception

Having launched our carefully composed RF carrier signals into the air, it's now time to examine the means of recapturing them at the receiving site and turning them back into an electrical signal for application to the TV set. Like any radio receiver the TV tuner has the job of selecting one channel from the many presented to it, and rejecting others which are out of channel. The conventional UHF aerial seen on millions of rooftops is the *Yagi* type, named after its Japanese inventor, through other types are sometimes used.

Aerial Bandwidth

Aerial elements are usually referred to in terms of wavelength, and a dipole is said to be 'tuned' when its length corresponds to one-half of the wavelength of the carrier it is to receive. Because the wave velocity reduces slightly when it is captured by the aerial element, the aerial (physical) half-wavelength is about 95 per cent of the signal half-wavelength.

At the high frequencies of the UHF channels, the active element may not be much more than 200 mm. However, to avoid attenuating the sidebands of the wanted colour signal, the aerial must have a sufficient bandwidth to pass not only the signals of one channel, but also those of the other three in the local group. As we have seen, the local group of channels may extend over 88 MHz or more. If the aerial bandwidth is insufficient or if the aerial is of an incorrect channel grouping for the area, chroma sidebands may be seriously attenuated to give severe desaturation; more often one or more channels of the group may have a poor S/N ratio due to uneven response of the aerial. The standard colour-codes for receiving aerial groups are given in Table 5.1.

The Yagi array

The most popular type of TV aerial to date is the Yagi design, developed in the 1920s and much used since then for its good gain and directional properties. It consists basically of a half-wave dipole, one or more reflectors and a large number of directors, as shown in Figure 5.4. When the array is used for reception the signal energy is 'focused' on the dipole (which of course is the active element) by the directors, and any which disperses beyond it is collected and re-radiated back to the dipole by the reflector behind it. The net result is an array with high gain (compared to a single dipole) due to its highly directional property which also confers an ability to reject signals which arrive off-beam, useful in areas like the UK where many transmitters are present, working on the same or adjacent channels, and where buildings and natural topography can cause signal reflections.

Whether an element is to act as a reflector or director depends on its length relative to the dipole. When it is longer it behaves as a reflector and when shorter

Table 5.1. Groupings and colour codes for UHF receiving aerials

Group	A	B	C/D	E	K	W
Channels	21–37	35–53	48–68	35–68	21–48	21–68
Colour code	Red	Yellow	Green	Brown	Grey	Black

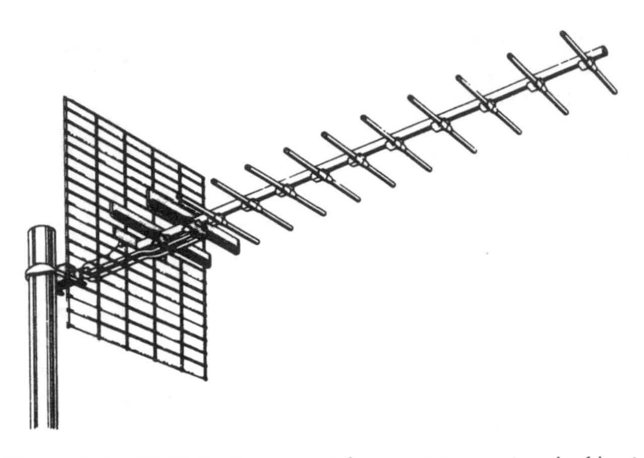

Figure 5.4 *UHF Yagi array set for receiving a signal of horizontal polarisation*

as a director. The element spacing on the boom in terms of wavelength is also of importance in determining the behaviour of an array. The gain tends to increase as more directors are added, but the law of diminishing returns eventually comes into play, and 18 elements is the normal maximum in domestic aerials. If more gain is required, two or more aerials can be used together, linked to each other, and the receiver, by a *phasing harness.* The characteristics of this, and the relative spacing of the aerial arrays, can be engineered to achieve specific requirements of gain and directivity.

The reflector and director elements are termed 'parasitic', meaning that they are not receiving dipoles as such. As more elements are added, so the centre impedance of the dipole falls, and in order to maintain a reasonable match to the feeder the dipole is folded double as shown in Figure 5.5. This artifice steps up its centre impedance by four times and thus compensates for the presence of multiple parasitic elements.

Aerial installation

While set-top and loft aerials may give passable reception in clear zones within the primary service area of a transmitter, the only reliable installation is a roof or chimney-mounted aerial with a low-loss feeder to the TV receiver. The elements must agree with the polarisation of the transmitter to be used, which generally (but not always!) means that they must be horizontal for main transmitters, vertical for relays. In situations of low signal strength, a slight upward tilt of the front 'firing' end of the array may improve results by a dB or two, though a *masthead amplifier* may be necessary to achieve an adequate S/N ratio in the picture in very poor reception areas. This consists of a small wideband transistor amplifier mounted very near the dipole to amplify the signal before further degradation occurs in the feeder cable, which now carries a DC power supply for the masthead amplifier, as well as the UHF signals on their way to the receiver.

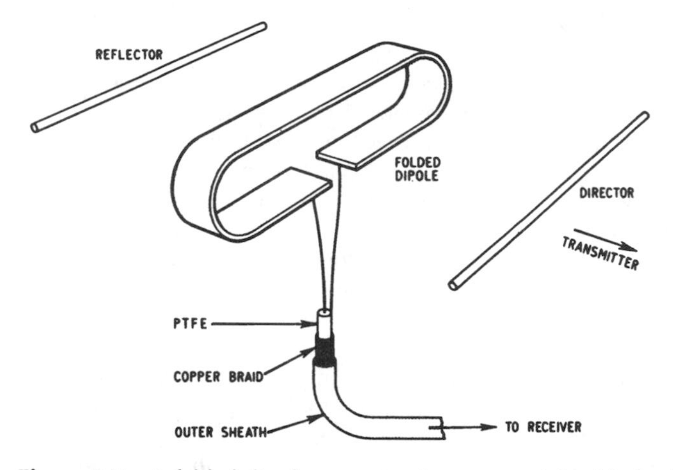

Figure 5.5 *A folded dipole connected to a co-axial cable feeder. Notice the reflector behind and the director in front, relative to the direction of the transmitter*

Normally the aerial needs to be 'looking' directly at the transmitting mast, and this is generally achieved by swinging it while monitoring signal strength on a suitable hand-held meter. Sometimes, due to multipath reception and the resulting 'ghosting' effect on pictures, an orientation must be chosen for minimum spurious pickup rather than maximum signal strength.

Log-periodic aerials

An alternative to the Yagi array is the log-periodic type, which works on a different principle, although it looks superficially similar. It consists of a pair of parallel booms mounted a few centimetres apart with separate dipoles mounted at intervals along its length, their connections being transposed between adjacent elements. At various frequencies in the UHF band different dipoles come into tune due to their differing lengths, while the twin beams on which they are supported act as a transmission line, at the end of which the signal is matched into a co-axial cable. The elements adjacent to the active dipole (whichever it may be at the frequency in use) act as directors and reflector. Thus the log-periodic aerial has some directivity, but not the very narrow beamwidth of a Yagi array of comparable size; its gain is also comparatively low. The main advantage of the log-periodic type is its superior bandwidth and absence of spurious resonance effects. As such, it is useful as a measuring aerial (the broadcasters use them to plot field-strength contour maps), and in areas where a wide spread of grouped channels is encountered. They are also popular for use on caravans and boats where (hopefully!) reflections and signal strength problems will not be encountered, and one aerial can be expected to operate over the whole of UHF bands IV and V.

Down-lead and matching

The characteristic impedance of the dipole of an aerial system is designed to appear as close as possible to a 75R resistor. This matches the impedance of the transmission co-axial cable which conveys the signal to the receiver's aerial socket, ensuring least loss from standing waves due to mismatching. At UHF the attenuation in the down-lead is significant and low-loss feeder is recommended to minimise this, installed to avoid sharp bends or kinks. The co-axial cable itself consists of a flexible outer sheath and an inner conductor supported through the centre of the sheath by low-loss insulating material, called the dielectric. This type of feeder has a characteristic impedance of about 75R. The half-wave dipole in a Yagi array has its ends terminated in a weatherproof box, wherein the feeder is connected, sometimes via a small *balun* or matching transformer. In the absence of this the inner core of the co-ax cable connects to the upper dipole end while the outer sheath is bonded to the lower dipole end.

For further information on aerials, siting and transmitters in the UK consult the BBC Engineering Information department on 08700 100123 for analogue broadcasts; or ONdigital (08702 410392) for digital reception enquiries.

6 *Colour decoding*

Since we discussed the PAL encoding system we have strayed into RF propagation and amongst the chimney pots. Readers who feel it necessary are advised to run over Chapters 3 and 4 again before continuing here with the decoding or restoring process.

Let's first define what we are seeking to achieve. We left the chroma signal, in PAL form, riding on the luminance waveform as a modulation signal for the transmission system. This modulation signal will be rendered in baseband form at the receiver's vision detector, and the job of the decoder is to process this signal to derive R, G and B drive signals to drive the display device.

Taking first the reference (burst) signal, we need to use this to establish within the decoder a reference timing 'clock' against which to measure the timing variations of the chroma signal itself – it will be recalled that the timing of the subcarrier is indicative of the hue being described. We need to provide an inverting switch to restore normality to the V signal on 'PAL' lines; and control it by means of the swinging alternations of the burst signal. In the chroma signal chain is provided the delay line matrix which overcomes the effects of differential phase distortion. During demodulation the subcarrier is discarded, then the baseband V and U signals are de-weighted and matrixed to derive a G – Y signal. Finally, all three colour-difference signals are added to the Y component to give low-level primary-colour signals. Further amplification brings these up to the level required by the picture tube or other display device.

Chroma channel

The chroma channel is contained in the block diagram of Figure 6.1. It consists of the two chroma amplifier stages, with the ACC (automatic colour control) and colour-killer connections, the delay-line driver, the PAL delay line and the PAL matrix. The other sections, though associated with the chroma channel, will be considered later.

Now the composite signal from the vision detector, assuming that the receiver is correctly responding to a colour signal, is passed to chroma amplifier stage 1 through a high-pass filter. Since the chroma information is centred on 4.43 MHz, the high-pass filter accepts all this information while attenuating the

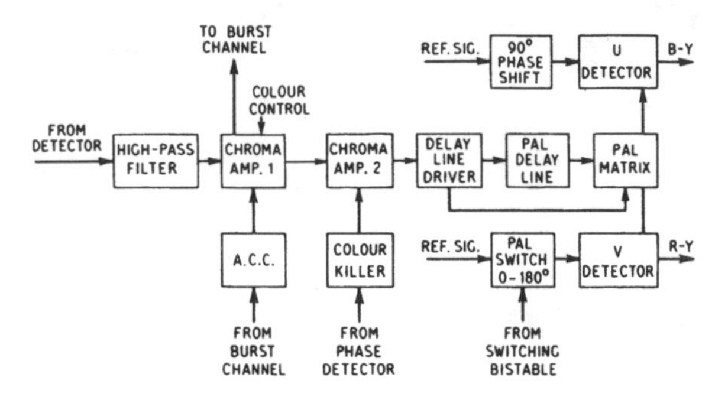

Figure 6.1 *Block diagram of chroma section of colour decoder. Each section is fully explained in the text*

lower-order components constituting the luminance information. Thus in essence only chroma information arrives at the input of the first chroma stage. The signal is then amplified by the second chroma stage and by the delay-line driver, these having a bandpass characteristic of about ±1.2 MHz centred on 4.43 MHz. The skirts of the response show a fairly steep roll-off at either side, determined by the filtering in the chroma signal path to the input. Manual control of the level of the chroma signal is provided by the user colour control, while automatic control (ACC) is also provided at the first chroma stage. The second stage in the block diagram is controlled by the colour killer.

The ACC is generally applied to the stage as a potential which regulates its biasing and hence gain depending on the level of the chrominance signal, as indicated by the amplitude of the colour-bursts (the only part of the chroma signal transmitted at a constant, known level). This reference level is picked up from the burst channel (described later) and then rectified to yield the control potential.

The second chroma stage in Figure 6.1 is deliberately designed to be non-conducting (e.g. biased off) in the presence of monochrome signals. This allows the receiver to work in black-and-white without the colouring circuits being active. If they were open during this time colour interference and noise might mar the monochrome display due to spurious colouring signals getting through the chroma channel. Now, when the colour killer detects a colour transmission by the presence of bursts, it produces a 'turn-on' bias which opens up the chroma channel, thereby letting the legitimate colouring signals through. The colour killer generally receives ripple signal (at half-line frequency; see later) from the burst-phase detector and this, or a part of it, is rectified and smoothed to produce the switch-on bias for the controlled chroma stage. A final function of the second chroma stage is that of *burst blanking*. We do not need the burst signal in the chroma chain beyond this point (it has already been diverted to the reference signal channel) and a line-rate pulse, suitably delayed, is fed to the second chroma stage in time-coincidence with the burst signal to blank out the latter and prevent it from upsetting functions further downstream in the signal path, such as *clamping*.

The delay-line driver thus delivers chroma signal to both the PAL delay line and the PAL matrix, and it is this part of the chroma channel which deletes the 'phase sensitivity' of the chroma signal, and neatly separates the V and U signals for feeding to their respective demodulators.

The PAL function

We have already dealt with the PAL function from the signal point of view in Chapter 4, but since it is particularly important no apology is made for referring again to it here, but this time more specifically from the point of view of the receiver. The relevant circuit section of the PAL delay line and matrix is given in Figure 6.2. Some sets employ an arrangement just like this, while others differ in detail; nevertheless the net result is the same in all cases.

Chroma signal is driven into the delay-line input transducer which converts the electrical signal into an acoustic one, whose rate of propagation through the glass block is relatively slow. Mechanical vibrations travel along a defined path within the glass block, whose edge surfaces are arranged to fold the path by multiple reflections; see Figure 6.2. The delay suffered by the acoustic wave in this long folded path is just under the period of one TV line, after which it arrives at the output transducer and is reconverted into an electrical signal by what amounts to a piezo microphone on the receiving face of the glass block. Chroma signal is also picked up from preset P1 and from this fed to the output of the line. The output transformer T2 thus receives two lots of chroma signal, one delayed in the line, and one direct. It is the job of T2 to add and subtract (i.e. matrix) direct and delayed lines of chroma signal. These processes, it will be recalled, effectively remove the 'phase modulation' from the chroma signal.

It works like this. We get addition and subtraction because the transformer T2 is bifilar wound and centre-tapped. The effect is that the signal from the delay line across winding A is exactly 180° out of phase with the signal also across that winding from preset P1, while the signal from the delay line induced into winding B is exactly in phase with the signal also across that winding from preset P1. The antiphase signals across winding A thus cancel out, assuming equal strengths (one is subtracted from the other), while the in-phase signals across winding B add together.

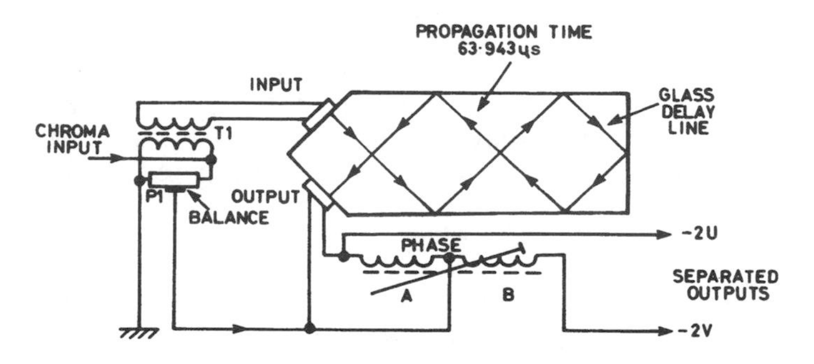

Figure 6.2 *The delay line and its matrix. The folded accoustic path within the delay line is arranged to take 63.943 µs to traverse*

Let us first consider the action on the U components of the direct and delayed signals. The subcarrier upon which the U signal is based is not phase-reversed at any time in transmission, so it has the same base phase on every line. Since the delay is engineered to be 63.943 μs it holds $283\frac{1}{2}$ cycles of reference subcarrier, and the emerging signal is half a cycle out of phase with the direct one – this represents a phase relationship of 180° so that when direct and delayed U signals are *added* they will cancel out thus: – U + U = 0. From the subtracting function, we get: – U – U = – 2U, assuming equal strength direct and delayed signals of unity value. The left-hand side of winding A on T2 in which subtraction occurs is thus endowed with a 'pure' U signal (albeit negative) and no V signal at all.

The V components of the direct and delayed signals are similarly processed, but the net result is different because (see Chapter 4) the transmitted subcarrier upon which the V chroma is based is reversed in phase during alternate lines, which means that the V components themselves are similarly processed. The adding part of the matrix thus 'sees' V signal of, say, 'positive phase' via the delay line and V signal of 'positive phase' via the direct route. If this is not clear, remember that $283\frac{1}{2}$ cycles in the delay line gives a phase inversion between input and output. With the transmitted signal having also undergone a phase inversion since the last line (PAL feature), the two inversions cancel out *on every line* so that the adder (T2 section B) will produce V + V = 2V on every line, while the subtractor, section A, will always produce V – V = 0. So it is that the right-hand side of T1 section B will produce pure V signal with no cross-talk from the U channel.

To summarise, therefore, the U detector receives only U chroma signal because the alternate-line phase reversals of the V chroma signal result in cancellation of the V chroma components in the subtractor section, while the V detector receives only V chroma signal because the in-phase V components reinforce in the adding section while the antiphase U components are added and thus cancelled. In other words, the composite chroma signal is very neatly separated into its two V and U constituents, these now carrying information in terms of amplitude *and not phase*. This means that phase distortion (which is tantamount to 'timing errors') in the system can no longer introduce hue errors on the display. If differential phase errors are fairly large there is a mild by-product in the form of reduced saturation, but this is more subjectively tolerable than a change of hue!

We still need to introduce the reference signals to the V and U chroma detectors (their operation was described in Chapter 3 and illustrated in Figure 3.17), in phase quadrature of course, to recover the quadrature-modulated baseband V and U signals. We must also neutralise the effect of the ±V chroma signal at some stage. In Figure 6.2 P1 adjusts the matrix balance so that the direct and delayed signals have equal amplitude. Imbalance here can give rise to Hanover bar effects as described in Chapter 4.

We can now return to Figure 6.1 again and see that the reference signal to the U detector undergoes a 90° phase shift which provides the quadrature condition just noted. It may be recalled that in our simple illustration of the delay line matrix system we showed the U output as being in *negative* form (– 2U at subtractor output) and to correct this we would need to either invert the U signal before detection or *introduce a further 180°* phase shift in the reference signal to the U demodulator. Either will achieve the desired effect of rendering, at the U detector output, a correct-polarity signal. In the case of the V-detector we have to neutralise the phase inversions of the transmitted signal, and as described for U above, this can be achieved by either inverting the V signal itself, or the subcarrier feed to its detector, but this time on a line-by-line rather than a permanent basis. The switch thus operates effectively at line timebase frequency, but remember that this is the 'toggling' rate; Figure 6.1 shows the switch in the reference signal path.

The PAL switch commonly consists of a pair of diodes or transistors within an IC, switched alternately to opposite phases of the local reference signal, the actual switching being governed by a bistable generator triggered by pulses from the line timebase or sync separator. It thus switches state line by line, each state having a frequency equal to half the TV-line repetition frequency.

Provided the switching count corresponds to the plus and minus phases of the V chroma signal, therefore, the V detector will work as though it is receiving a V chroma signal of constant base phase; but if the switching count is in error the V detector will fail to work correctly and the displayed hues will be in error. To ensure that the switching is correctly synchronised to the transmitted V chroma phase inversions, line-identification pulses are also fed to the bistable from the phase detector responding to the swinging bursts, as we shall see soon.

Burst and reference signal channels

Another section of the PAL decoder is given by the block diagram of Figure 6.3. Here the burst gate and amplifier receives composite chroma signal from the chroma channel. The gate part is a transistor or diode which is switched at line frequency such that it conducts only during the periods of the bursts; during the rest of the television line it is non-conducting, thus performing the opposite function to the burst-blanking switch in the chroma amplifier described earlier. This means, then, that only the bursts of the chroma signal are amplified and that the output from this 'block' consists only of a series of bursts with no picture content.

Note that the bursts are also fed to the ACC circuit in Figure 6.1 which, after rectification and smoothing, provides the control bias for the chroma amplifier, already considered. There are many possible causes of chroma-signal level

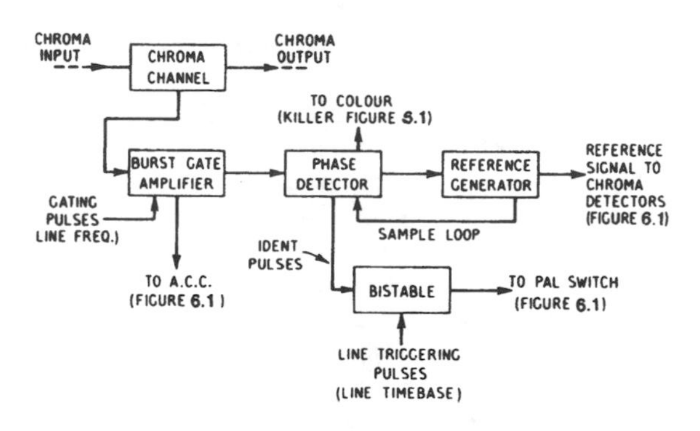

Figure 6.3 *Block diagram of reference and burst channels*

variation, but the ones we are most concerned with are those where the chroma signal is affected to a different degree than the luminance component. The resulting imbalance would upset saturation if it went uncorrected by some form of ACC; the most common causes of this sort of imbalance are RF mistuning and aerial or propagation defects which unduly emphasise one section of the RF channel bandwidth.

Returning to Figure 6.3 and the bursts, however, they are primarily fed to a phase detector which compares their average phase (remembering that their 'swinging' characteristic takes them alternately ±45° relative to the – U axis; see Figure 4.3) with the phase of the locally-generated reference signal. This reference signal for such phase comparison is fed via a 'sample loop.'

The reference generator is crystal-controlled but slight phase variation of the reference signal is made possible by a varicap diode effectively in shunt with the crystal. The varicap diode is widely used in applications where a DC-controlled reactance effect is required. It consists of a p-n semiconductor junction operated in reverse-bias which, under this condition, exhibits a capacitance effect; the effective value decreasing as the reverse bias is increased. This happens because the *depletion layer* between the heavily doped p and n regions acts as a dielectric and its width is a function of applied reverse bias. Thus we can vary the crystal's shunt capacitance, and hence its phasing, by simply applying a DC voltage to the varicap device.

Now, for correct chroma detection it is essential that the phase of the reference signal applied to the detectors corresponds exactly with the phase of the subcarrier which was suppressed at the transmitter. The 90° phase shift required between the two detectors is provided by the phase shifter in Figure 6.1, which has already been mentioned.

The phase detector

It is the job of the phase detector to provide a control potential for the varicap diode such that the phase of the reference signal is always held constant. This is easily possible because the average phase of the bursts corresponds exactly to the average phase of the subcarrier at the transmitter. The phase detector, in fact, produces a DC output of a positive or negative value corresponding to the phase error between the bursts and the locally-generated reference signal. This is called the *error* voltage, and it is superimposed upon the standing reverse bias applied to the varicap diode so that the phase of the reference signal is continuously held at the correct value, with any tendency to drift being immediately corrected. Thus is set up a phase-locked loop (PLL) in which the output signal is locked or *slaved* to the incoming reference, in this case the colour bursts. Modern electronic equipment makes much use of PLLs.

The correctly-phased reference signal is then fed to the V and U chroma detectors via the sections shown in Figure 6.1. Some designs will have a DC amplifier between the phase detector and the varicap diode, while there is always a 'buffer' stage (probably an *emitter-follower*) between the reference generator and the chroma detector feeds.

Ident pulses

We have seen (Chapter 4) that the bursts are swinging in phase line by line such that on one line the phase relative to the – U chroma axis is + 45° and on the next line – 45° and so on. These swings are geared to the phase reversals of the V chroma signal to provide a facility for the identification of the + V and – V lines (odd and even lines, Chapter 4) of the chroma signal, thereby allowing the PAL switch to be correctly synchronised.

So far as the phase detector is concerned the swings of phase are processed as phase modulation, which means that the phase detector's output contains, superimposed on the error voltage, a ripple at half-line frequency (about 7.8 kHz) – *half* line frequency because the phase is the same on every *other* line.

This ripple can be amplified and squared-off to produce ident (short for identification) pulses which are fed to the PAL switch bistable. It is arranged that when the line-frequency triggered bistable is correctly synchronised (and there is only a 50/50 chance at the outset!) the ident pulses will have no effect, but if the bistable is out of agreement with the transmitted V chroma the ident pulse will inhibit its operation for one count, so that it 'misses a beat', as it were. Thereafter the bistable phase will be correct and the ident signal will not be required until a new channel is selected or the set switched off and on again.

Video section

Having arrived at colour-difference signals once more (at the output terminals of the V and U detectors), we are now ready to process them towards the production of RGB signals. The first step is amplification in the R – Y and B – Y preamplifiers whose gains are adjusted to 'de-weight' the signals as described in Chapter 3, and the relative gains are 1.140 for R – Y and 2.028 for B – Y. Next comes the matrixing section for G – Y (see Figure 6.4) and a *clamping* process. This is similar to that described earlier for TV signals on arrival at the transmitter, and ensures that all three signals, R – Y, G – Y and B – Y are sitting at the same reference level so far as their black level is concerned. This is essential for the next stage, which

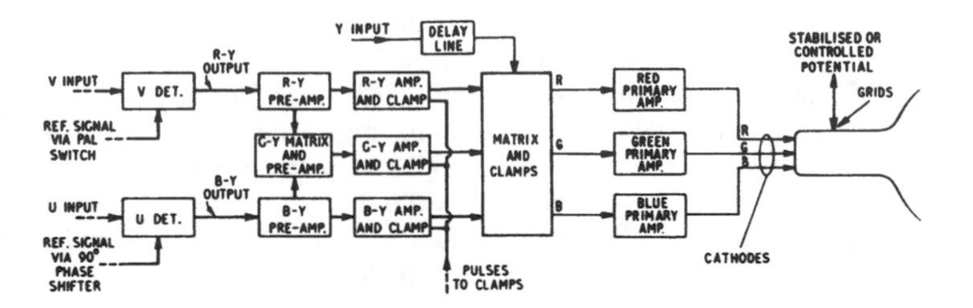

Figure 6.4 *The 'back-end' of the decoder showing the stages of signal processing between the chroma detectors and the picture tube*

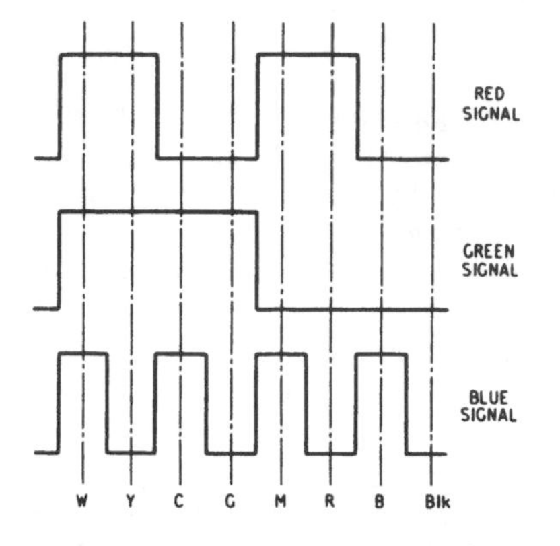

Figure 6.5 *RGB waveforms for a standard colour-bar signal. These result from matrixing Y and individual colour-difference signals*

is the coming-together of the luminance and colour difference signals to form primary-colour signals.

Let us see how this final process is done. Again it is a simple algebraic function in which the Y signal is *added* to each colour-difference signal, having itself undergone a similar clamping process. So (R – Y) + Y = R, (G – Y) + Y = G and (B – Y) + Y = B. When the difference and luminance signals are added, then, we are back to primary-colour signals. For the standard colour bars of our Figure 3.16 the primary-colour waveforms are shown in Figure 6.5. In Chapter 4 we saw (Figure 4.5) that to achieve time-coincidence between narrowband chroma signal and wideband luminance signal at the transmitter a delay line was inserted in the Y channel. At the receiver's decoder a similar delay is required to prevent the chrominance being printed slightly later (and hence to the right, as TV picture scanning is from left to right) than the corresponding luminance picture.

The delay required is of the order of 300–800 ns depending on circuit design, and the wideband delay line can be seen in Figure 6.4. It is made of series L and parallel C components, rather like a π filter; its cut-off frequency, however, is way above the highest luminance frequency in the video signal. Later decoder designs use a solid-state delay line in the form of a 'bucket-brigade' device. At its input the chroma signal is chopped into 'segments' which are then clocked by switches along a chain of capacitors, typically 192 in this application. The technique is similar to the charge-transfer system shown in Figure 1.3.

At the point of emergence from the matrix the RGB signals are at low level, perhaps 4–5 V black to 'white' and we shall wish to drive a colour picture tube with them. To provide the 90 V or so drive required by the picture-tube guns, each primary colour signal is passed through a wideband amplifier, very often a complementary type using two or more transistors in each channel. The characteristics of the three primary-colour output amplifiers must be closely matched in terms of gain, black level and bandwidth to ensure that the primary colours agree or 'track' at all brightness levels and frequencies. This will achieve correct relative and combined colour fidelity at all brightness, hue and saturation levels.

An IC decoder

The entire PAL decoder function is contained within a single IC (integrated circuit) or 'chip' in a TV set. These use a mixture of analogue and digital functions and require a minimum of external components. Typical of such a decoder chip is the device illustrated in Figure 6.6. In broad outline, it follows the block diagram and description already given, but several detail differences and new features will be noticed, and these will be described in turn.

Luminance chain

The delayed luminance input (pin 8) is AC-coupled to permit clamping within the IC. An amplifier and line-rate black-level clamp is first encountered, then, followed by insertion of a reference black level generated within the chip. The insertion period (3L) occurs once per TV *field* for reasons which will become clear later. Luminance signal then passes directly to the R, G and B matrices, and there we will leave it for the moment.

Chroma path

Chrominance signals enter the chip on pin 4, sourced from the same point (vision detector) as the luminance signal described above; in place of the latter's wideband delay line, they will have come via a high-pass filter, and so will consist of the baseband signal spectrum between about 3.3 and 5.5 MHz to embrace the chroma signal bandwidth. The first step is the ACC function, carried out in the controlled chroma amplifier – controlled, that is, by the potential derived from a peak detector and stored in the capacitor on pin 3. Next comes a gated saturation control stage under the influence of a DC control voltage from the user saturation control. The 'gated' function is puzzling here. In fact the gating prevents the saturation control from having any effect on the burst signal, which thus passes, at a constant level, though the chroma chain, delay line and all. The gated chrominance amplifier increases the chroma/burst amplitude ratio by conferring about 12 dB extra gain during the 'picture chrominance' period of 52 μs. An impedance buffer follows, whence the signal leaves the chip on pin 28 en route for the delay line.

Emerging from the delay line, the separated U and V signals now enter the chip on pins 22 and 23 respectively and are passed to the demodulators. The burst is also present, remember, and this is extracted from the delay line outputs for use in the gated burst detector at the bottom of the diagram.

Reference chain

The reference oscillator in this design runs at exactly twice subcarrier rate – 4.433619 × 2 = 8.867238 MHz, and this is the nominal frequency of the crystal connected to pin 26. Let us explore the reason for this double-frequency technique. We know that the two chrominance detectors require switching pulses with the quadrature relationship, and ordinarily this would require a 90°

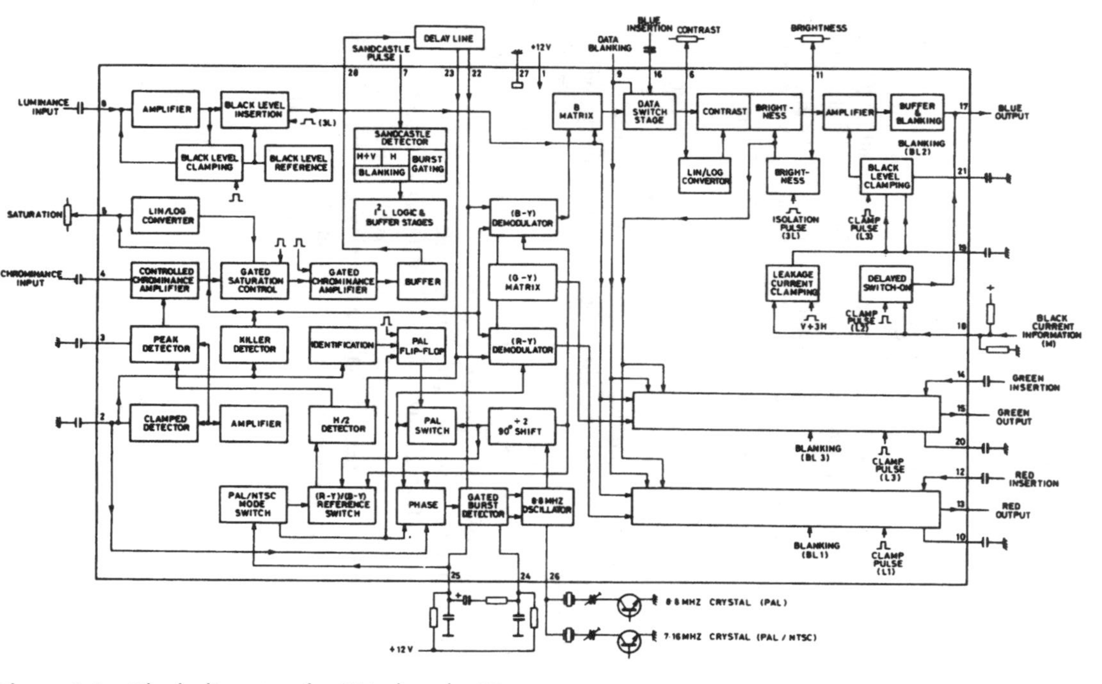

Figure 6.6 *Block diagram of a PAL decoder IC*

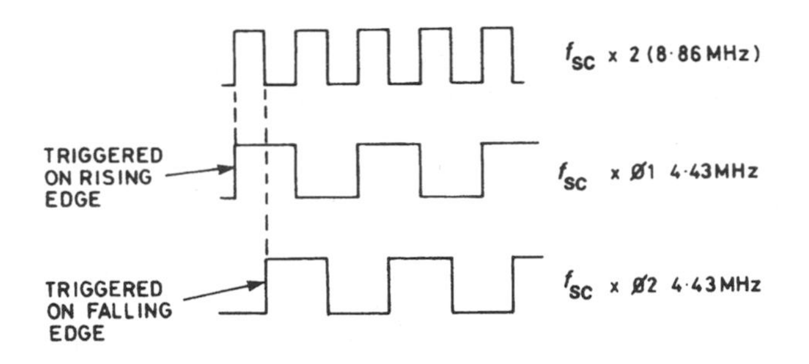

Figure 6.7 *Derivation of quadrature drive signals for the chroma demodulators. A crystal oscillator generates a squarewave at 8.86 MHz (2 f_{sc}) and a pair of ÷2 bistables, respectively triggered from rising and falling edges, produce two f_{sc} drive signals in quadrature*

phase-shift network of some kind in the reference path to the U detector. If we have a reference signal at $2f_{sc}$ we can halve its frequency by use of a bistable counter rather like the ident flip-flop described earlier. If we arrange two such flip-flops, one counting on negative flanks, the other counting on positive flanks, we shall have two 4.43 MHz squarewaves as outputs from them; essentially these outputs will have a 90° (quadrature) phase relationship as required for the synchronous detectors. Figure 6.7 illustrates the technique.

Back now to the bottom of Figure 6.6. The ÷2 90° shift block contains the bistables just described and one output operates the B – Y demodulator. The second output is inverted on a line-by-line basis in the PAL switch on its way to the R – Y demodulator whose baseband output is thus restored to + (R – Y) on both 'PAL' and 'normal' lines. The gated burst detector operates only during the back porch, comparing the R – Y axis reference signal at 4.43 MHz with incoming burst to render an error voltage at pins 24 and 25, between which are connected a time-constant RC circuit to act as 'flywheel', smoothing out the ripple due to the swinging burst. The smoothed error voltage is applied to an internal varicap diode to 'pull' the crystal oscillator into lock as described earlier. Thus is the PLL set up, and the demodulator feeds are now slaved to incoming burst and held rigidly on the correct phase vector axes for correctly-timed detection and sampling.

Ident circuit

The operation of the ident detector is a little different in this chip from the 'basic' practice. Here we have an H/2 detector to compare the subcarrier output of the PAL switch with the R – Y axis component of the burst signal. Let us look a little deeper into this. We have seen that in a properly set-up delay-line matrix the V and U outputs contain only *amplitude* information, with phase variations already taken care of (page 64). Considering then the effect of the burst signal on the V (R – Y) signal, we can see (Figure 4.3) that on a 'normal' line a positive voltage will appear in the V channel during the burst period, then during an inverted or 'PAL' line a negative voltage will be present during the burst period.

When the H/2 detector discovers a discrepancy between R – Y signal phase and R – Y reference phase, the PAL flip-flop gets a 'reset' pulse from the H/2 detector via the ident stage; this corrects the switching sense and ensures agreement between the PAL switch and the transmitter.

ACC and colour killer

Once the PAL switching phase is correct, the output voltage of the H/2 detector is directly related to burst amplitude and so becomes the feedback signal in the ACC loop, finding its way back to the controlled chroma amplifier in the form of a DC control voltage. In the absence of a burst signal killer action is evoked with the chip, taking place at *two* points in the signal path to ensure that spurious chroma suppression is adequate. The saturation control input is taken low and the synchronous detectors are simultaneously cut off.

Control pulses

For economy of lead-out pins, all control pulses enter the chip on its pin 7, and three basic pulse streams are required. A field pulse is present at 20 ms intervals along with a 12 μs line blanking pulse at 64 μs intervals. During the latter, and coinciding with the back porch, a further pulse, about 4 μs wide, is added at a higher level (hence the *sandcastle* shape of the composite pulse). These are separated within the chip by 'level slicers,' then timed and processed for many different functions as the chip diagram makes clear.

NTSC capability

In view of the international market intended for such devices, the chip shown in Figure 6.6 has a dual-standard feature in that it can also work with NTSC signals. By a clever arrangement of mode-switching and 'dual-personality' of some lead-out pins, the PAL switch can be disabled and a tint control brought into operation by DC mode switching at pins 24 and 25 of the chip. If the NTSC subcarrier frequency differs from 4.43 MHz (as it does in the USA, but not in certain 'hybrid' signal formats used in the UK in connection with special VTR playback modes etc.) the transistor switches at the bottom of the diagram can be changed over to bring in a suitable reference crystal – the one shown is 7.16 MHz, twice the 3.58 MHz USA standard subcarrier frequency.

In combination with another chip, the device can also handle SECAM signals. Many colour decoder ICs can deal with PAL, SECAM and NTSC signals.

RGB video stages

It is in the post-demodulator stages that modern IC designs differ most from the traditional model of signal-processing practice. In Figure 6.6 only the blue channel is shown in its entirety – the R and G stages are identical. We left the Y signal at the RGB matrices, and it's time now to take up the story from there. The output from the matrix is of course in primary-colour form and it first encounters the *data switch*. For each primary-colour channel this forms a double-pole switch able to select either broadcast or locally-generated signals under the influence of the data blanking input on pin 9. When pin 9 is raised to logic level 1 the switch operates and locally-generated video data (digital signals, typically from a teletext decoder or a computer) is passed from the 'insertion' pins (12, 14 and 16 for RGB respectively) to the primary-colour outputs. The data blanking input can operate at video rate if necessary, enabling superimposition of text or graphics on analogue-derived pictures; a good example is the 'mix' mode in text-equipped receivers. As with luminance and chrominance signal inputs to this IC, the data is AC-coupled in, then clamped within the chip.

Next comes the contrast control stage, and here again we have a DC control input feeding a voltage-controlled amplifier, facilitating remote cordless command of the set's functions. Because the voltage-controlled attenuator within the chip has an essentially anti-log control input/signal output relationship, and in view of the fact that linear control functions are easier to arrange from panel controls or remote-command decoders, the chip contains at pin 6 a lin/log converter – a similar device is present at pin 5, saturation control input. Because contrast is independently but simultaneously controlled in all three primary colour channels, matching between them must be very good; it can be seen that the same applies to the next (brightness control) stage, and that inserted RGB signals in the data blanking mode also come under the influence of contrast and brightness controls.

The brightness control works on a clamping basis, restoring the three black levels to a point set by the DC control voltage on pin 11. This is carried out during the '3L' period of black level insertion at the luminance amplifier, and described earlier.

Automatic grey-scale tracking

As we shall see in the next chapter, a colour picture tube contains three separate electron guns, one for each primary colour. For correct picture reproduction in black-and-white or colour it is essential that these three guns have exactly the same emission characteristics. If the cut-off points (beam-extinguish voltage) vary between them, for instance, the same signal applied to all three will give different results on each, so that we may find red-tinted low lights because the red gun has a higher (cathode) cut-off voltage than its companions. Alternatively, perhaps, the green gun will cut off at a lower cathode voltage than red or blue, leaving a magenta tint in place of dark grey in the display. These differences, initially due to manufacturing tolerances, can be easily compensated for by pre-set adjustment of black-level voltage for each individual gun. As the tube ages, however, differential drift of the gun characteristics takes place; this necessitates occasional resetting of the internal black-level controls during the life of the tube to maintain tint-free reproduction.

The IC illustrated in Figure 6.6 can automatically adjust the black-level point for each gun in the picture tube. We shall now examine the operation of this automatic grey-scale correction system.

During TV lines 22, 23 and 24 and their companions 334, 335 and 336 on the twin-interlaced field, the scanning spot is at the top of the screen and out of sight,

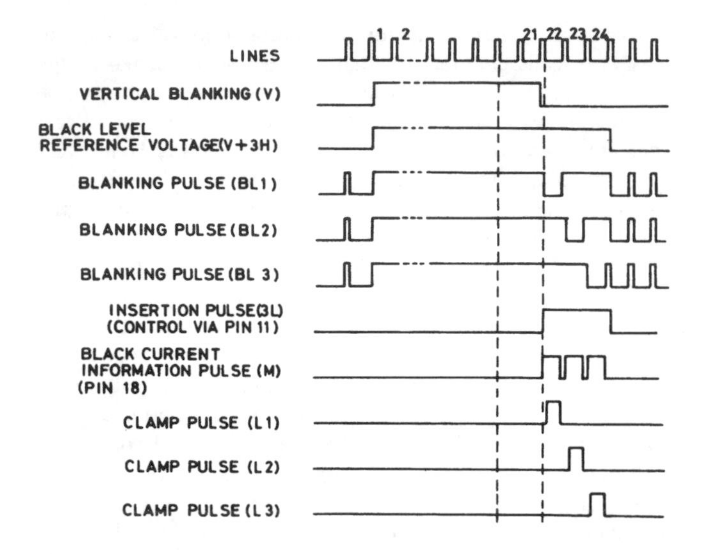

Figure 6.8 *Time-related auto-grey-scale-shift and blanking waveforms for the auto-grey-scale correction. This diagram is centred on the field blanking interval*

being just about to 'light up' to start describing the first active TV lines of the field. It is at this time that the auto-grey-scale compensation takes place. For the duration of these three lines the 3L insertion-pulse blanks all video information in the RGB channels; see the time-related waveforms of Figure 6.8. A 1-line duration clamp pulse is now inserted into each of the colour channels in turn; red, blue then green corresponding to L1, L2 and L3 in Figures 6.8 and 6.6.

This clamp pulse drives each gun of the picture tube just to the cut-off point and the beam current in the tube (nominally zero at black level, in practice a few microamps for measurement purposes) is measured and passed into the chip on its pin 18. Thus pin 18 (row 8 in Figure 6.8) receives three pulses in quick succession, sequentially giving information on the cut-off points of the R, B and G guns of the tube in turn. These reference levels are gated to the three storage capacitors connected to pins 10, 20 and 21 respectively, and the charge stored in each of them is proportional to the cut-off point of the corresponding tube gun, which will vary with temperature and aging of the picture tube.

The charge on the storage capacitor is made to modify the amplitude of that gun's clamp pulse so that a separate feedback loop is maintained for each gun; with all three guns thus controlled, automatic grey-scale correction is achieved, and the need for a black-level adjustment preset in each channel is eliminated.

Further integration

As TV design has progressed, many more functions have been integrated into the video processor IC, including the IF, audio and signal-demodulator sections and timebase generators. Even so the system of Figure 6.6 has been retained here because it gives a better insight into the techniques used than the diagrams associated with more highly integrated chips.

RGB output amplifiers

If the display device is a picture tube, either a shadow-mask device or three separate display tubes in a projection set-up, we need three matched-voltage amplifiers to raise the RGB signals to a level sufficient to drive the modulating electrodes, usually the cathodes. The bandwidth of these needs (if data and other video sources are to be handled) to be in excess of the 5 MHz or so of normal Y broadcast signals because rise and fall time (the time taken for a video signal to raise itself from black level to white, and vice-versa) is inversely proportional to bandwidth, as we saw in connection with Y/C registration and Y delay lines. Where data is being displayed the sharpness of the vertical edges of the characters depends on RGB bandwidth, the greater the better! As previously mentioned, the three amplifiers need to be closely matched in all respects as they separately handle components which are reblended in the display to form a composite and very critical image.

A typical video drive amplifier is shown in Figure 6.9. Here the video amplifier is Tr1 working in common-emitter mode with a gain of about 25. The video input

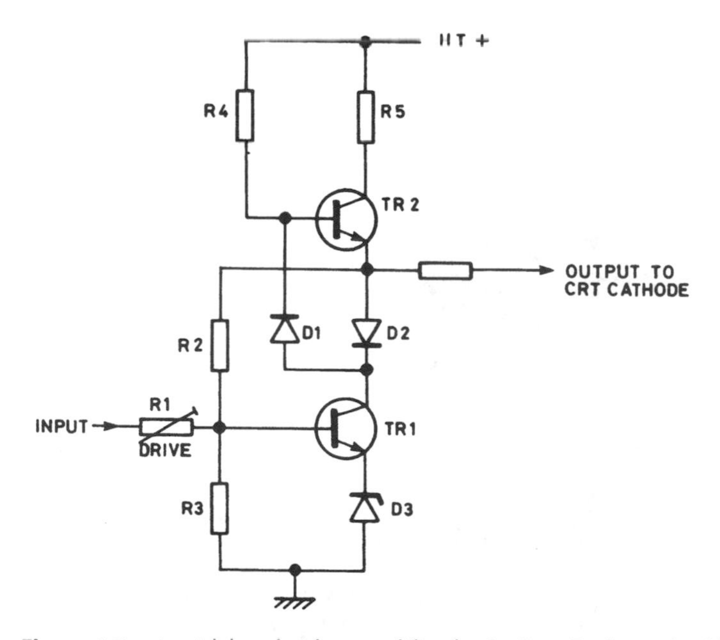

Figure 6.9 *A wideband video amplifier for R, G or B channels. Circuit operation is explained in the text*

signals to this stage are positive-going for 'white' and the polarity inversion in the stage renders negative-white output signals suitable for cathode drive of the picture tube. The difficulty in arranging the output amplifier is that it has to drive a wideband signal into the considerable capacitance of the tube cathode and its associated wiring; this calls for low output impedance. On transitions from black to white Tr1 is being driven on and its low impedance quickly discharges the stray capacitance to give a fast fall-time to the output waveform and a sharp leading vertical edge to the televised object, be it a data character or picture highlight.

When the transistor turns off, however (trailing vertical edge of picture object), the stray capacitance can only charge relatively slowly via the amplifier's load resistor R4 and because this RC time constant is long a trailing smear effect will be manifest on the picture. This is where Tr2 comes into operation, acting as a low-impedance emitter-follower with its base bias coming from R4. The shunt capacitance quickly charges via Tr2 and low-value R5, then, and crisp pictures are ensured by the low output impedance, and hence great bandwidth, of this video output stage. Two further identical amplifiers are needed in a practical TV, of course, and they share the emitter-potential reference diode D3 and work from a common HT supply of about 200 V.

One other aspect of the RGB stages must be mentioned, and this is the necessity for drive adjustment. In Figure 6.9 R1 is made variable and acts as a 'contrast' control for the primary colour in question. The presence of similar control in the two other channels enables us to adjust the *drive* for reasons which will become plain in the next chapter. Where no auto-grey-scale correction is provided by the decoder chip, a further preset is required, this time to set the black-level for each gun. It effectively alters the standing voltage at the picture tube cathode to facilitate manual matching of the cut-off points of the three guns.

Figure 6.10 *Picture-tube base panels usually incorporate the final RGB amplifiers to reduce stray capacitance in the cathode wiring. Here a high-voltage IC (top left-hand corner) is used for RGB signal drive. The signals enter the board at the socket top right*

Decoder summary

In this chapter we have seen the 'chroma unscrambling' procedure in some detail, both from a theoretical point of view and in terms of practical hardware. A photograph of an RGB amplifier mounted on the picture-tube base panel is shown in Figure 6.10.

7 TV display systems

A brief account of the operation of a monochrome picture tube was given in Chapter 1, and here we shall enlarge on that description to take in colour tubes and finally other types of colour display. A great deal of research and development has gone into the technology of colour picture tubes over the years, and direct-view types are now available with screen diagonals up to 89 cm. Improvements have been made in screen materials, gun technology and energy demand. The colour tube uses all the techniques of its simpler counterpart, and by extension and refinement of these is able to present a full colour picture on a single 'integrated' screen – no mean achievement! These direct-viewing colour tubes work on the shadowmask principle originally brought to fruition by Dr A. N. Goldsmith and his research team in the American laboratories of RCA Ltd in 1950. The original device was rather clumsy and cumbersome by today's standards, but it embodied all the fundamental features of the current generation of colour tubes, and was amenable to mass production with all the cost advantages that could bring. The practical realisation of a relatively compact and decidedly cheap domestic colour TV receiver was (and is) dependent on the direct-viewing shadowmask tube concept.

The three-gun picture tube

Colour television picture tubes have to carry three different screen phosphors in order to produce three displays in perfect registration, one each in red, green and blue. If the face of such a tube (when operating) is examined under a magnifying glass the phosphor-dot formation might be seen, as shown in Figure 7.1. The phosphor make-up of the dots, of course, looks the same for each group of colour, and it is only when the phosphors are stimulated by the electron beams that the different colours are produced.

There thus exists a multiplicity of 'groups' of the three colour phosphors, and each group, detailed in Figure 7.1, is called a triad (*group of three*). In a typical tube there are almost 500 000 triads, together forming an interwoven pattern over the entire screen area.

The three guns are arranged in the tube neck so that the electron beam from one energises only the red phosphors, that from the second only the green phosphors

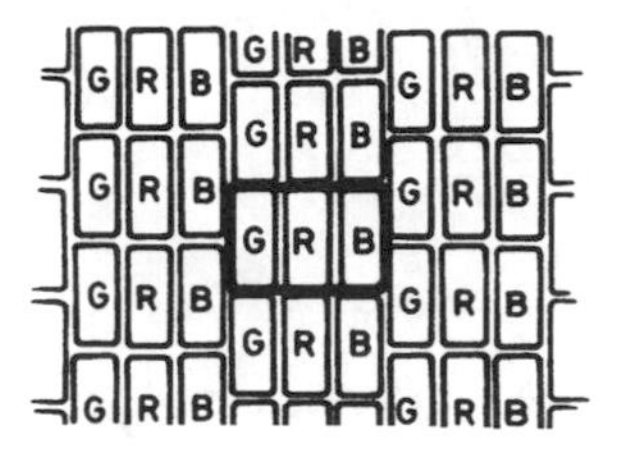

Figure 7.1 *The formation of the red-, green- and blue-glowing phosphor dots on the screen of the in-line shadowmask picture tube. The three-dot formation is called a triad*

and that from the third only the blue phosphors. For this reason the guns and beams are often referred to as red, green and blue. They are not coloured, of course! When a group of triads is caused to glow at proportioned colour intensities, the eye is deceived into perceiving that the three primary colours occur at the same point and so it discerns a spot or pixel of a colour corresponding to the mix of the three lights. When the mix is correctly proportioned for white light, the spot or pixel appears without colour as a 'shade' between white and black – i.e. a shade of grey.

Remembering our description in Chapter 2 of additive mixing by lights, it is easy to see how the spot or pixel can have any of the three primary colours red, green or blue; a complementary colour, yellow, cyan or magenta; or indeed any colour of intermediate value that can be produced by an admixture of the three primaries available, and falling within the 'television triangle' of the chromaticity diagram, Figure 2.3. This embraces a wide range of colours which make the colour TV display subjectively equal to a good-quality colour photographic print.

The diagrams of Figure 7.2 give a basic impression of the working of the shadowmask three-gun picture tube. It is, in fact, three 'tubes' in one glass envelope and sharing the same screen. The electron-gun assembly has three cathodes which work quite independently to produce the three separate electron beams, each of a different strength according to the primary-colour signals by which they are being controlled.

Shadowmask

In addition to the components of the tube already mentioned, there is what is known as a shadowmask, from which the tube gets its name. In appearance this is something like a metal gauze pierced by a large number of holes. In fact there is one hole, or slot, for each triad and the purpose of the mask is to ensure that the phosphor dots of the different colours are hit only by the electron beams from the guns of corresponding colours. This will be made clearer shortly.

The shadowmask is located about 12 mm behind the phosphor screen, and the three beams are made to converge on a single hole and cross over within it to

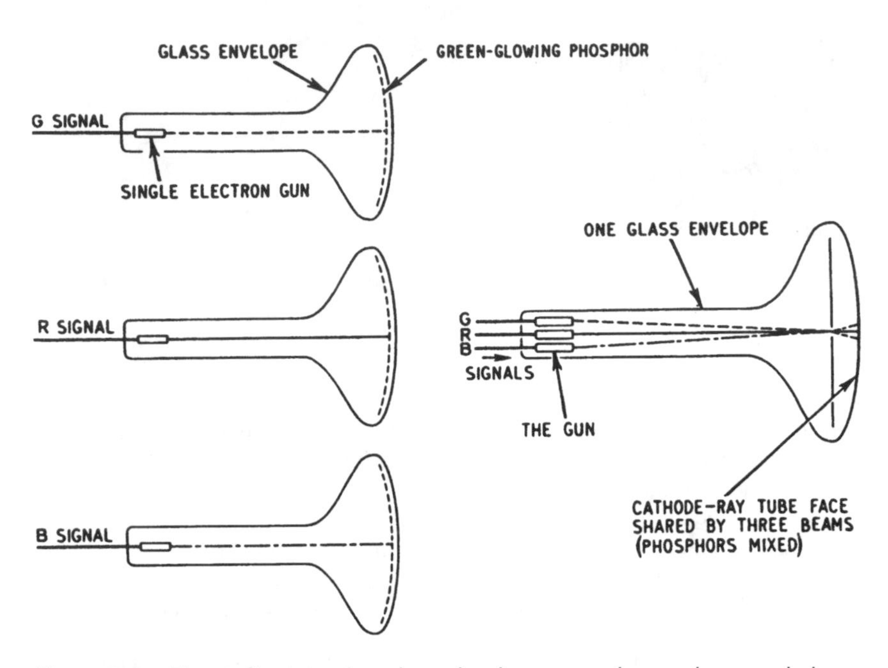

Figure 7.2 *These diagrams show how the three-gun tube can be regarded as 'three-tubes-in-one'*

diverge slightly beyond so that the red, green and blue beams strike their corresponding fluorescent colour phosphor dots.

Each dot is then activated by the electroluminescent process described in Chapter 1 to yield an amount of coloured light proportional to the intensity of the electron beam striking it. As already mentioned, the closeness of the dots and the integrating action of the human eye merges the three individual colours into a single spot whose colour and intensity are closely controlled by the video signal. A broad impression of the total TV illusion is given in Figure 7.3 which illustrates the simultaneous analysis, passage, reassembly and perception of a single pixel, of which many millions are sent every second!

The shadowmask is made of a thin sheet of metal identical in size, shape and contour to the glass faceplate of the tube. It is manufactured by a photochemical etching process similar to that used for the production of lithographic half-tone plates for printing. The positional and etch-timing accuracy has to be several orders higher for the shadowmask, however!

Arranging the phosphor dots

To produce the phosphor pattern on the inner surface of the glass faceplate a very ingenious method is used. The green glowing phosphor material is first sprayed onto the screen so that it covers the whole area, and then the shadowmask is

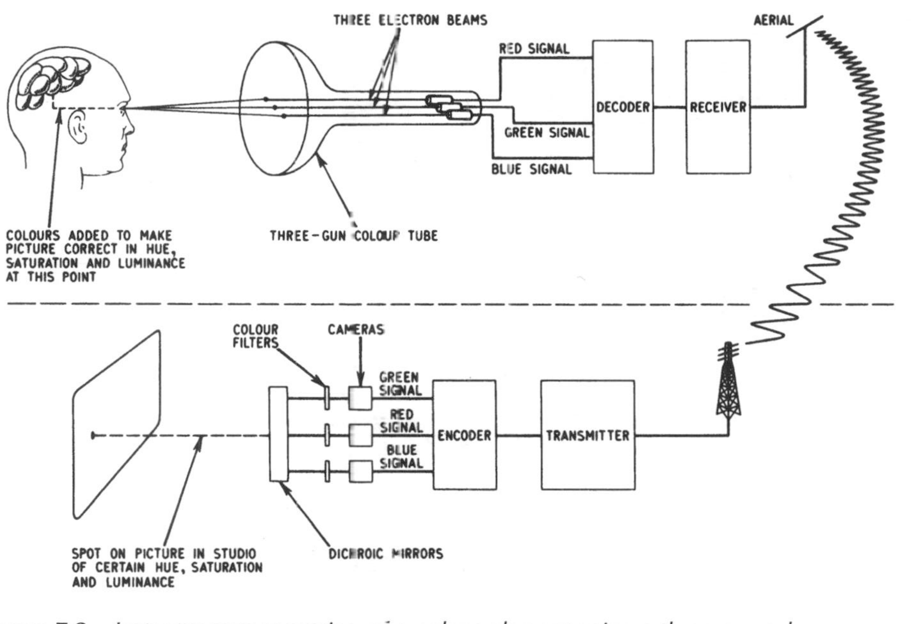

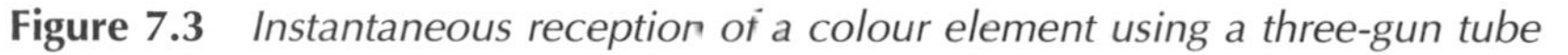

Figure 7.3 *Instantaneous reception of a colour element using a three-gun tube*

mounted accurately in the faceplate. To 'fix' the green phosphor material on the screen in the dot-pattern required, ultra-violet light is arranged to pass through the shadowmask holes at an angle corresponding to that of the electron beam from the green gun when the tube is correctly active. The effective origin of the green beam, so far as the shadowmask is concerned, is the *deflection centre*, a point near the physical centre of the scan coil's field. It is here that the ultra-violet light source is placed during the phosphor-fixing stage in manufacture.

When the green phosphor has been selectively hardened in the required dot pattern, the shadowmask is removed and the screen is washed to remove the unexposed phosphor material which was in the shadow of the mask. We now have a uniform pattern of green-glowing phosphor dot 'islands' in the 'sea' of the glass inner-screen surface.

This process is repeated separately for the red- and blue-glowing phosphor materials with the ultra-violet light in these cases positioned in the deflection-centre point corresponding to the colour being fixed. At the end of the operation, then, we have a regular pattern or mosaic of dot triads aligned with the shadowmask and with the incident angle of the electron beams from the right guns. The pattern is shown in Figure 7.1.

There are several refinements carried out on the basic dot-mosaic screen for the sake of good contrast ratio and high efficiency. We saw in Chapter 2 that the darkest part of the picture can only be as dark as the unenergised screen itself, and the phosphor materials used are pigmented and dark-coloured to absorb ambient light falling on the screen and prevent reflectance. Additionally the small space between the phosphor dots is filled with a similar light-absorbing black pigment; these artifices increase the contrast range. Like a monochrome tube the shadowmask type is aluminised and this highly reflective screen backing increases brightness by reflecting phosphor-light forward instead of wasting it within the internal bowl cavity. The aluminium layer also acts as final anode and forms a protective ion-barrier for the screen.

Tube construction

Before we go on to examine the operation of the shadowmask tube let us look at some of its physical characteristics. The heater's power consumption is low, and a quick (4 second) warm-up is achieved by concentrating the heating element just behind the front surface of the cathode. The neck diameter varies with tube type, but 29 mm is common in PIL tubes where a single 'gun' is used (still producing three beams of course!).

Deflection angles are generally 90° or 110°. Very small colour tubes used in portable colour TVs have deflection angles down to around 50°, while at the other end of the scale some have a deflection angle of 114°, which represents about the limit of current technology. The conductive inner and outer coatings of the 'bowl' section of the tube are resistive rather than highly conductive in many colour tube designs; these form an EHT reservoir capacitor with the glass wall of the tube as dielectric. The resistive coatings absorb suddenly-released energy in the event of internal flashover within the tube, and harmlessly dissipate it over a large surface area. Coming now to the shadowmask itself, we have seen that correct alignment of the beams on their own phosphor dots is essential for correct operation and good purity of colour. The correct alignment of the shadowmask is crucial to this. Because only about 25 per cent of the mask area is 'holes' and the rest is 'wall', most of the energy in the electron beams is dissipated within the mask rather than at the phosphors, and the total current flowing in the mask can exceed 1 mA. With an EHT voltage in the region of 25 kV the dissipation in the shadowmask can thus approach 30 W, and if this is sustained for long appreciable heating of the mask will take place. The resulting expansion must be taken up in a defined and controlled way if buckling and distortion are to be avoided – these would upset beam landing and purity. The mask, then, is either fitted under tension (Trinitron vertical aperture-grille) or in a special frame incorporating bi-metal support clips to permit an axial movement of the mask to take up expansion while maintaining hole-triad registration. As we shall see, Trinitron tubes use a mask consisting of continuous vertical slots, and this is more vulnerable to heating, especially local heating at a point of high brightness in the display. To give lateral support to the vertical grille members in large-screen Trinitron tubes, three very fine horizontal platinum tie-bars are micro-welded across the grille at intervals. In many large picture tubes the shadowmask is made not of steel but of *invar*, an expensive metal with a low coefficient of expansion. This reduces the problem of 'hot-doming' at high levels of brightness and contrast.

Regarding the screen itself, the size quoted used to be the diagonal measurement from corner to corner of the *glass faceplate,* larger than the picture. Today, screen sizes are quoted in V (visible) sizes of the picture diagonal. This may vary from 10 cm to 89 cm, though most domestic sets are fitted with tubes whose diagonal ranges from 50 to 68 cm. The faceplate has two layers, the outer of which acts as an implosion guard in the event of sudden collapse of the glass envelope. These two front layers together are typically 2 cm thick in a 68 cm tube, and this largely accounts for the weight of a colour receiver! The atmospheric pressure on the envelope of an evacuated tube is very great.

Gun and beam path

Having set the stage, as it were, we shall now look at the basic operation of the shadowmask tube. The single electron gun of the tube is illustrated in Figure 7.4, where it can be seen that a common accelerating anode system and electron lens is shared by all three beams which travel through the gun assembly in very close proximity (5 mm) to each other. The three heated cathodes are in-line abreast at the left of the diagram, and their electron beams emerge through three small holes in the common control grid. They next encounter the first anode (A1) which accelerates them towards A2; the latter, in conjunction with the potentials on nearby A1 and A3 forms an electron lens for beam focusing. On emergence from the gun assembly the beams are on converging paths, as shown on the right of Figure 7.4.

In fact the beams' trajectory is such that they cross over in the shadowmask hole and diverge just beyond to strike their appointed phosphor dots – an idea of this is

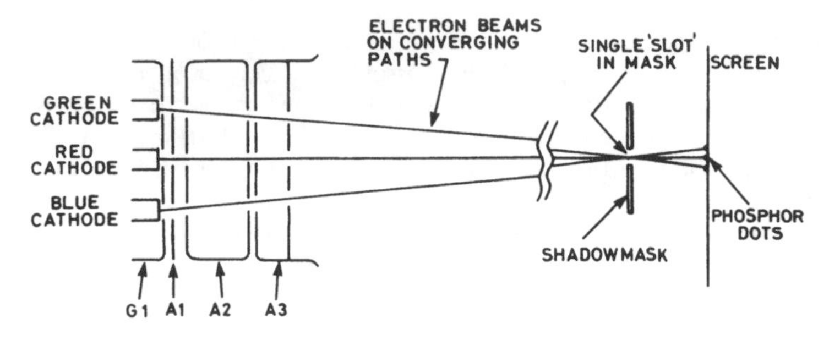

Figure 7.4 *The internal arrangements of a shadowmask tube, viewed from above*

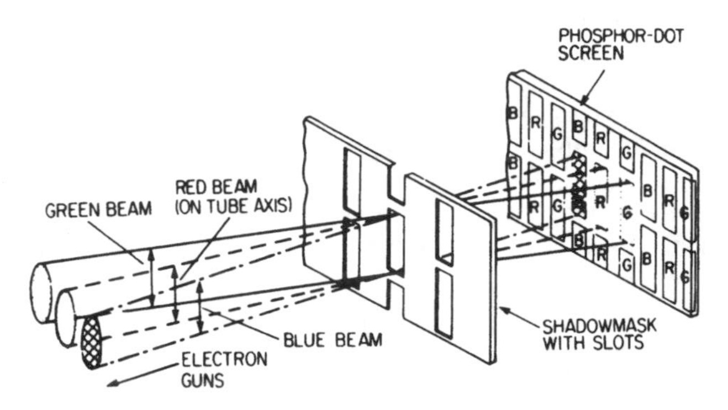

Figure 7.5 *Working principle of the shadowmask and phosphor screen. For each beam, all but the intended phosphor dot is in the shadow of the mask*

conveyed in Figure 7.5, where the centre beam (red in our example) is on the axis of the tube with the outer beams, green and blue, having an approach angle of around 1° with respect to red.

Purity

Clearly, the approach angle of the three beams to the mask must be spot-on to get the 'shadowing' effect right. If the effective origin of the beams does not exactly coincide with the position of the light source used for dot fixing during manufacture, one or more of the beams will spill over onto part of a neighbouring phosphor dot which will be the 'wrong' colour for that beam. If this happens to red, say, and it spills onto the blue dots the red gun will give rise to some blue light from the screen and a magenta 'stain' will mar what should be a pure red field. To prevent this a pair of two-pole ring magnets is provided around the tube neck in the gun area, and adjusted to achieve an on-axis beam path for good purity at screen centre. Secondly, the entire scan yoke is set axially (by sliding it along the tube neck) to establish coincidence between deflection centre and original dot-fixing light-source position. When this is correct, purity is established over the entire screen area for all three colours.

We have now achieved our aim, which is the establishment of three superimposed images, one in each primary colour and independently controlled for brightness. The triple beam may be regarded as a tri-coloured light pencil, with each colour winking and flashing as the picture is scanned out by the common deflection fields to describe a full colour picture. There is another factor to be considered, however, and this is the necessity to *register* the three images so that they exactly overlay at all points on the display; this is called *convergence*.

Converging the beams

What Figure 7.5 does not convey is that each beam is larger than one shadow-mask hole, and in fact overlaps the slot edges. This means that not only must the beam approach angle be correct, but the spatial position of each beam must be adjustable to take up manufacturing tolerances and ensure that the three rasters are exactly superimposed; any error here will cause colour fringes on borders and outlines in the reproduced picture. What is required, then, is a means of independently shifting the beams in the tube neck, and this is arranged by further ring magnets around the tube neck to provide well-defined and adjustable magnetic fields embracing *individual* beam paths within the electron gun. These fields are set up by a combination of four-pole and six-pole ring magnets.

There is a *pair* of ring magnets to set up the four-pole field, and a further pair to provide the six-pole field. Two are provided in each case so that by relative- and co-adjustment the *strength and direction* of the resulting magnetic field can be manually set to give complete control of beam positioning. None of these fields affect the central beam which travels straight down the axis of the tube after the fashion of that of a monochrome tube – this implies that the rasters due to the outer beams (green and blue in our case) are made to conform to the position of the raster traced out by the centre (on-axis) beam. Let us look first at the effect of the four-pole field. It is shown in Figure 7.6 and can be seen to move the outer beams *differentially* in either vertical or horizontal planes, the direction of shift being determined by field direction, and the amount of shift by field strength. This permits us to register the blue and green images to give a cyan overlaid image, but this will not necessarily coincide with the red image, and this is where the six-pole field comes into play.

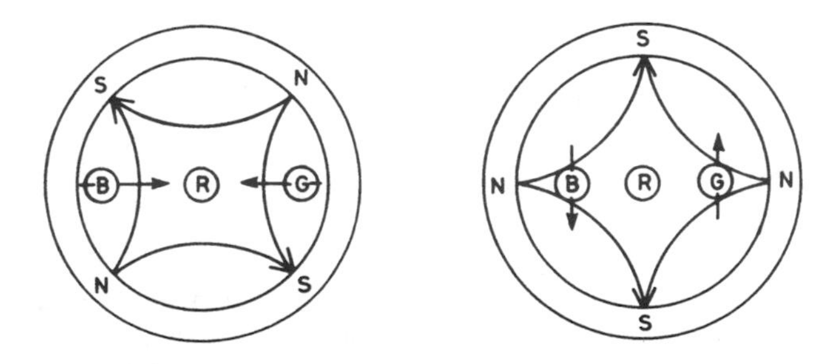

Figure 7.6 *Four-pole static convergence rings with field pattern*

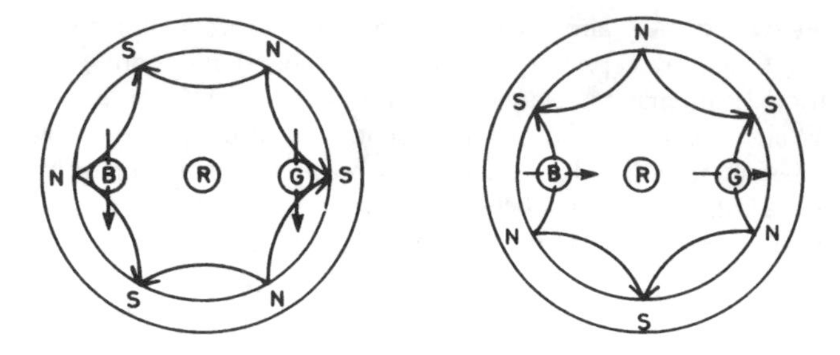

Figure 7.7 *Six-pole static convergence rings with field pattern*

The six-pole field is shown in Figure 7.7. Here the positions of the outer beams are moved together in the same direction, which may be vertical or horizontal in compliance with the orientation of the lines of flux. Again the flux strength (and hence the distance the beam moves) is controlled by the relative movement of the ring magnets, while the flux direction is varied by moving the six-pole rings *together*. Thus we can position our cyan raster exactly on top of the existing (and unvarying!) red raster.

On plain mixed-colour fields misregistration of the colours is not visible, so that convergence adjustment is carried out by observing the display of a crosshatch or dot-matrix pattern from a special signal generator.

Dynamic convergence

The convergence process described so far will only overlay the images in an area corresponding to a small circle at the centre of the screen. As the beams are deflected to left and right in the line-scanning process, the fact that the outer guns are off the tube axis means that the point of convergence will fall progressively short of the screen unless the latter is cylindrical with cylinder and deflection centres coincident as represented in Figure 7.8*a*. Because practical TV screens are much flatter than this, the three rasters become horizontally displaced (Figure 7.8*b*) and arrangements must be made to compensate for this. While a dynamic convergence system could be devised along the lines of the static magnets already described, but with line-rate flux variations (i.e. a strap-on convergence yoke fed from the line timebase), a much neater system has been evolved, based on the concept of a special deflection yoke whose magnetic field is tailored to achieve correct dynamic convergence.

Deflection yoke

The degree of deflection suffered by an electron beam under the influence of a magnetic field depends on the strength of that field, i.e. how many lines of force it cuts. In a homogenous (uniform) field, all three beams will be 'bent' through the same angle as in our Figure 7.8*a*. Because the beams are horizontally separated

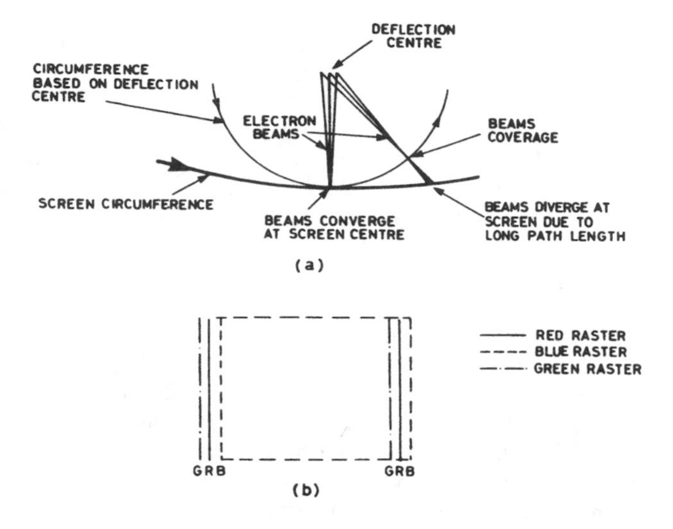

Figure 7.8 *The causes of misconvergence: (a) shows how the beams diverge in areas away from screen centre, while (b) indicates the resulting misregistration*

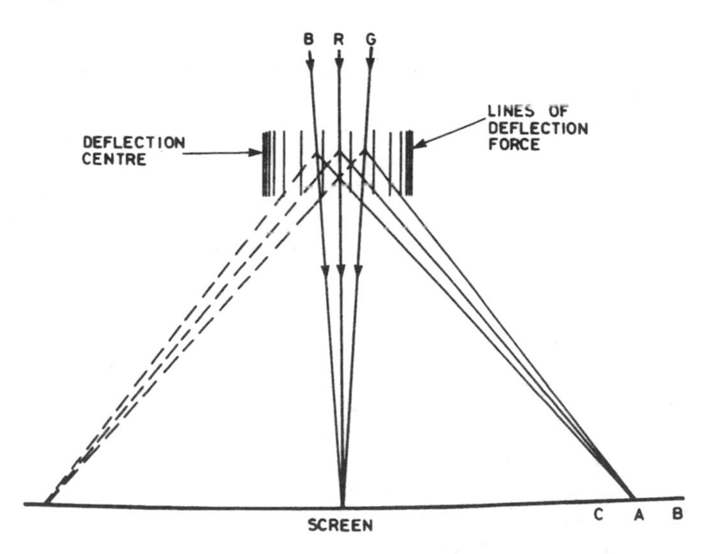

Figure 7.9 *Beam trajectories in the deflection field. The latter is non-uniform to achieve correct convergence, as explained in the text*

by a few millimetres on their journey through the deflection field, however, we can arrange for them to encounter different flux levels by carefully 'grading' the magnetic deflection field across the tube-neck diameter; see Figure 7.9. Here the lines of force are widely separated in the area of the tube axis, and become progressively closer towards the tube's outer walls. This means that each of the outer beams will, during horizontal deflection, be turned into an area of stronger or weaker flux than its fellows, and so suffer a greater or lesser deflecting force. Provided the flux gradient across the tube cross-section is just right, complete convergence of the rasters can be achieved.

This effect is clarified by the beam trajectories in Figure 7.9. This is drawn from above and taking the case (solid line) of deflection to the right, we can see that the central red beam passes through an angle determined largely by the magnetic flux at the centre of the field to strike the fluorescent screen at point A. The blue beam, if it were similarly deflected, would cross the red beam some way short of the shadowmask and land at point B causing misconvergence. However, the blue beam can be seen to be (at the deflection point) turning into a weaker field, causing it to be deflected through a lesser angle than red and 'aiming' it precisely at screen point A. Now consider the green beam. In a uniform deflection field it would cross the red beam too early and diverge to strike the screen around point C. This time, then, we need to increase the green beam's deflection angle to achieve screen registration; as the green beam passes through the deflection field it is turned into a stronger deflecting field to achieve this. For line-scanning deflection to the left the opposite happens; in all cases the beam on the inside of the deflection-centre 'bend' is bent through a greater angle than the norm, and the beam on the outside is bent through a lesser angle.

What happens during vertical deflection? Again the yoke-to-screen beam-path gets progressively longer as we move upwards or downwards from screen centre, so that *horizontal* displacement of the three images will take place with increasing severity towards screen top and bottom. Again an astigmatic deflection field will correct this in similar fashion to that described for horizontal dynamic convergence. The vertical deflection field is barrel-shaped (Figure 7.10) so that the required horizontal lines of force have an increasing vertical component away from the tube axis. This vertical component has the effect of reducing the *horizontal* divergence of the outer beams to achieve horizontal register at screen top and bottom.

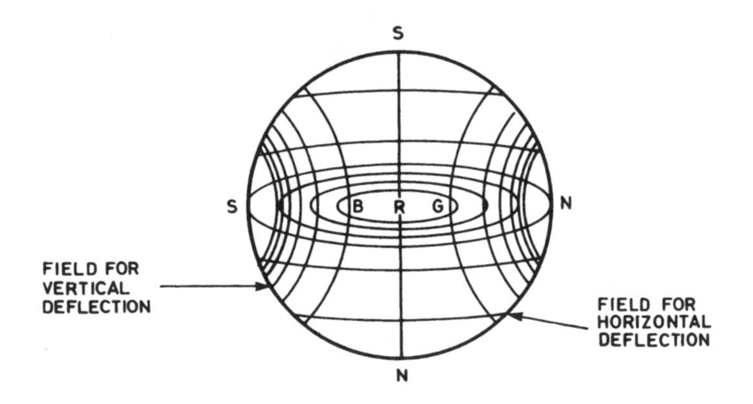

Figure 7.10 *The astigmatic deflection fields required for a self-converging in-line picture tube. The field shapes shown are achieved by precision construction of the deflection yoke*

Thus correct convergence is maintained all over the screen area by a precisely-tailored deflection yoke. The positioning of the yoke along the tube axis is dictated, as we have seen, by the requirements of purity setting, and the gun end of the yoke is concentric with the tube neck due to their common diameter. To take up manufacturing tolerances, however, the flared front end of the scan yoke assembly may be tilted in both vertical and horizontal planes to 'trim' the relative trajectories of the electron beams and thus form a final adjustment of screen-edge convergence; this adjustment is carried out during manufacture or as a servicing routine, after which the yoke is wedged and sealed in position.

The necessary very precise field shaping was achieved in early designs by a toroidal deflection yoke in which the positioning and distribution of the individual turns of wire are defined by notches cast in the supporting frame. The fields are further modified by magnetic 'shunts' and 'enhancers' fitted inside the tube or the scan yoke which respectively disperse or concentrate local magnetic flux to create exactly the field shape required for self-convergence.

Later and current picture-tube designs work in just the same way, usually having the green gun central. Further development in deflection-yoke technology has achieved a saddle-shaped winding pattern whose appearance is broadly similar to that of an ordinary monochrome deflection yoke. However, the wires are grouped into 'streams' to create the necessary magnetic field pattern in some types. More precise manufacturing tolerances have eliminated the need to pan and tilt the deflection yoke to achieve screen-edge registration of R, G and B images in certain designs, and even the need for the neck-mounted magnetic rings has been done away with in tubes which incorporate a single *Internal* magnetic ring mounted at the top of the gun assembly. At the time of manufacture it is 'printed' with a composite magnetic field to fully correct convergence, purity and geometry.

Faceplate

As picture-tube design has evolved the viewing screen has progressively become flatter and its corners squarer. The effects of a flat screen are to reduce reflection of ambient light; to eliminate geometrical picture distortion; and to give a more pleasing appearance to the set.

The Trinitron

Another approach to in-line tube design is represented by the Trinitron type which in fact was the first in the field, having been introduced by Sony in 1969. The main difference between this and the types so far described is in the form of the shadowmask. Instead of the discrete phosphor-dot triads of the conventional tubes, the screen of the Trinitron is composed of several hundred vertical stripes of red-, green- and blue-emitting phosphors. The mask consists of a metal aperture grille which has one continuous vertical slot for each three (R, G, B) phosphor stripes. The general principles of the design are illustrated in Figure 7.11. It can be

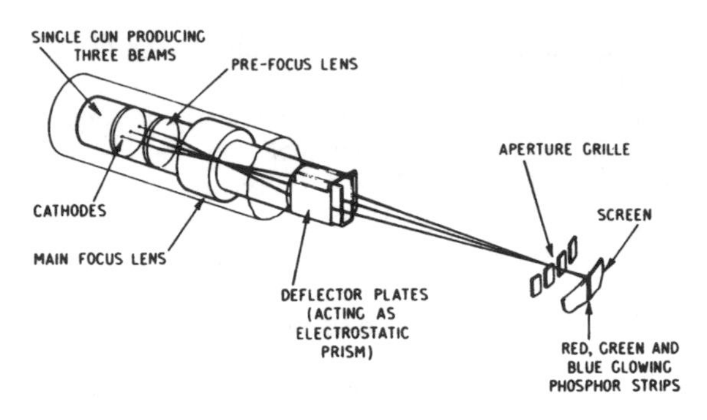

Figure 7.11 *The main features of the Trinitron single-gun, three-beam aperture-grille picture tube*

seen that (unusually amongst colour tubes) the beams cross over within the single gun assembly before crossing again at the aperture-grille slot.

The RGB beam crossover within the focus field confers the advantage on this design of having all three beams pass through the centre of a relatively large electron lens. The analogy to light beams in optical lenses is obvious, and focal aberrations in the 'outer' beams are thus minimised. A further 'electron-optic' function is carried out further downstream by the deflector-plate system shown in the middle of Figure 7.11. These plates act as a prism due to a PD (potential difference) between them of a few hundred volts. By adjusting this PD the outer beams (red and blue) can be shifted laterally in opposite directions to converge them with the central green beam. The static convergence control, then, is associated with the EHT voltage source and connects to the prism plates via a coaxial EHT cable; in other designs a potential-divider resistor is built into the tube to provide static-convergence and other electrode potentials.

In other respects the Trinitron is similar to the PIL tube already described. Vertical static convergence is catered for by adjustable neck-mounted permanent-magnet systems, and such dynamic convergence trimming as is required is carried out by tilting the scan yoke. A simple four-pole dynamic convergence coil is present and if necessary (depending on manufacturing tolerances) energised by a line-rate (and in large screen types, a field-rate) adjustable sawtooth current derived from the timebase section.

Two advantages which the Trinitron offers over the opposition are the increased mask 'transparency', leading to higher picture brightness, and the enhanced vertical resolution, both due to the absence of a vertical matrix component in the shadowmask design.

Degaussing

Any external magnetic field will upset the beam path in a picture tube as a result of which beam landing accuracy – and hence purity – will be affected. Even the vertical component of the Earth's magnetic field can have an influence (the horizontal component cannot upset operation of an in-line gun tube) and nearby magnetic objects such as loudspeakers, radiators and steel joists can give rise to sufficient field strength to spoil the display.

To prevent these effects, the flare of the tube has a built-in magnetic shield, and to neutralise any magnetic field it may itself collect, a built-in degaussing system is provided in the receiver. It takes the form of a pair of coils embracing the tube cone which are briefly energised at switch-on by a decaying burst of 50 Hz mains power. This comes from a double PTC (positive-temperature-coefficient) device whose resistance rapidly increases with the heating effect of the mains current so that the degauss energy decays to zero after two or three seconds, and remains so due to the presence of a 'bleed' resistor which keeps the PTC combination warm. To prevent pickup and distribution of line scan-yoke energy the degauss coils are bypassed at TV line frequency by a suitable capacitor.

Geometric distortion

In a monochrome picture tube, especially a wide-angle type, the flatness of the screen leads to a form of raster distortion known as pincushion, in which the raster borders are bowed in. This is corrected by permanent magnets which neutralise the effect by applying an equal and opposite pre-distortion of the scanning field. Colour tubes have similarly flat faces and so the same inherent defect is present, but any permanent-magnetic correction system is not practical for reasons which should now be obvious! Instead the pincushion correction is carried out dynamically, as it were, by amplitude-modulating the line-scanning current with a field rate *parabolic* wave-form, and by superimposing a line-rate sawtooth wave on the field-scanning current.

The degree of pincushion distortion which has to be catered for depends on the type and vintage of the tube in question; early types needed considerable correction on both axes, whereas later designs eliminated the need for vertical correction and reduced the horizontal requirement. Some tubes, by further refinement of the deflection yoke design, eliminate the need for pincushion correction altogether, but wide-angle types need external correction of the horizontal scanning current. It's done by a diode modulator in the line-scan stage which alters the scan amplitude on a field-rate basis.

Grey-scale tracking

A very important requirement is for the beam current versus cathode voltage characteristic of the three guns to match over the full range of brightness. Since it is impossible for guns to be manufactured with these parameters exactly matching (and to expect them to wear at equal rates), the tube has to be equipped with various adjustments to provide accurate *grey-scale tracking* as it is called.

We have seen that white light is produced when the mix of red, green and blue lights is correctly proportioned. This must occur at all drive levels from black to full brightness for each gun, for a drift out of alignment as drive is changed would alter the relative intensities of the three colours and thus change the proportions of the mixture, which would result in the white (or grey) display becoming coloured. The effect on a colour scene would be to alter the hue or saturation with changing luminance level. The black cut-off points are matched by varying the clamp potential on the individual cathodes by manual or automatic means, as we saw in the previous chapter. To match the brightness of the primary colours at high drive levels, the 'slope' of the video amplifier signal is adjusted by drive controls ('primary-colour contrast pre-sets') with which the highlights of a monochrome display are set to eliminate any coloured tint.

Focusing

We have talked much in this chapter of converging the beams for the purposes of picture registration, and this may be regarded as a focusing of the three beams inasmuch as we aim to get them to cross over at the shadowmask holes. Each individual beam may be likened to that in a monochrome tube, however, and the necessity to focus the beam itself is as important as in a black-and-white set; if good focus is not achieved, the colour image, while remaining converged, will have poor definition and render a blurred picture.

Each electron beam consists of a multitude of negative 'charge packets' – electrons – jostling and falling over each other in their headlong dash towards the highly-positive final anode. The fact that they are all *negative* charges makes them tend to repel one another, 'spreading' the beam. To correct this they pass through an electrostatic lens system (Figures 7.4 and 7.11) in which the *outer* electrons in *each* beam are sent on an inward trajectory such that all the electrons in the beam come to a sharp focus at the phosphor screen. Adjustment of the focal length of this electron lens is carried out by varying the voltage on the focus anode(s) – usually A2; A3 in Trinitron types – by means of a potentiometer associated with the diode-split EHT transformer. Typically, the potential required is about 18 per cent of final anode voltage in *bi-potential* tube type, and around 27 per cent of final anode voltage in *high bi-potential* tubes. With a 25 kV EHT voltage in both cases, this calls for around 4.5 kV and 6.8 kV respectively, with a focus control range of about ±15 per cent. This potential is applied to one of the CRT base pins, and to avoid insulation and sparking problems the adjacent pins are widely spaced from it.

Beam current limiting

In our earlier description of the shadowmask, mention was made of the heating effect of the beam current flowing in the mask, and its special expansion arrangements to compensate. This works up to a limit, but at around the 1.2 mA mark (depending on screen size) the mask is at the end of its tether, metaphorically and literally! Beyond this point, purity suffers and convergence and focus become degraded. The *average* beam current in the tube can be measured indirectly by a monitor in the 'cold' side of the EHT supply, or directly at the cathodes. When the beam current reaches the upper limit, tube drive is automatically pulled back to prevent picture distortion. This 'pull-back' effect is carried out via the decoder (or luminance) IC by reducing the voltage available to the contrast, brightness and colour control-potential inputs.

Colour-tube performance

We have seen that there is one hole in the shadowmask per dot-triad, and because a TV pixel can be no smaller than a single triad, the definition capability of the tube is determined by the number of triads on the screen surface, assuming all other conditions of scanning, focus and convergence are optimum. In general the mask and dot structure are the limiting factors for resolution overall in a receiver, especially in the smaller screen sizes where the mask/hole matrix is not reduced in size in proportion to the screen area. Current medium- and large-screen direct viewing tubes are adequate for 625/50 colour reception, where the limitations on broadcast bandwidth offer a picture definition which can just be resolved by these tubes.

For higher-definition pictures and particularly high-resolution data and graphics displays, fine-matrix tubes are available with more triads per square cm, leading to an inherently better resolution capability. This puts further demands on the convergence system, which in any tube should ideally converge all three rasters to within one pixel over virtually all the screen area.

Screen shape

The shape of the viewing screen evolved gradually, starting with a circular faceplate embraced by a quasi-rectangular mask. For many years the aspect ratio (width/height) was 4:3, while the 'bulge' of the faceplate was gradually reduced. Current tubes have optically-flat faces and many of them are widescreen types with an aspect ratio of 16:9. The principle of operation of widescreen and flat-faced picture tubes is the same as those described above, the yoke and gun technology having been stretched, as it were, to accommodate the much more stringent requirements of these types.

Projection TV

For a large picture the direct-viewing system we have been describing is not practical for many reasons, foremost of which are weight, bulk and the effect of atmospheric pressure on a large evacuated glass envelope! Until a practical large flat-screen device is developed (no easy matter, as we shall see) the only alternative to direct-viewing CRT (cathode ray tube) displays is some form of projection system. The two basic approaches to this are to project the coloured image from a conventional-format tube onto a large screen by means of a lens; or to project separate RGB images onto a common viewing screen, rather like the set-up we used in Chapter 3 to illustrate the basic concept of colour TV and light mixing. The former approach has been used in domestic applications, but its light output is pitifully low since the limited CRT phosphor-light becomes spread over a large area. RGB systems are more complex and expensive, but are capable of much brighter picture reproduction for reasons which will become clear, and it is on these that we shall concentrate now.

In a projection system many configurations are possible, and much depends on the required screen size and viewing area. To make a (relatively) compact set, the light path is often folded by an intermediate mirror and the image formed on an integral screen. This system may use *back-projection*, where the picture is built up on the rear of a ground-glass screen which (for domestic and small-audience commercial use) is fitted into a free-standing cabinet which thus resembles a huge conventional console TV set. An alternative approach is the front-projection system where projector and viewers are on the same side of a reflective screen. This offers the great advantage of being able to give the screen a 'gain' factor by building a directional property into it – the concave screen throws a lot of light over a small viewing area and a useful increase in brightness is realised in the limited viewing area. Off-beam the brightness is very low indeed. A light 'gain' of three to five times is possible with a directional screen.

In the sort of RGB projection system we are discussing here, the light sources are three separate picture tubes with different phosphor screens, one each of R-, G- and B-emitting – corresponding to the three TV primary colours. The chief requirement is high brightness and with no shadowmask to intercept beam energy, and a high accelerating (EHT) voltage, the limiting factor for brightness is the heating effect of the beam dissipation on the glass faceplate. Very expensive projection tubes are made with sapphire faceplates to withstand high temperatures, but a more homely arrangement is the use of a forced-air cooling system for three conventional small cathode ray tubes. They are mounted in line abreast like the electron guns inside a conventional picture tube.

The next operation is to gather as much light as possible from each tube and project it onto the viewing screen, along a straight or folded light path. Either conventional lenses or a *Schmidt* optical system can be used; the basis of the Schmidt system is a highly-polished spherical mirror which gathers the scattering light rays from the tube and concentrates them into a parallel beam, the reverse process to that used in astronomical telescopes, or indeed the satellite receiving dish to be described in Chapter 13. The parallel beam passes through a corrector lens on its way to the viewing screen. To prevent contamination of the highly-polished reflector and to ensure perfect and permanent positioning, the entire Schmidt system is built into the tube envelope in some designs – Figure 7.12 represents the LightGuide tube by Advent Corp of the USA. The internal screen is about 9 mm diagonal, and made of aluminium, coated with R, G or B phosphor.

Whether simple lenses or Schmidt units are used, the fact that the three projection lights are not coincident in position means that the light beams strike the viewing screen at different angles and, as with our shadowmask picture tubes, this causes misregistration of the three primary colour images. To correct this, a convergence system is incorporated in the deflection system of two or more tubes to introduce an equal-and-opposite distortion to each individual raster, and these are finely adjusted to overlay the three images on the viewing screen. Where the screen (or the incident light beams) is at an angle to 'normal', keystone distortion will also be present, and this is corrected in the same way.

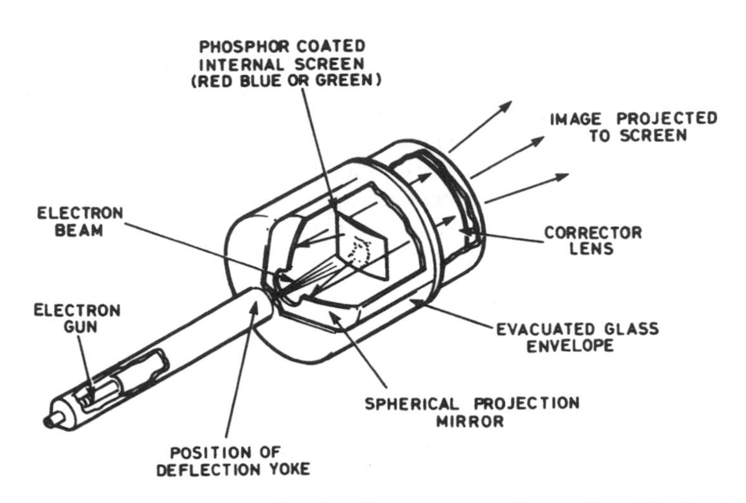

Figure 7.12 *An internal-Schmidt projection tube (LightGuide) by Advent Corp of USA*

Light valves

In all the systems discussed above for TV display, the basic source of illumination is phosphor-light, and for very bright or very large displays this technology is pushed to the limit. If we were able to emulate ciné practice, and use the picture information to *modulate* a separate, external light source on its way to the viewing screen, overall illumination would be limited only by the brightness of the 'light-house.' The first system of this sort used the Eidophor principle, wherein the TV raster is traced out on a spherical mirror in an evacuated glass envelope. The face of the mirror is coated with a special oil whose surface is deformed by the electron beam. The TV lines are formed as grooves in the oil film, and the depth and wall-angle of these grooves depends on beam intensity – picture brightness. A carefully-formed and precisely-angled light is focused onto the oil film which then passes or blocks the light so that the spherical mirror's reflectance is controlled by the TV signal. Emerging light is focused on the viewing screen by a lens, and the whole shooting-match is triplicated for the three primaries of colour.

LCD displays

The most common alternative to CRT technology for producing TV pictures is the LCD (liquid crystal display) system, consisting of a matrix of crystal cells. The basic difference between a picture tube and an LCD panel is that the former generates point-sources of coloured light while the LCD controls, at each picture element, the amount of light passing through it. The basic property of a liquid crystal is that its molecules are arranged in the form of thin bars or flat plates, and of many configurations; the *nematic* type is used for TV and has long, thin molecules, all aligned in the same direction. During manufacture the molecular structure is twisted through 90° (see Figure 7.13) so that the polarisation of light passing through the molecules is similarly twisted. The degree of twist of the molecules can be controlled by applying an electrical field at right-angles to their axes, as shown in Figure 7.14*a*. An LCD element has a ten-micron (1 micron = 10^{-6} metre) liquid-crystal layer sandwiched between two transparent electrodes, and when the voltage applied to the electrodes is varied, the rotational polarisation of the molecules in the crystal can be varied between zero and 90°; the effect is illustrated in Figure 7.14*b*.

The basic action of an LCD is shown in Figure 7.15. The polarisation of natural light is random: its photons vibrate in all directions. A polarising filter can be used to select light which is polarised in only one plane. The LCD element is

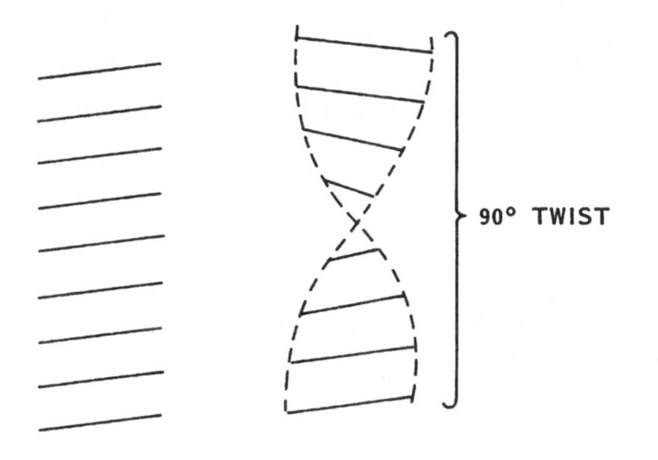

Figure 7.13 *Molecular structure of a nematic liquid crystal: (a) natural state; (b) twisted for use in LCD display panel*

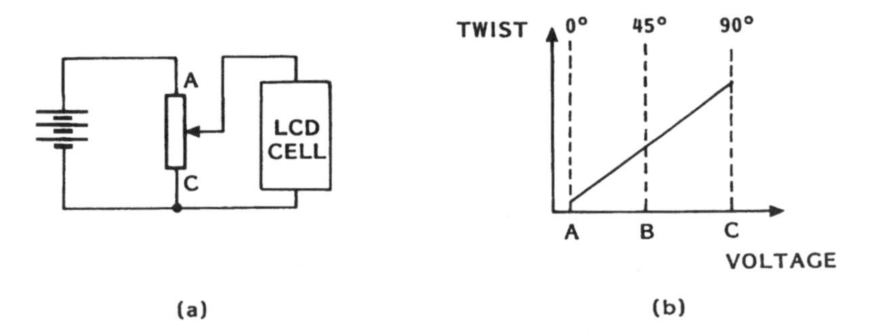

Figure 7.14 *(a) As the potentiometer is turned up, progressively more voltage is applied to the LCD cell, increasing its molecular twist. (b) Twist is in proportion to applied voltage*

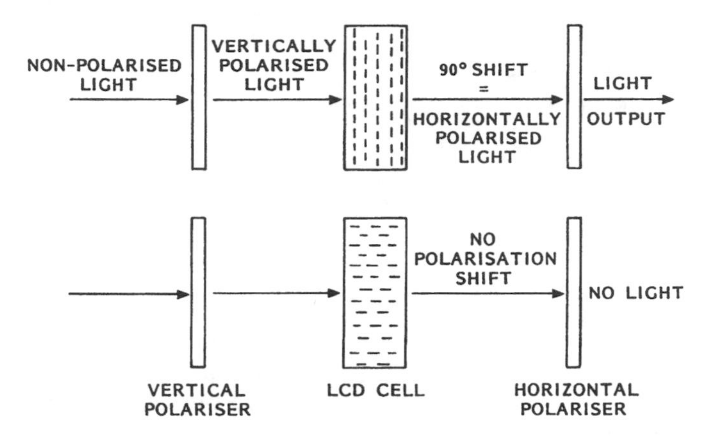

Figure 7.15 *Sandwiched between opposing polarisers, the crystal layer can be made to act as a light cell*

sandwiched between two polarising filters with opposite characteristics, one passing vertically-polarised light and the other horizontally-polarised light. The 'transparency' of the assembly depends, then, on the crystal twist and hence the voltage applied. As the drive voltage is increased, the light passing through the crystal becomes progressively more at odds with the filter on the output side until a point is reached where all the incoming light is blocked and the screen is black. In some types of LCD the two polarising filters have identical characteristics so that the pixel is black for zero drive voltage and white (transparent) when the drive voltage is maximum. The voltage/opacity characteristic of the cell is almost linear (Figure 7.14*b*) and the drive voltage can be applied in either direction; in fact, to prevent ionisation and consequent deterioration, AC drive is used.

In a practical display panel, between 100 000 and 400 000 cells are arranged in a rectangular matrix, each cell being controlled by its own built-in thin-film field-effect transistor (TFT). By individually controlling the conduction of each TFT on the basis of a rows-and-columns drive signal, each picture element can be opened, closed or driven to any point between the two extremes to build up a pattern of light and shade, corresponding to a black and white TV picture whose definition is governed by the number of elements in the matrix. The drive circuit is integral with the LCD display signal, and consists of a series of shift registers, sample-and-hold stages and output buffers. When a TFT is switched on, once per TV frame, the display data for the pixel it represents is stored as an electric charge in the capacitor formed by the LCD cell. The charge remains, and the cell's light transmittance is held at the same level, until the pixel is updated when the next frame is scanned. This gives the best possible picture contrast and brightness.

Colour LCD displays

The panel described above gives a monochrome picture, not very popular with TV viewers! To produce a colour display each individual cell is given its own

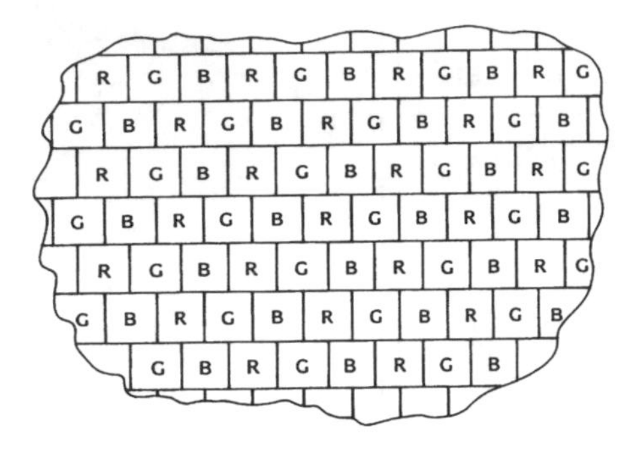

Figure 7.16 *An offset RGB matrix gives best possible definition in a colour LCD display panel. The same applies to CCD image sensors, which generally use* complementary *colour-matrix overlays*

colour filter, producing an RGB matrix like that shown in Figure 7.16, with the RGB video drive signals routed to the appropriate LCD cells. Because we now have three cells per pixel, the definition capability is reduced – it's overcome by using more and smaller pixels, though the complexity of the drive circuits is increased in proportion.

LCD back-lighting

As we've seen, the LCD panel operates basically as a multi-point light valve, and needs a light source to produce a picture. It's generally provided in the form of one or more fluorescent tubes with reflector and diffuser to give bright and even illumination over the entire screen surface. In pocket and portable TV sets and LCD camcorder viewfinders, the fluorescent lamp is very tiny.

LCD projectors

As we have seen, the brightness produced by any sort of cathode ray tube is limited by the energy that can be imparted to its electron beam and dissipated at its screen. When (as with a three-tube projector) the available light has to be spread over a large picture area the result, even with a directional viewing screen, may be barely adequate. Because an LCD panel works on the principle of a light-valve there are no such constraints when it is made the active element in a TV projector: a very bright lamp can be used as the light source, its output being intensity- and colour-modulated by the LCD matrix between the lamp and the screen, which latter can now be made flat, cinema style, needing no 'gain' by virtue of directional properties – indeed a white wall can be satisfactorily used as a reflective screen.

Some LCD projectors use a single RGB panel of the sort described above, with a powerful backlight and a lens system, but it's difficult to achieve sufficient definition in a large picture projected from an LCD panel in which three cells are required for each pixel. Commercially-produced LCD projectors for home and industrial use have three separate LCD panels; the beam coming from the back-light is separated into its R, G and B components, separately modulated then recombined into a single beam for projection.

A typical three-LCD panel projector offers a horizontal picture resolution of 300–450 lines, using three separate (R, G, B) panels each with a diagonal measurement of about 8 cm, and each containing over 100 000 cells/pixels. Thus the display carries 300 000 or more transistors. A high-power metal-halide lamp is provided as light source, capable of giving a contrast ratio of 100:1 via an f4.5 projection lens with a 'zoom' facility of about 2:1. The internal arrangement of the projector is shown in Figure 7.17. The lamp is backed up by a parabolic reflector (1) which absorbs IR radiation (heat) while reflecting visible light forward for reflection from a cold mirror (2) which absorbs IR and UV (ultra-violet) energy. Next the light passes through an IR/UV cut filter, (3), after which it is 'cold' with its energy confined to the visible part of the spectrum. Now the beam reaches the first (red) dichroic lens, (4) which has the property of reflecting red light while allowing blue and green to pass through. The extracted red light is deflected through 90° by a reflector (5), and passed through a condenser lens which makes the light rays parallel on their way to the red LCD for modulation by the red picture information. Meanwhile, the green and blue components of the light beam emerging from mirror (4) are intercepted and separated by dichroic lens (6) for separate passage to their own condenser lenses and LCD modulator panels. Two further dichroic lenses (7) and (8) are used to recombine the three modulated beams on their way to the projection lens and hence the screen.

The running temperature of the LCD panels is held below 50°C by a forced-fan ventilation system which sucks in air at the back of the projector housing and blows it out of the top. Internal temperature is monitored by sensors hooked to a control system capable of shutting down the lamp if necessary.

Study of Figure 7.17 shows that the image produced by the R-LCD is not reflected at all on its way to the screen, while that from the B-LCD is reflected once and from the G-LCD, twice. The result would be that the blue component of the picture would be horizontally reversed compared to the G and R images. It's compensated for by electrically scanning the B-LCD in the opposite direction to the other two.

Projector lamp

The metal-halide lamp used as light source is very powerful and efficient. It consists of a mercury-arc lamp with its electrodes 1–2 mm apart; initially an arc is struck between them by application of a 12 kV pulse. The resulting discharge vaporises a drop of liquid mercury which radiates an intense glow, and establishes, when warm, a low resistance path between the electrodes. At this point an AC feed voltage of about 80 V is sufficient to maintain conduction and light emission: it's regulated by a control circuit which holds the lamp dissipation at about 150 W.

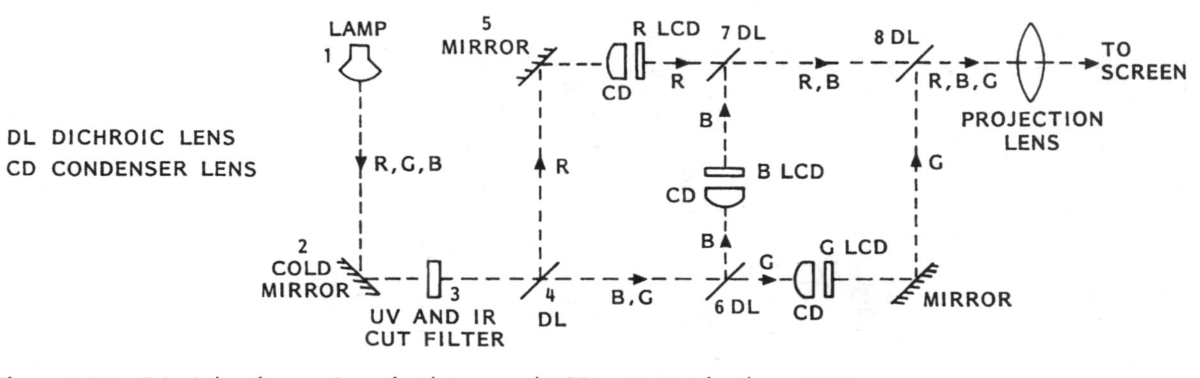

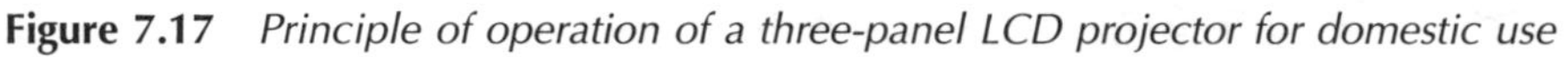

Figure 7.17 *Principle of operation of a three-panel LCD projector for domestic use*

It's important that the spectral emission of the lamp is fairly evenly divided across the visible spectrum, and a standard mercury-arc lamp has a very uneven distribution, with its major component in the blue and UV region. To overcome this, various metal halides, e.g. caesium, are added to the mercury in the quartz glass envelope, which also contains an atmosphere of argon gas. The result is an even spectral distribution from 380–760 nm. The lamp's operating life is about 1000 hours, monitored by the system-control section of the projector.

Plasma display systems

Because it does not require separate electronic circuits for each pixel, the cathode ray tube is simple in concept and relatively cheap to mass-produce. We have seen that an LCD display requires at least one transistor per pixel, making it difficult to reliably manufacture, especially in large screen sizes. For large flat screens with good performance – and where cost is a secondary consideration – plasma displays are available.

The most widely used plasma technology is PALC (plasma-addressed liquid-crystal), typically in a 46"/115 cm-diagonal screen size with 10 cm depth, 1 000 000 pixels, 500:1 contrast ratio and 160° viewing angle. PALC screens use LCD pixels, switched on not by a TFT transistor but by a gas discharge triggered by a pulse voltage of about – 300 V. When this takes place the video signal is effectively applied to the LCD section which glows in proportion to its amplitude. Each plasma channel controls a complete TV display line in which individual pixels are addressed and brightness-controlled, retaining their state until updated by the next discharge pulse: lines are fired in sequence in a non-interlaced progression. Video signals are stored electronically until their turn comes for application to the display system. The colours are produced by filters over individual cells, three to each effective pixel.

Another application of plasma technique uses coloured phosphors to produce the display. Here each phosphor-coated cell is filled with neon gas, into which is projected an electric discharge to generate ultraviolet rays. These strike and activate the phosphor to produce the picture directly. Until and unless the price of plasma screens come down they will not challenge tube- and LCD-type displays for general domestic use: currently plasma systems have four- and five-figure price tags.

100 Hz scanning

With large display screens, and especially on bright picture features, the 50 Hz refresh-rate gives rise to picture flicker, even though the viewer may not be conscious of it. The use of electronic field storage enables the scanning rate to be doubled to 100 Hz, eliminating all traces of flicker; more details are given on page 169.

8 *The TV receiver*

A colour TV set can be regarded as a monochrome receiver with refinements, to which has been added the colour display device and a decoder. Figure 8.1 shows the block diagram of a colour receiver with the 'extra' items required for colour in heavy line.

The colour display device is shown in heavy line because this is specific to a colour receiver, but as we have seen, a colour set can display very satisfactory monochrome pictures, whether or not the transmission is in colour. Thus the main and essential item in the colour set, apart from the display device itself, is the decoder, which has been fully dealt with in Chapter 6. In most other respects, monochrome and colour receivers are similar or identical – each contains a tuner with some means of viewer control; an IF amplifier with carefully-shaped frequency response; a vision detector to produce video and sound signals; an 'intercarrier' audio IF and detector stage; one or more audio amplifiers and loudspeakers; a sync-pulse separator; line and field timebase; and a power supply unit. The degree of sophistication of many of these circuit blocks needs to be greater for colour TV, and much depends on the price of the model in question, especially in

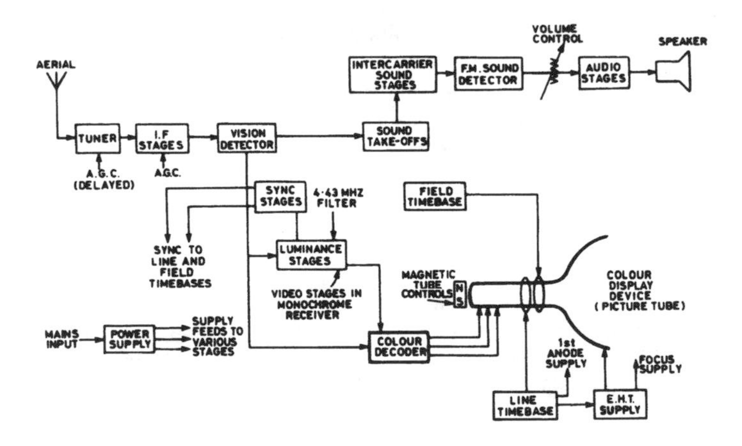

Figure 8.1 *Block diagram of colour receiver. The heavy-line blocks represent the extra stages required for colour*

such areas as remote-control, sound quality and interfacing ability. In recent years the production volume of black-and-white sets has been very low, consisting mainly of portable, small-screen types.

The function and operation of the various 'common' circuit sections are outlined in this chapter. More details are given in other reference books, e.g. *TV and Video Engineers Pocket Book* by the same author.

The tuner

The tuner's job is to amplify the weak signals picked up by the aerial, to select or 'tune' the required one, then to amplify and translate it, complete with all the information that it carries, to a lower (intermediate) frequency called the 'IF'. It will be recalled that the aerial itself is a tuned circuit, so the aerial and tuner should be regarded as a 'matched pair', a point perhaps which should be borne in mind by the aerial-erecting fraternity!

It should be remembered that the selected signals are in a specific channel and consists of both the modulated sound and vision carriers. Reasonable balance between all components of the transmitted signal on any one channel is required, though FM sound limiting and the ACC function make the set reasonably tolerant of 'tilt' in the response of the aerial or tuner section. Most of the gain and selectivity of a TV receiver is designed into the IF amplifier, and the main requirements of the tuner are a low inherent noise figure and an ability to suppress the 'image' frequency, which the IF amplifier is quite unable to reject. Normally the local oscillator (these tuners are, of course, superhet devices) runs 'high' with respect to incoming frequency, and the IF output from the tuner is the *difference* between the two, normally 39.5 MHz for vision. There are two input frequencies which will give rise to this 'beat' frequency, one 39.5 MHz above the oscillator frequency, one 39.5 MHz below. The former is the so-called 'image' and any signals arriving at the tuner on this wavelength must be strongly rejected.

The tuned circuits in the tuner unit must keep in step with each other as different frequencies are selected, and this calls for good matching in manufacture. The variable tuning is carried out by means of fixed Lecher bars (inductors), in conjunction with varicap diodes which we met in our discussion on the colour decoders. Tuning control, then, is carried out by varying the DC potential applied to them. The control voltage source may be as simple as a single potentiometer or as complex as a frequency-synthesis, self-seeking ensemble. Perhaps these last terms deserve a little more explanation! A self-seeking system, when initiated, sweeps up the TV transmission band(s) by itself, stopping each time it encounters a station for committing (manually or automatically) to its solid-state memory. Frequency-synthesis tuning offers a self-seek and memory facility along with a 'direct addressing' feature in which a required channel *number* (21 to 68 for UHF) can be requested by the viewer and automatically tuned. This involves a very stable crystal oscillator in a PLL embracing the tuner's local oscillator. The PLL includes a *programmable divider* to set the ratio between crystal and local oscillator and this division ratio governs the tuning point of the TV. All this is achieved by special digital ICs.

IC tuner

A modern integrated-circuit alternative to the conventional TV tuner is shown in block-diagram form in Figure 8.2. It uses no conventional tuned circuits and thus needs no tuning control voltage: all its tuning instructions come from the TV's control microcomputer via the SDA and SCL lines of the I^2C control bus which we shall meet later in this chapter.

This IC tunes from 50–860 MHz and uses a double-superhet technique. The RF amplifier stage consists of a wideband gain-controlled low-noise amplifier suitable for use with both aerial and cable input signals. It feeds mixer 1 where the first frequency conversion takes place; in fact it is an *up*-conversion to a first IF frequency well above 1 GHz. This passes through an external filter, a simple inexpensive two-pole ceramic resonator with a bandwidth of about 15 MHz: this defines the initial passband and provides image-frequency rejection. In conjunction with on-chip image-rejecting mixer design, image-signal suppression of 65 dB is achieved over the entire tuning range. The signal then undergoes a second, *down*-conversion in mixer 2, whose local oscillator 2 runs at a frequency such that the second IF frequency is about 40 MHz (vision), a standard for terrestrial TV. The second mixer stage is again a special image-rejecting type to provide suppression of 65 dB to the image signal produced in the first mixer stage.

The two local oscillators are fully integrated into the IC, and generate the required frequencies with reference to a single external crystal. Tuning resolution is 62.5 kHz, giving 128 steps in the (typical) 8 MHz channel width. The system uses a complex frequency-synthesis circuit whose components – varactor diodes, voltage-controlled oscillators, phase/frequency detectors, programmable dividers and charge pumps – are all on the chip. The IF amplifier is gain-controlled in the same way as for a conventional tuner/IF ensemble (see below) to optimise the

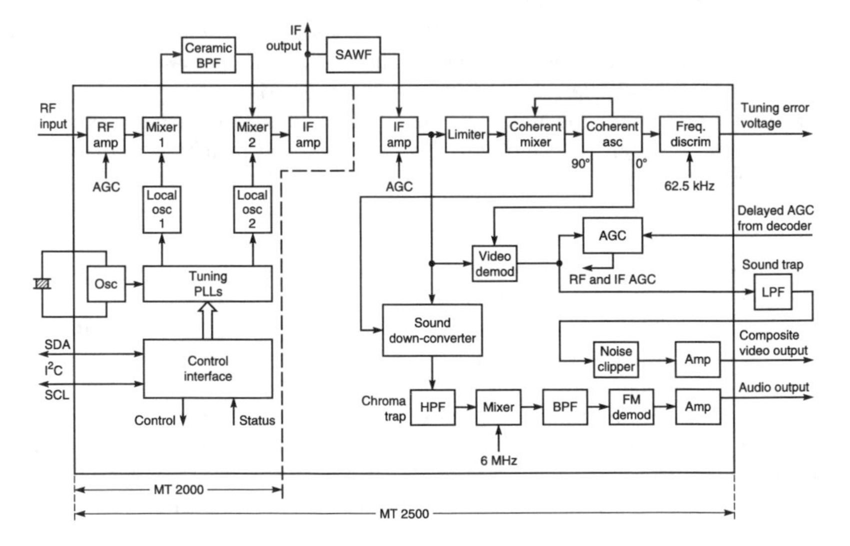

Figure 8.2 *IC tuner architecture. Type MT2000 contains only the components to the left of the dotted line (Microtune Inc)*

noise performance and minimise cross-modulation when large input signals are present. This device has an AGC range of 96 dB, a noise figure of 8 dB, image rejection of more than 57 dB, and cross-modulation performance of less than 1 dB in the presence of 30 mV input signal.

The 'tuner-only' version of the IC in Figure 8.2 ends at the dotted line to the left of diagram centre. The 'complete-receiver' IC incorporates further processing which will now be described; either chip can be used for digital TV by taking the IF output signal (top centre) to a suitable demodulator.

Continuing to the right, the 40 MHz IF signal is 'shaped' in a SAW filter – see below – for passage to a further IF amplifier, after which analogue demodulation takes place. This involves a phase-locked loop (coherent oscillator and mixer) locked to the vision carrier frequency to provide in-phase (I) and quadrature (Q) feeds for the vision demodulator and a sound down-convertor respectively. The demodulated video signal passes through an external trap (RHS of diagram) to take out the sound carrier, then back through a chip-internal noise clipper to remove impulse interference.

Down-converted intercarrier sound (see below) has an FM carrier frequency of 6 MHz for the UK, and is fed through a chroma trap to remove colour subcarriers, then an on-chip self-tuned filter on its way to the FM demodulator. After filtering and de-emphasis the baseband signal is ready for amplification and passage to the speaker(s). Take-off of a Nicam stereo signal feed is also possible with this IC.

The tuner chip is controlled by an industry-standard two-wire serial control bus which allows interrogation and readout of the contents of all the status registers on the chip, and enables the device to be programmed in software. Data registers in the tuning PLLs are loaded to tune in a specific channel.

IF amplifier

Returning to conventional circuits, and looking again at Figure 8.1, the level of IF signals (both sound and vision) from the tuner are raised by the IF amplifier. The response characteristic is carefully tailored to selectively amplify sound, chroma and vision signals, and reject out-of-band products, by a SAW (surface-acoustic-wave) filter, which may almost be regarded as a cross between a crystal and a glass delay line! With selectivity provided by this filter, and gain by the following amplifier, the IF signal is brought to a condition suitable for demodulation, or detection.

Vision AGC

Automatic gain control is applied to both the tuner RF amplifier and the IF amplifier. The vision signal is *sampled* (usually at line rate) to measure its amplitude, and the sample is produced in DC form as an *error* voltage or current whose level is proportional to signal strength. This is fed back to an early stage in a way that varies its gain, thus setting up a feedback loop which maintains a constant output level from the vision detector.

While the *level* of signal can be held constant by AGC over a tremendous range of RF input levels, the *noise* (snow/grain effect) is proportional to received signal strength. Below about 300 μV (for modern UHF tuners) this noise becomes obtrusive, and to achieve the lowest possible noise level in weak signals the gain of the tuner is held at maximum unless the signal becomes so strong (2–3 mV) that overloading of the tuner could occur, when its gain is automatically reduced. This is called *delayed* AGC.

Vision detection

The IF signal is demodulated within an IC by a synchronous detector working on a sampling system similar in essence to that employed for colour demodulation in a decoder. In place of the local oscillator we have a 'tank' circuit which is energised by the incoming IF signal itself, and critically tuned for linear demodulation. This is followed by a filter to remove residual IF carrier components. The vision signal is now ready for application to the decoder block as described in Chapter 6.

Intercarrier sound

At the vision detector input both vision (39.5 MHz) and FM sound (33.5 MHz) signals are present, and it is arranged that intermodulation takes place between these two. As a result, a *difference* frequency of 6 MHz is produced, containing modulation components of both contributing signals. Its AM (vision) modulation is minimised by a 'swamp' technique of depressing the 33.5 MHz sound carrier (in the early vision IF stages) to a level below that of the lowest (i.e. peak white) excursion of the vision signal. A tuned acceptor circuit picks off the post-detector 6 MHz beat signal for processing in an IC containing a series of amplifiers and limiters which remove any residual vision components, leaving a pure FM signal at 6 MHz containing sound information. After detection – again a synchronous, *quadrature*, detector, is used – the audio signal is back in baseband form, without any vision buzz if the limiters are doing their work! Amplification and processing in volume and tone controls render the audio signal in a form suitable for driving a loudspeaker.

In the MAC and Nicam receivers to be described later in the book, the audio signals, which as we know are in stereo, are produced in a totally different way. Here we have a strobed data-slicer which produces numbers corresponding to the quantised audio signal levels; they become recognisable after their passage through a D-A converter.

Sync separator

The detected video signal also feeds the sync separator to produce the all-important timing pulses for scan synchronisation, described in Chapter 1. This is a relatively simple circuit, internal to an IC, which discriminates between vision and sync signals on an amplitude basis, a sort of electronic bacon slicer, as it were. Usually a noise-cancelling circuit is provided in the chip which disables the separator during interference bursts, preventing picture judder and tearing. The sync separator is in the same package as the line oscillator, and TV receiver chips incorporate sync separator, line and field oscillators and several associated functions in a single IC package.

Signal-processing IC

All the low-level signal-processing stages in a colour TV can easily be integrated on a single chip, an example of which is given in Figure 8.3. At the top LH corner the IF signal from the SAW filter enters the vision IF amplifier and PLL (phase-locked-loop) demodulator. From here is produced an AFC (automatic frequency control) potential for the tuner – it counteracts drift of the input carrier or the tuner itself. Baseband video signal, complete with intercarrier sound, passes down to the video amplifier block. Taking first the 'auxiliary' outputs from here, a sample goes up to the AGC section, whose link to the tuner feeds back information on signal strength at high input levels. A video mute block prevents 'snow' being displayed in the absence of a transmission, and from this the video signal comes out of the IC on pin 6. The sound carrier is selected by a ceramic filter ('sound bypass') to re-enter the IC on pin 1, where it encounters a limiter stage to clip off any amplitude modulation. A PLL FM demodulator follows, from which the baseband (mono) audio signal passes through a mute/pre-amp/de-emphasis stage on its way out of the IC on pin 15. It is intercepted by a switch – under I^2C bus control – to enable other sources of audio, entering on pin 2, to be selected (along with external video on pin 17) when the set is switched to AV mode.

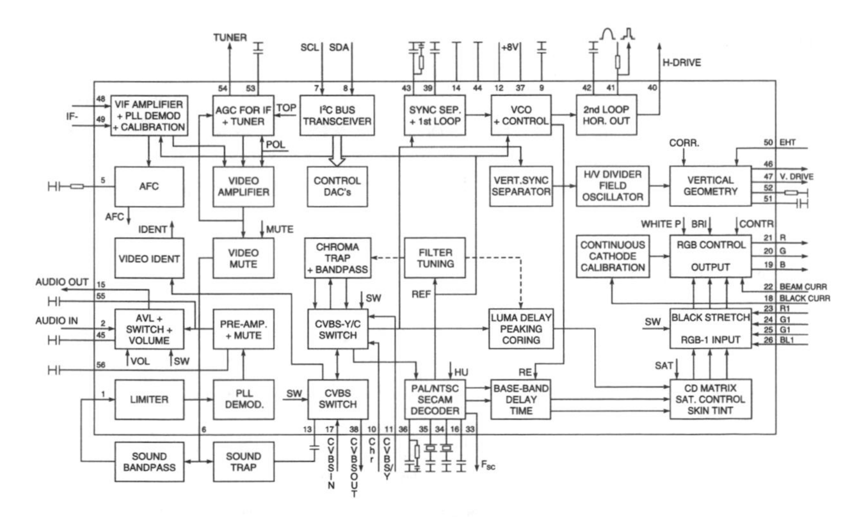

Figure 8.3 *Internal architecture of 'jungle' IC for a TV receiver. Operation is explained in the text*

The main path for the video signal is out of IC pin 6 and through a sound trap, sharply tuned to 6 MHz, into chip pin 13. Here it encounters two routing switches, both working under I^2C-bus instructions. The first of them selects between internal and external (coming in on pin 17) composite video; the second selects composite or Y/C signals, the latter entering on IC pins 11 (Y) and 10 (C). We shall meet Y/C signals later. Pin 38 gives a route out for the composite signal, typically via an AV socket. The chroma trap and bandpass block is used to separate the video and colour subcarrier signals contained in a composite video signal. The entire PAL decoder, also capable of NTSC and SECAM operation, is contained in the bottom centre block in the diagram. It has no need of an external delay line: this is catered for by on-chip components. The components at IC pin 36 provide time-constants for the subcarrier flywheel filter, while 4.43 MHz and 3.58 MHz crystals at pins 35 and 34 respectively regenerate colour subcarriers for PAL and NTSC signals. In the latter case hue control is effected via the I^2C control bus. R-Y and B-Y are separated in the internal baseband delay line on their way to a colour difference/RGB matrix in the bottom RHS of the diagram. This also incorporates the saturation (colour) control. Luminance signal on its way to the RGB matrix is optimised in the peaking/'coring' block, and delayed to coincide with the lower-bandwidth, slower-rising chrominance signals.

RGB signals from the chip-internal decoder pass into the primary-colour switch, where selection is made between them and signals entering on pins 23/24/25, usually from a teletext decoder. 'Black stretch' provides level-dependent amplification to low-level (near black) video signals if required. It's claimed to enhance detail in dark grey picture areas, and can be switched off if required. There's little doubt that broadcasters and studio engineers would switch it off! The final picture-processing block in the chip is centre-right, RGB output. As well as providing buffered feeds out to pins 21/20/19, this provides brightness and contrast control for each of the three channels: separately and in parallel, so that external RGB input signals, when selected, can also be controlled. Brightness and contrast settings are pulled back when the voltage entering on IC pin 22 indicates that the beam current in the picture tube is near the point at which its shadowmask will hot-bulge. Also present here is an auto grey-scale correction system of the sort we met in Figures 6.6 and 6.8. Here the black current sample enters the chip on pin 18 for entry to the cathode calibration block. RGB signals pass through a two-stage amplifier, usually mounted on the tube base panel, on their way to the cathodes of the picture tube.

It remains to trace the path of the synchronisation pulses. At top centre of the diagram of Figure 8.3, the sync separator stage is fed with composite video signal, from which it strips out line sync pulses and feeds them to a comparison stage (details below) which keeps the TV line frequency in step with the broadcast picture. Line drive pulses from IC pin 40 switch the line output stage at 15.625 kHz. A second feed from the composite video signal operates the on-chip vertical sync separator which picks up on the long-duration pulses in the field blanking interval – see Figure 1.5 and associated text. The result is a single 'clean' pulse once per field to reset an H/V divider, a counter circuit which triggers field flyback every $312\frac{1}{2}$ scanning lines. Thus synchronised to the broadcast signal, the field oscillator's drive waveform passes through a shaping stage (vertical geometry) on its way out of the IC on pins 46 and 47. There are two influences on the vertical geometry corrector: a feed coming into IC pin 50 contains a sample of EHT voltage (see later) to prevent the picture size changing with beam current; and, from within the chip, a correction factor by which – under I^2C bus control – the vertical scanning current can be adjusted in amplitude and shape to render good picture geometry and linearity.

Control of all the functions of the IC depends on the I^2C bus transceiver near the top centre of the diagram; it governs all the switches in the chip, plus the D-A convertors which set parameters like colour, brightness, contrast and sound level as well as factory/technician settings such as picture-white points and vertical scan correction. Towards the end of this chapter we shall see the origin of the SCL and SDA control pulse-trains which enter this chip on pins 7 and 8. To round up some of the pins not so far mentioned, nos 12 and 37 provide a +8V operating voltage; no. 33 provides a sample of locked colour subcarrier; no. 26 takes a picture-blanking pulse from the text decoder for use when text is superimposed on a picture; and no. 41 has the dual purpose of dispensing a sandcastle pulse output and receiving a flyback input from the line scan stage. There are other pins and functions of this 56-pin package which are not absolutely essential to the plot, as it were!

Timebases

The vertical and horizontal scans are produced respectively by the field and line timebases subjecting the electron beams in the picture tube to deflection forces by electromagnetic means, as explained in Chapters 1 and 7. Considerable magnetic force is needed to provide full deflection of the beams vertically and horizontally, and this force is generated in the field of the scanning coils on the neck of the tube. Figure 8.4 illustrates a typical deflection yoke and associated line output transformer for use with a colour picture tube.

Line timebase

The line timebase (Figure 8.5) consists of a group of building blocks; those which work at low voltage and current levels are incorporated in an IC. The 'pacemaker' is a pulse-generating oscillator whose frequency and phase can be controlled by an externally-derived voltage. It runs at line-scan rate, which is 15 625 Hz for 625-line 50-field systems and 15 750 Hz for 525-line 60-field systems. For a locked and stable picture display it must be synchronised with the broadcast line sync pulses (Figure 1.5, Chapter 1) not only in frequency, but in phase too. This is achieved by a *flywheel* synchronisation loop, whose function is to 'average' the effect of the sync pulses over a relatively long period so that impulsive noise and interference in the TV signal cannot upset scan timing to give a ragged effect on vertical picture features.

Incoming pulses from the sync separator are checked against line oscillator pulses in a *phase comparator*, whose output voltage indicates by its polarity whether the line oscillator is running faster or slower than the broadcast sync pulses; and by its amplitude the difference in speed between the two. This error voltage is used to control the line oscillator frequency by 'steering' it to the point

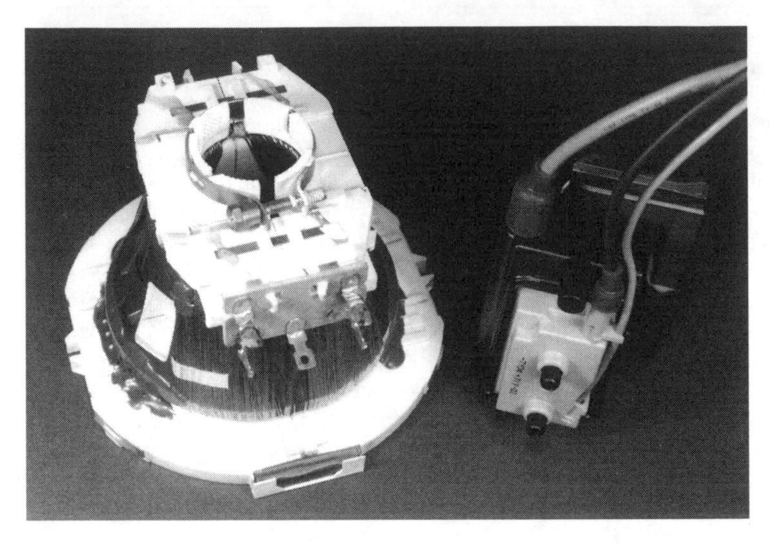

Figure 8.4 *Accessories for use with a 90° PIL tube: the deflection yoke and diode-split line output transformer*

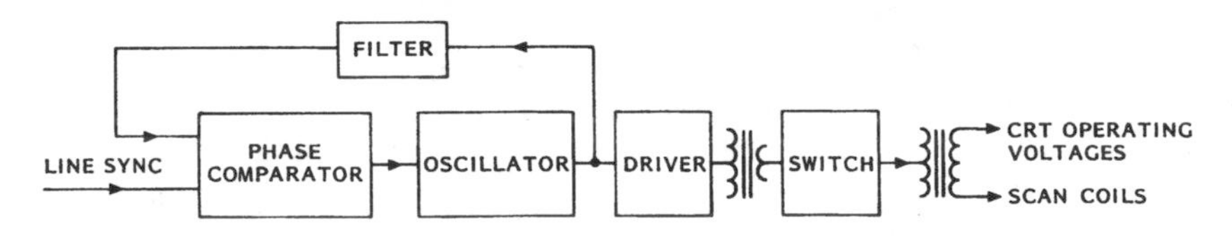

Figure 8.5 *Block diagram of the line timebase and scan-current generator of a TV set or monitor*

where there is zero frequency or phase difference between the two pulse chains, at which point the line oscillator is locked to the broadcast scan rate. The flywheel effect is provided by a low-pass filter – with a relatively long time constant – in the path of the DC control voltage feed to the line oscillator. In practice there are two loops, a coarse one to quickly bring the oscillator onto correct frequency, and a fine one to establish precision phasing of the line oscillator to the vision signal; adjustment of this phase loop, by manual or automatic means, determines the horizontal position of the picture on screen. Both loops are clearly shown at the top right of the chip block diagram of Figure 8.3. For use with a tape or disc video source the time-constant of the flywheel loop filter is shortened, often by a sampling/detector circuit within the line oscillator IC; this prevents horizontal instability and twitch of the picture as a result of the timing jitter which is inevitably superimposed on the video signal by a mechanically-scanned readout system. In this circumstance the timing jitter affects both vision signal and sync pulse, so too long a time constant in the flywheel filter would smooth out the variations in scanning phase while the still-jittering picture information would give rise to a ragged and wavering image. A suitably short flywheel time constant permits the individual line scans to follow the random changes in pixel timing, thus straightening up the vertical features of the picture.

The line oscillator – whose basic timing element may be a quartz crystal, a ceramic resonator or a C/R combination – is designed to free-run in the absence of a synchronising signal so that the timebase continues to run (and generate EHT voltage) in the absence of an input signal or while tuning. The output from the oscillator section consists of a series of pulses with a repetition rate of 64 μs for a 625-line TV set or monitor. Throughout the rest of this chapter we'll assume that we're dealing with a 625-line 50-field system.

Having generated and synchronised a drive pulse for the line output stage, we need to provide impedance matching and drive power for the line output transistor, which in this role acts purely as a switch, as we shall see. It's usually provided in the form of a line driver stage, consisting of a medium-power transistor driving a step-down transformer as shown right of centre in Figure 8.5. On each line drive pulse the transistor saturates, inducing a pulse in the low-impedance secondary winding of the drive transformer. It is used to inject current into the base of the line output transistor for precise control of its conduction.

Line output stage

Although some practical line output circuits seem complex, their basic operation is very simple, and their essential elements are just four components as shown in Figure 8.6. Starting with zero energy in the coil and the picture-tube's scanning spot at screen centre, the switch S is closed, shorting out D and C, and applying the supply voltage across the inductor L. It's a basic characteristic of an inductor under these circumstances that the current in it builds up from zero in a linear fashion. After about 35 μs, and with the current in L still increasing, the switch S is suddenly opened, whereupon the magnetic energy now stored in the ferrite coil of L is released, reversing the current flow and charging tuning capacitor C via large capacitor Cres, which is part of the power supply section, and may be regarded here as a short-circuit to the fast-moving currents involved in a line-scanning circuit. At the point where S was opened, the flyback of the scanning spots was initiated, and now they are back at screen centre and halfway through their rapid

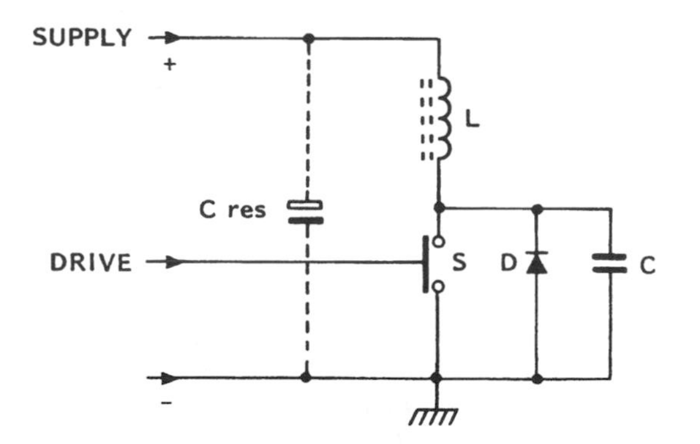

Figure 8.6 *Operating principle of a line output stage. In practice S is a transistor*

'flyback' phase, during which the picture is blanked. The energy in L and its core is once more at zero, all having been transferred to C.

Continuing the action of an LC tuned circuit, the charge in C is fed back into L, and this exchange of energy between L and C would continue until it is all dissipated by the circuit losses were it not for the presence of clamp diode D, which comes into conduction after one half-cyle of LC oscillation, at the point where the bottom end of L attempts to move negative of ground potential. The magnetic energy stored in L now decays linearly to zero, feeding energy back into the supply line, and providing the first half of the next horizontal scanning stroke. When the scanning spots reach the centre of the screen once more there is zero energy in both L and C, at which point switch S closes again, initiated by the precision-timed line drive pulse, to start the complete cycle anew. It is repeated at intervals of 64 μs to give continuous horizontal scanning of the electron beams across the face of the picture tube in synchronism with the broadcast picture. In practice, L is not the scanning coil; it is a board-mounted transformer serving many purposes. The scan-coils proper are connected in parallel with it via a DC isolating capacitor which also provides *S-correction* to compensate for the non-linearity of scanning speed inherent in the use of a flat-faced picture tube; and usually a *saturable* inductor (linearity coil) to correct for the resistive component of the copper-wound scan coils. Switch S is in practice an NPN or field-effect power transistor, sometimes incorporating diode D in its envelope. It has to withstand a very high positive collector voltage during the line flyback stroke, whose duration is about 12 μs, determined by the resonant frequency of LC (approx. 42 kHz) and corresponding to the line blanking interval in the picture transmission shown in Figure 1.8.

The line output transformer (LOPT) is a useful source of energy for other parts of the TV set or monitor. Secondary windings feed the picture-tube heater, the first anode and, by means of voltage multipliers, the focus and final anodes, the latter at a typical DC potential of 25 kV. This is called EHT (extra high tension) and is produced by a *diode-split* extension to the LOPT wherein several separate EHT secondary windings are used, each forming a 'cell' in association with a rectifier diode and a reservoir formed by inter-winding layer capacitance. Each of these EHT cells, buried within the LOPT, typically generates 6 kV DC and they are connected in series to yield an EHT potential appropriate to the tube size in use. Except for very small tubes this will be between 20 and 28 kV. The focus voltage for the picture tube is also generated in this area by 'tapping off' the output of the first or second EHT cell for application to the *thick-film* focus and A1 control potentiometers, which are also built onto the LOPT. These supplies, along with other LT and HT lines derived from the scan transformer, have good *regulation*, benefiting from the close stabilisation which must necessarily be applied to the operation of the line output stage to ensure constant scan amplitude (picture width) in the face of changing EHT current demand with changes of picture brightness and mains supply voltage. Stabilisation is largely the function of the power supply section (PSU) to be described later in this chapter.

Thus the basic diagram of Figure 8.6 has grown up into that of Figure 8.7, which though still simplified is more representative of commercial practice. The component designations of the previous diagram have been retained to help in identification. For use with a wide-angle picture tube (110° deflection, generally in

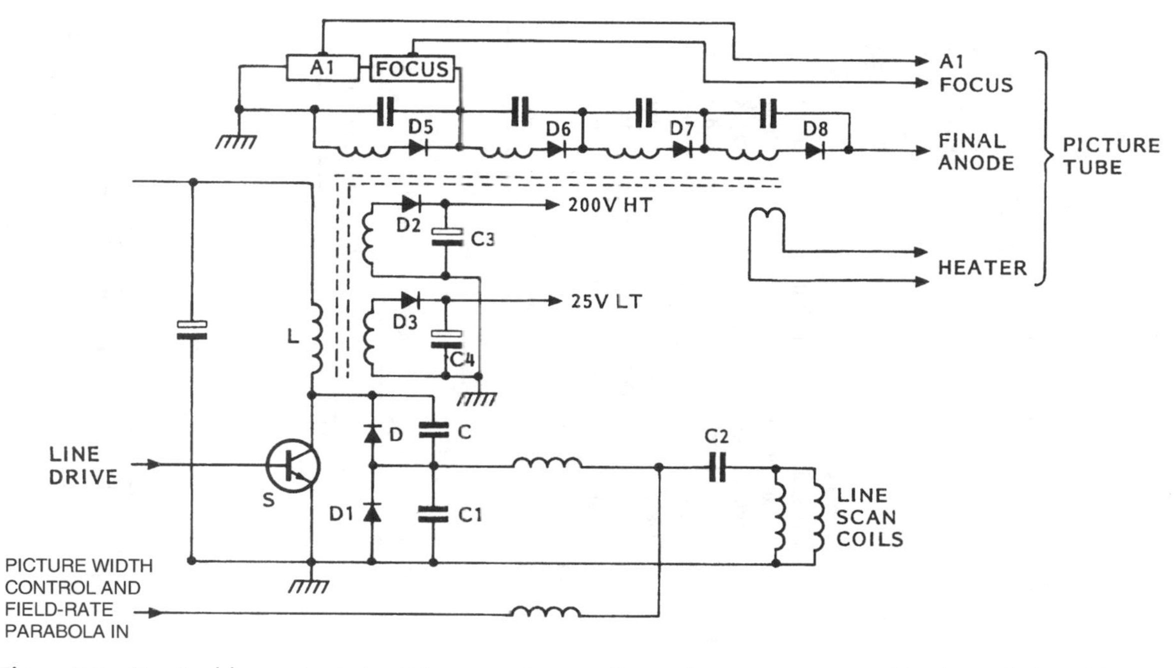

Figure 8.7 *Practical line output circuit incorporating auxiliary voltage generators and horizontal pincushion distortion correction*

screen sizes above 53 cm/21″) it is necessary to amplitude-modulate the line scanning current at field rate with a parabolic waveform. It's done by varying the resonant tuning of the scan-coil circuit – not the LOPT – at 50 Hz frequency by means of a *diode modulator* drive from a medium power transistor.

The diode modulator, driven by a linear DC amplifier, can operate at the field rate called for by the pincushion-correction, and also at any required DC level, permitting picture-width adjustment. Widescreen TV sets have scan-amplitude switching to facilitate various 'zoom' modes, and it is here that these are implemented. In different zoom/widescreen modes horizontal overscan may typically range up to 8.5%: see Table 8.1. Auto widescreen switching commands can come from Scart pin 8 (e.g. from a VCR or DVD player) or from the teletext datastream, called WSS – see the next chapter.

Field timebase

The function of the field timebase is simpler than that of its line-scanning counterpart. Its job is to deflect the scanning spots relatively slowly (cycle time 20 ms) from top to bottom of the viewing screen, then rush them quickly back to the top during the field blanking interval. During the downward stroke the line timebase draws over 300 scanning lines on the screen, and the timing is such that they are distributed evenly over the viewing area. The lines of one field are traced out in the spaces between those of the previous field to satisfy the interlace requirement explained in Chapter 1. The triggering of the field timebase by the broadcast sync pulses is critical for good interlacing performance.

Figure 8.8 is a representative block diagram of a field timebase system. It starts with a free-running oscillator capable of being triggered by the separated

Table 8.1 *Tube-scanning excesses for wide screen TV display*

Mode	Vertical Overscan	Horizontal Overscan
Wide	8%	8.5%
Superlive	15%	4.7%
Cinema	34%	8.5%

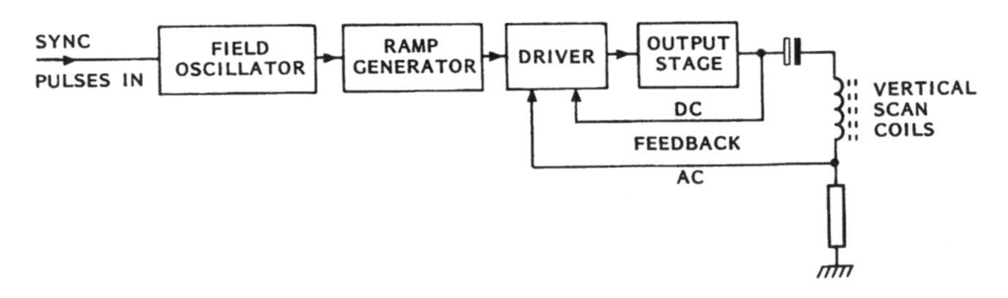

Figure 8.8 *Block diagram of field timebase section of a TV or monitor*

broadcast field sync pulses at 50 Hz rate. The timing pulses produced by the oscillator trigger a precision ramp generator whose output is a sawtooth waveform with a period of 20 ms and excellent linearity. It is amplified by a driver stage for application to the power output stage and thence to the field coils in the deflection yoke. AC and DC feedback are used respectively for linearity correction and stabilisation of the working conditions.

Field oscillators do not need a flywheel synchronisation system; they are triggered directly from the separated sync or make use of a digital countdown circuit, working from line rate, inside the timebase IC: see Figure 8.3. As with the line oscillator, all that is required from the timing generator is a pulse, once per scan, to initiate flyback: the generation and shaping of the scanning waveform is carried out further downstream in the circuit, though not by the components associated with the scan coil itself at this low frequency of 50 Hz. As we have seen, the required scanning current is linear, with a rapid reverse of flow at the bottom of the screen, corresponding in time with the field blanking interval. The next stage in the field timebase circuit, then, is a sawtooth voltage generator. In practice it consists of a capacitor charged from a constant current source: this gives rise to a perfectly linear voltage ramp across the capacitor, terminated after about 19 ms by a shorting switch driven by the field oscillator. After about 1 ms, during which field flyback takes place, the short is removed and the ramp capacitor permitted to begin charging again. To avoid loading the capacitor its waveform is buffered by an amplifier with very high input impedance on its way to the driver and output stages, which are very similar to those of a medium-power audio amplifier.

The power amplifier, now invariably incorporated into a heat-sinked IC, is made up of a Class B complementary-symmetry transistor pair whose mid-point feeds the deflection coils via a large DC-blocking electrolytic coupling capacitor. The DC feedback loop around the output stage stabilises operation in the face of temperature and supply-voltage variations, while AC feedback is used to set the gain of the amplifier and provide scan-linearity correction. It is taken from a low-value resistor in the ground-return path, which puts the deflection yoke itself within the loop and prevents picture height shrinkage as the coils warm up in use – the copper wire with which they are wound has a positive temperature coefficient.

Most modern practice uses separate ICs for generation/synchronisation of the sawtooth drive voltage and the driver/output stage. Figure 8.9 shows a skeleton circuit of an IC-based field output stage. The IC is powered from a 25 V supply rail, derived from the LOPT, which enters at pin 6. During the scanning stroke C1 is allowed to charge to 25 V via D1, but for the duration of field flyback its negative plate (IC pin 3) is internally switched to IC pin 2. The effect is to reverse-bias D1 and 'jack-up' the chip's supply voltage at pin 6 to about 50 V to rapidly reverse the current in the deflection coils for a fast flyback stroke. This removes the need to continuously operate the IC from an (e.g.) 50 V supply and thus reduces internal heat dissipation and improves efficiency. Sawtooth current flows through the scan coils via coupling capacitor C2 and low-value (typically 1 Ω) resistor R2 to ground. R2 acts as a sampler, developing a sawtooth voltage which corresponds to coil current. It is fed back to the driver stage via IC pin 1, having passed through an RC network incorporating height and linearity correction, often including S-correction to compensate for the necessity of scanning a flat-faced picture

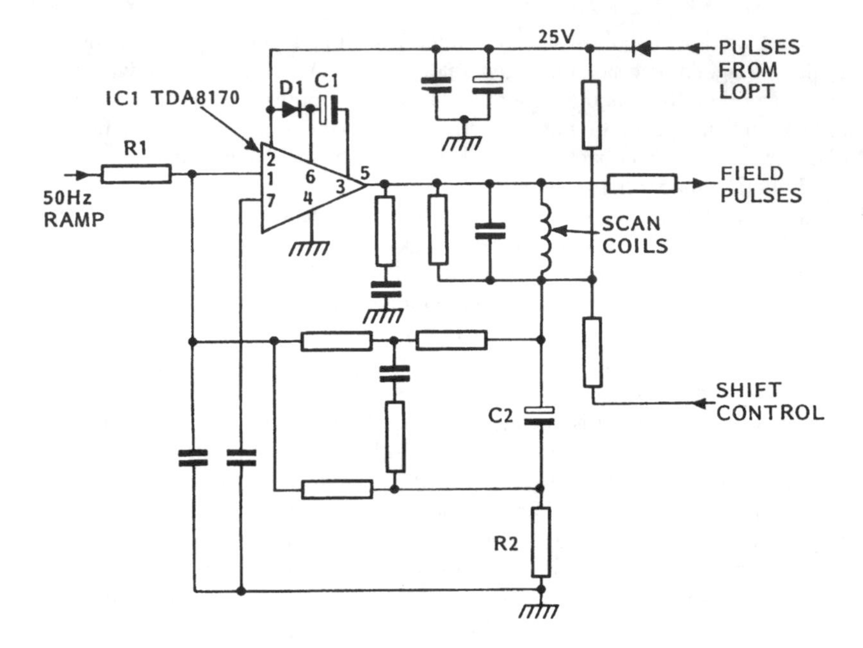

Figure 8.9 *Complete circuit of commercial field output stage using a single IC type TDA8170*

tube as described above for line scanning. In many sets provision is made for cutting off the tube's electron beams in the event of failure of the field deflection so that the screen phosphors do not get burned by the very bright horizontal line which would otherwise result.

Power supply section

A TV set or monitor requires a stabilised and closely regulated supply for its line scanning circuit to ensure that picture size and brightness remain constant as beam current, mains supply voltage and internal temperature vary. Other sections of the set also require closely-stabilised supply voltages, which may be derived from the LOPT itself or from separate feeds from the power stabiliser section: usually a mixture of both is used. The power supply unit (PSU) of a TV set, and indeed most other types of electronic equipment including VCRs, generally works from the domestic mains supply and provides one or more isolated supply lines whose voltage is held steady in spite of wide variations in the current drawn from them and the mains input voltage. Also required are protection circuits to prevent damage in the events of excessive current loading and excessive output voltage due to a malfunction in the stabiliser circuit.

Most PSUs use the 'chopper' technique, in which the regulating element is an electronic switch, as shown in Figure 8.10. The incoming AC mains voltage is

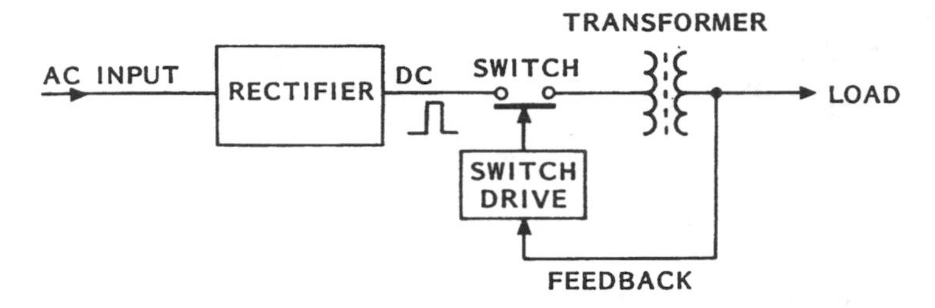

Figure 8.10 *Operating principle of a switch-mode power regulator*

rectified and smoothed in a mains-live area of the set to produce a DC operating supply of about 330 V for application via a high-speed switch transistor to the primary winding of a ferrite-cored transformer. Because the switching transistor is fully off or fully on at any one time, it dissipates very little energy and so runs coolly and efficiently. The ratio of on-time to off-time of the switch (*duty-cycle*) is varied to suit the load demand. When little energy is required the switch dwells in the on position for only a short time during each cycle of operation, briefly topping up an inductive and/or capacitive reservoir which supplies the load during the switch's off period. When the load's energy demand increases, for instance when the picture goes bright and pulls more beam current, a feedback/monitoring circuit automatically increases the duty-cycle of the switch so that it remains on for a longer period in each operating cycle and increases the energy fed to the reservoir. In this way the energy drawn from the primary power source (usually AC mains) almost exactly matches that demanded by the set from instant to instant, with minimal wastage within the PSU itself. The energy reservoir, whether in inductive or capacitive form, need not be large or bulky: the chopper-switch operates at frequencies in the region of 30 kHz so that small and light inductors and capacitors can be used.

In Figure 8.11 the incoming mains voltage is full-wave rectified by bridge rectifier D1–D4 to produce an unregulated supply of about 330 V across reservoir capacitor C1. This is switched to the primary winding of transformer T1 by electronic switch Tr1. Each time the switch closes a current ramp builds up in the primary winding of T1 and magnetic energy is stored in its ferrite core. When the switch is opened, say 30 μs later, the magnetic field collapses, transferring energy to the secondary side of the circuit: rectifier diodes D5 and D6 conduct the current flowing in the transformer's secondary windings to charge C2 and C3. The phasing of the windings is arranged to prevent conduction on the secondary side of the transformer when Tr1 is conducting. The loads are continuously supplied by reservoir capacitors C2 and C3.

All the secondary-side supplies are activated when the transformer's magnetic field collapses, so it's only necessary to monitor one of them for control and regulation purposes – this is normally the one that supplies the major proportion of the load. The output voltage here is monitored to provide feedback to the chopper control circuit. A comparator checks the feedback voltage against a reference level, generally derived from a zener diode, to produce an error/control voltage proportional to the load. It controls a pulse-width modulator whose product is a pulse train of varying duty-cycle which governs the conduction

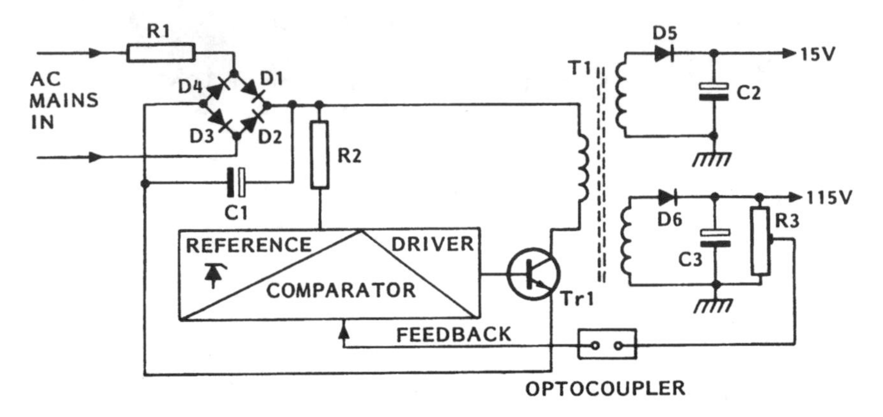

Figure 8.11 *Skeleton circuit of IC-based switch-mode power supply with mains isolation in transformer T1*

period of the switch to suit the current demand. Thus is the stabilisation loop completed.

In most modern circuits the frequency as well as the duty-cycle is varied to suit different load demands. Many circuits are self-oscillating, with feedback from the transformer to the transistor to provide oscillation. Others (Figure 8.11) use a purpose-designed IC as the control element to provide current drive to the base of a bipolar transistor, or voltage drive to the gate of a power field-effect transistor (FET) which provides the switching action. The power transistor may be incorporated in the IC, resulting in a simple circuit with few peripheral components. The three resistors in Figure 8.11 merit a few words of explanation: R1 is a *surge-limiter* to restrain the inrush current when C1 charges at mains switch-on, protecting the rectifier diodes and mains-switch contacts. R2 initially powers the control IC, while pre-set potentiometer R3 is used to set up the stabilised output voltage: it would typically be used to adjust the supply voltage to the line output stage to a level of 115 V. Electrical isolation between live mains and the circuits of the TV is provided by T1 whose primary and secondary windings are separated by insulating layers capable of withstanding over 2 kV. This demands an isolating break in the feedback path, too, often provided by an *optocoupler* whose two highly-insulated halves are linked by an infra-red light beam, generated and monitored by a light-emitting diode (LED) and a photodiode respectively. Another form of power-supply circuit is described in chapter 22.

Some late designs of TV and monitor use a 'hot-coil' system, in which the line scan coils and other timebase components are mains-live. Special care in respect of electric shock and safety/insulation standards should be taken in servicing and adjusting these.

TV control systems

All but the simplest of modern TV sets have remote control and an internal system-control section, usually based on a microprocessor. It governs the tuning;

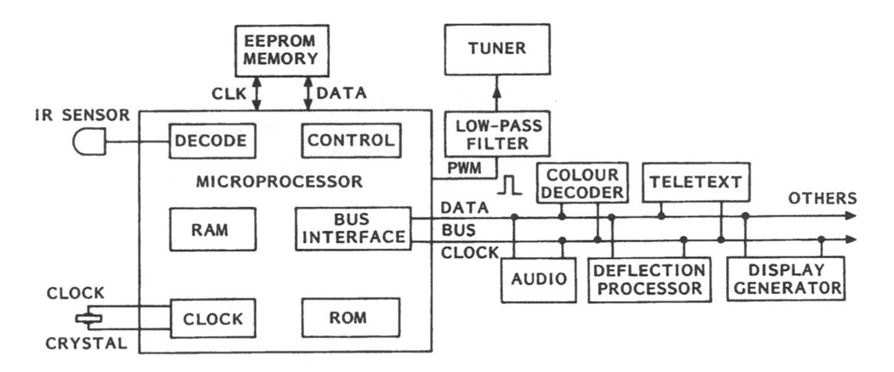

Figure 8.12 *Microprocessor control of TV set functions. Communication is via the two-wire serial data bus shown on the right*

programme selection; 'analogue functions' like volume, brightness and colour; standby switching; teletext functions; and internal signal routing. A non-volatile memory IC (EEPROM) holds user-programmed data like tuning points, preferred analogue settings and, in some designs, factory- or installation-settings such as picture geometry, line phase and decoder/RGB operating points.

An overall view of a microprocessor-based TV control system is given in Figure 8.12. The microprocessor IC is an eight-bit type, custom-designed and mask-programmed for the application and for the specific TV model to which it's fitted. In common with all micros, its heart is a central processor unit (CPU) which consists of gates, inverters, adders and registers to store the results of calculations. Its operating program/instructions are held in a read-only memory (ROM) within the IC, while some random access memory (RAM) is provided for temporary storage of data and instructions. Data enters and leaves the IC by way of *ports*, an important one of which is the interrupt (INT) port, whereby the remote control instructions are passed to an internal serial-to-parallel decoder for instruction processing. Tuning is controlled by pulse-width-modulation (PWM) square wave output for integration into a DC control voltage: it's passed to the tuning control pin of the RF tuner. There is more on microprocessors on page 334.

Most of the commands to and from the microprocessor pass via a two-way *serial data bus*, I^2C, in which one conductor carries clock timing/synchronisation pulses, and the other a string of eight-bit serial data words consisting in order of address and command bytes. The address bytes indicate which of the several ICs on the bus is being called up, and then which of its internal functions requires to be accessed. The following data/instruction byte is fed into the appropriate register within the 'slave' IC, then if appropriate converted into an analogue level to conform to the command. At the end of each byte the slave IC acknowledges its receipt to the master, which is usually the control microprocessor. Because the bus is bi-directional and accessible to all the devices connected to it, provision is made for any of them to assume the role of master and call upon any other (including the microprocessor itself) to be slave – each IC has its own unique address in the form of an eight-bit identification code, the first part of the address sequence.

The same two-wire data bus is used for communication between the microprocessor and the memory IC in both reading and writing modes. The data (which for programme tuning consists of 13 consecutive bits) is passed after the memory chip and location addresses have been specified, confirmed and acknowledged between processor and memory ICs.

Remote control

The cordless control handset contains a pulse generator/encoder IC which generates a string of serial pulse data; each command on the handset has its own unique code, recognisable by the control IC at the receiving end. The pulse train is fed, in the handset, to a small power amplifier feeding one or more LEDs whose radiation is in the (invisible) infra-red spectrum. The chopped infra-red light pulses are intercepted by a photodiode on the TV's front fascia panel for reconversion to electrical pulses and amplification in a very high gain AGC-controlled amplifier before clipping and passage to the INT port of the microprocessor.

To round up this thumbnail sketch of the TV control system, see also page 332, let's see what happens when two consecutive commands are sent by the viewer via the infra-red remote control. First, say, BBC2 is keyed. As the key is stroked the remote control IC comes to life, its clock oscillator starts up, and it strobes and samples the key matrix lines to ascertain which has been closed. Now it generates the appropriate pulse train and switches on and off the LED to send 'chopped' light pulses at about 38 kHz rate. At the receiver they are picked up and amplified in the infra-red pre-amp, squared and passed into the microprocessor IC of our Figure 8.12. The arrival of the first pulse interrupts whatever the micro is doing and diverts the data into a decoder where the command is interpreted. In this case the instruction is to access the section of the memory in which the tuning data for channel 2 is stored. A memory address is generated and sent along the serial bus to the memory (EEPROM) IC which passes the tuning data back over the bus to the microprocessor. On its receipt the control IC changes the duty-cycle of its tuning-drive pulse train to generate the precise DC voltage required to line the tuner up with the local BBC2 transmitter as pre-programmed by the installer or viewer. It now changes the (e.g.) on-screen channel indicator to CH2 or 02, then settles down to monitor the local keyboard and infra-red sensor for any further commands.

Let's imagine that the next command is to decrease picture brightness. The command is generated and sent by the infra-red hand-unit as before, though with a different code of course. On its receipt by the microprocessor it is decoded, then loaded into a register for temporary storage while the address of the luminance processor IC is generated, sent down the data bus and acknowledged. Now the address of the brightness data register within the IC is generated, sent and acknowledged. Finally, the data itself is sent in the form of an eight-bit (only six of them are used for this particular function) byte and loaded into the register. They are translated by a D-A converter to a DC control voltage – lower than before, since we are *decreasing* the brightness – which governs the operating point of the

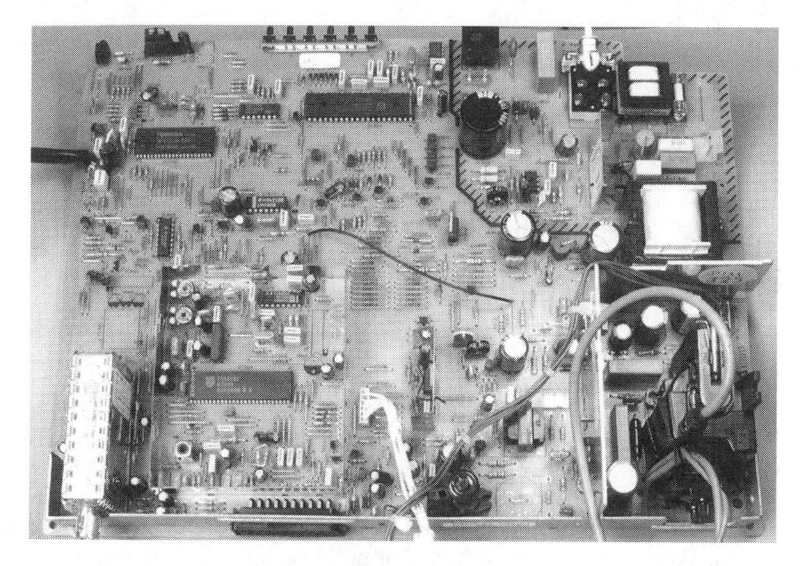

Figure 8.13 *All the electronics of this TV set are mounted on a single board. The power-supply section is at top right, with the line output transformer below it. At bottom left can be seen the tuner and receiver sections and the 21-pin AV (Scart) socket for interchange of signals and operating commands*

black-level clamp in the luminance channel. The picture brightness decreases accordingly.

Many TV sets are fabricated on a single printed-circuit board. An internal view of a typical TV set appears in Figure 8.13.

9 *Teletext*

Teletext is the oldest form of digital broadcasting technology, having been used in the UK for over twenty five years, during which the decoder (an 'optional extra' in a TV set) has steadily shrunk in size and cost, though not in the complexity of the internal workings of its ICs or 'chips'. Teletext uses 'spare' lines in the broadcast field blanking interval to carry a digital pulse stream which when decoded can type out data on the TV screen in the forms of text and simple graphics like maps and diagrams. TV lines 7 to 18 in even fields and 320 to 331 in odd fields are used to carry the data, which can be seen as lines of twinkling dots above the 'live' picture if its height is turned down.

Each text line carries a colour burst to maintain the colour decoder's subcarrier oscillator synchronisation, along with the usual line sync pulse and blanking interval. The line time given to text pulses is the same for that of picture signal: 52 μs. To prevent intercarrier buzz with FM sound systems the amplitude of the text pulses is limited to 66% of the peak white signal. In this two-state AM data transmission system, binary zero is signalled by the text signal hitting black level, and binary one by the signal rising to the 66% level, which represents 0.462 V with a standard 1 V peak-to-peak video signal. The pulses are not square: they are filtered before transmission and further rounded by the receiver's IF and demodulator circuits, and thus reduced to virtually sinusoidal shape by the time they reach the text decoder. The pulses are broadcast in NRZ (non return to zero) form, so that the signal remains high to indicate a string of binary ones and low when a string of binary zeros are being conveyed.

Text data format

The text data rate is 6.9375 Mbits per second, corresponding to 444 times the line frequency, though the pulses are not necessarily locked to line scan rate. A data-stream representing 01010101 etc. throughout the line will thus correspond to a quasi-sinewave at 3.46875 MHz. At this rate there is room for 360 data bits on each TV line used for text conveyance: these 360 bits produce a single row of graphics or characters in the teletext screen display.

The 360 bits per line are grouped into 45 eight-bit bytes, the first five of which synchronise and prime the decoder, and provide an 'address' to indicate where the row is to be placed in the text display. The remaining 40 bytes define the characters in the row, which can thus have a maximum of 40 letters, numbers or graphic blocks. Blanks or gaps, e.g. between words, have their own code so that data is always present.

The text display has 23 rows, permitting a maximum of 23 × 40 = 920 characters to be displayed on a single page to form simple maps, diagrams and alphanumeric text. Upper- and lower-case letters can be produced, as well as figures, symbols and 'building blocks' for graphics, in any of eight colours formed by the three primaries, plus black and white.

The teletext code table is shown in Figure 9.1. Starting at the top left-hand corner, the first four bits define the row number and the next three the column number. The codes are thus used to determine the character so that capital T for example would be called up by byte 0010101 (bits 1 to 7), while byte 1111101 produces the symbol #. Note that to change from a capital to a lower-case letter only bit no. 6 changes, and that the codes for numbers are made by their binary

b_7 b_6 b_5					000	001	010		011		100	101	110		111	
Bits b4	b3	b2	b1	Col → / Row ↓	0	1	2	2a	3	3a	4	5	6	6a	7	7a
0	0	0	0	0	NUL①	DLE①			0		@	P	—		p	
0	0	0	1	1	Alpha^n Red	Graphics Red	!		1		A	Q	a		q	
0	0	1	0	2	Alpha^n Green	Graphics Green	"		2		B	R	b		r	
0	0	1	1	3	Alpha^n Yellow	Graphics Yellow	£		3		C	S	c		s	
0	1	0	0	4	Alpha^n Blue	Graphics Blue	$		4		D	T	d		t	
0	1	0	1	5	Alpha^n Magenta	Graphics Magenta	%		5		E	U	e		u	
0	1	1	0	6	Alpha^n Cyan	Graphics Cyan	&		6		F	V	f		v	
0	1	1	1	7	Alpha^{n}② White	Graphics White	'		7		G	W	g		w	
1	0	0	0	8	Flash	Conceal Display	(		8		H	X	h		x	
1	0	0	1	9	Steady②	Contiguous② Graphics	)		9		I	Y	i		y	
1	0	1	0	10	End Box②	Separated Graphics	*		:		J	Z	j		z	
1	0	1	1	11	Start Box	ESC①	+		;		K	←	k		¼	
1	1	0	0	12	Normal② Height	Black② Background	,		<		L	½	l		‖	
1	1	0	1	13	Double Height	New Background	-		=		M	→	m		¾	
1	1	1	0	14	SO①	Hold Graphics	.		>		N	↑	n		÷	
1	1	1	1	15	SI①	Release② Graphics	/		?		O	#	o		■	

Figure 9.1 *Code table for broadcast Teletext. This is held in a ROM look-up table in the character generator IC of the text decoder*

equivalents followed by the code 011. The 32 codes in columns 0 and 1 are called control codes, and they determine whether the codes in the following columns represent alphanumeric characters or graphics and assign some 'attribute,' such as a colour, to the characters to be displayed. These control codes all have bits 6 and 7 at zero: they produce no display themselves, occupying a single blank space in the displayed row – usually in the background colour of the preceding character.

Protection

On its way from the transmitter to the decoder the text data is vulnerable to corruption by noise and interference, so a protection system is used, in the form of an odd-parity bit at the end of each of the seven-bit codes shown in Figure 9.1 so that each becomes an eight-bit byte. The parity bit added may be a binary one or a binary zero, such that there is always an odd number of one bits in the byte. This enables the decoder to check whether any bytes with an even number of ones have been received and reject them as being corrupt. This simple system has two limitations: it fails if an even number of bits in the byte are incorrect, and it cannot provide data *correction*. Even so, it is adequate for use with text symbols, where a minor error often goes unnoticed, and is corrected anyway when the page is cyclically updated.

Some of the text data requires greater protection than is afforded by this simple parity check. If the data conveying the page number is corrupt, the wrong page could appear in response to a viewer's request, while if the row address data is wrong the row will be printed out in the wrong place on the screen. To avoid this the bytes that carry this information and the real-time clock data are heavily protected by means of a *Hamming Code* in which every alternate bit is a parity bit: bits 2, 4, 6 and 8 convey data while bits 1, 3, 5 and 7 provide protection, as shown in Table 9.1. This permits single-bit errors to be detected and corrected by inversion. The penalty incurred is that the volume of information conveyed by the byte is reduced.

Row coding

As we have seen, the text page is organised in 23 rows. Of these the most important is row 0 (top row) because it contains the magazine and page numbers, the service name and the date and time – it is called the header row. To capture a page for display the viewer keys in the magazine and page numbers: when this keyed-in data corresponds with the transmitted code the required page is selected and displayed on the screen.

The data format for the header row is shown in Figure 9.2. First comes a clock run-in sequence consisting of two bytes carrying a series of ones and zeros, 101010 etc. It synchronises the decoder's bit-sampling clock oscillator. Next comes a framing-code byte whose purpose is to identify the start-point of the data bytes in the following pulse train. Its pattern is 11100100, chosen to give

Table 9.1 *Hamming coding table*

Message	Hamming coded bits							
	B1	B2	B3	B4	B5	B6	B7	B8
0	1	0	1	0	1	0	0	0
1	0	1	0	0	0	0	0	0
2	1	0	0	1	0	0	1	0
3	0	1	1	1	1	0	1	0
4	0	0	1	0	0	1	0	0
5	1	1	0	0	1	1	1	0
6	0	0	0	1	1	1	0	0
7	1	1	1	1	0	1	0	0
8	0	0	0	0	1	0	1	1
9	1	1	1	0	0	0	1	1
10	0	1	1	0	0	0	1	1
11	1	1	0	1	1	0	0	1
12	1	0	0	0	0	1	0	1
13	0	1	0	1	1	0	1	1
14	1	0	1	1	1	1	1	1
15	0	1	0	1	0	1	1	1

a reliable reset signal to the rest of the decoder and to be amenable to correction of single-bit errors by a simple check and correction circuit. The next two bytes, which are Hamming-protected, carry the magazine and row-address information. They are followed by page number and time-code bytes to enable the viewer to select a particular page at a specified time. Following these, the next two bytes provide information governing the whole page which is to follow: the control codes shown in columns 0 and 1 of Figure 9.1 for subtitle, update, news flash and similar functions, all with Hamming protection. Eight data bytes in the header row have thus been used, leaving 32 for the display of characters. Except for the page number, the header row text is the same for all the pages of any one transmission. It shows the name of the service (Ceefax, Skytext etc.) and the day, date and real time in hours, minutes and seconds.

The following rows have the same clock run-in and framing codes as the header row. These are followed by individual magazine and row-address bytes, which are the only ones to have Hamming protection in the 'ordinary' rows. Bytes 6 to 45 contain eight-bit codes for the text or graphics in the line. Each byte is transmitted with the LSB (least significant bit) first and the parity bit last.

Transmission sequence and access time

We've seen that each text line transmitted during the field blanking period can produce one row of text on the screen. 24 TV lines are required to produce the 24 rows of a single page of text data, then, and at the normal transmission rate of

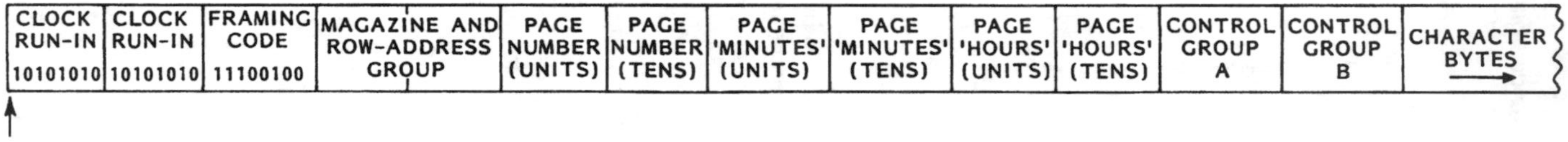

CLOCK RUN-IN 10101010	CLOCK RUN-IN 10101010	FRAMING CODE 11100100	MAGAZINE AND ROW-ADDRESS GROUP	PAGE NUMBER (UNITS)	PAGE NUMBER (TENS)	PAGE 'MINUTES' (UNITS)	PAGE 'MINUTES' (TENS)	PAGE 'HOURS' (UNITS)	PAGE 'HOURS' (TENS)	CONTROL GROUP A	CONTROL GROUP B	CHARACTER BYTES →

Figure 9.2 *Data format for Teletext header row. Each block represents an eight-bit byte*

12 text lines per TV field it takes two fields to transmit a single teletext page. At a field rate of 50 per second, 25 pages a second, or 1500 pages a minute can be transmitted. If 750 pages are being transmitted the worst-case access time will be 30 seconds and the average wait 15 seconds, assuming that the pages are transmitted repeatedly in sequence.

To maximise the user-friendliness of the system, high-priority pages such as indexes are transmitted more often than others. Also contributing to speed of access is a *row-adoptive* feature, in which the totally blank rows which occur in many text pages are omitted from the datastream. In addition to this, the text rows of several pages are interleaved, timewise, in the transmission to give a perceived reduction of the access time: this accounts for each row carrying its own magazine number. Subtitles, assigned to page 888, have the highest priority: they are slotted into the first available TV line after the cueing point. These techniques are limited in their effectiveness in that most of them shorten the waiting time for some pages while lengthening that for others. For the user the most effective reduction in page access time comes from the use of the *Fastext* system.

Text data memory and Fastext

There is no need to store the parity bits, so the RAM (random access memory) capacity required for a single page of text is 7 (bits) × 40 (characters) × 24 (rows), calling for 6720 bits, the minimum required in the decoder. Most decoders have a two-page storage capacity, in practice 2k × 7 bits, so that they can automatically capture the page following the one being displayed. If, for instance, page 183 is called up, page 184 is also captured and written into memory for immediate display. The action of the viewer stepping forward to 184 prompts the decoder to seek, acquire and store the data for the next page (185) in the now unused half of the memory, overwriting stored page 183 with the page 185 data.

Fastext decoders incorporate larger memories which can store the requested page and the next seven, automatically loaded in sequence by the decoder; as each page is discarded the next page in the sequence is written into the memory. This is based on the premise that viewers usually step through pages sequentially, and gives instantaneous new-page readout when they do. The broadcaster's text editor can enhance the Fastext function by anticipating the user's requirements and sending additional instructions to the decoder, based on the current page's contents, for extra page acquisition. It works thus: the bottom row of the text page has four colour-coded 'prompts' which may typically be news, sports, travel and weather. The remote control handset has four keys with corresponding colours. Pressing any one of them selects the magazine and page numbers simultaneously, loading the required pages into the memory for instant selection. If, for instance, travel is selected the on-screen coloured prompts change to air, sea, road and 'continue', the latter providing an escape from the travel menu.

A 'packet' system, in which additional data rows are transmitted but not necessarily displayed on screen, is used for Fastext data acquisition. The packets contain additional control data rather than characters and are sent in advance of the teletext pages (which now become packets 0 to 23) so that the control data arrives ahead of the page to which it relates. Packet 24 *is* displayed on screen as a row because it contains the Fastext prompts, while packet 27 provides the page-linking data. There are eight packets in the text specification, used for various purposes including linked character sets (primarily for other languages) and general-purpose data transmission. Packet 30 is assigned to PDC (programme delivery control) for videocassette recorders; the data it contains is captured and stored in a suitably equipped VCR, and used to initiate a recording at any required time on any channel which radiates PDC data. Its advantage is that it compensates for any unexpected alterations to the broadcast timing schedules.

Also contained in the teletext datastream packet is a WSS (widescreen switching) flag for application to the scan generators. It automatically selects the best scanning amplitudes for the programme being broadcast. This automatic function can be cancelled by the viewer if required.

Decoder operation

Figure 9.3 shows a text decoder in the simplest possible block diagram form. The first block gates the data pulses out of the video signal and decodes them. The next block selects the data requested by the viewer, according to instructions sent via the control logic block, and passes it to the page-memory RAM, whose write-in control signal also comes from the control block. Read-out from the RAM is again governed by the control block: it takes place at a slower rate, a complete field scan period being required for a page in the memory to be read out. It was written in during short 52 μs bursts at varying intervals.

The data from the RAM is fed to a character-generator ROM (read only memory) which uses it to produce the character graphics and symbols shown in columns 2 to 7a of Figure 9.1, and to give them the colours and attributes assigned by the control data stored with the character codes. The outputs from the ROM consist of RGB signals in serial data form, plus a blanking pulse train. They are fed to an RGB signal processing IC like that shown in Figure 6.6, where they would enter on pins 12, 14, 16 and 9 respectively for onward passage to the picture tube.

Data from the viewer's infra-red control handset is fed into the control logic section of Figure 9.3 in serial form, most often along the two-wire control

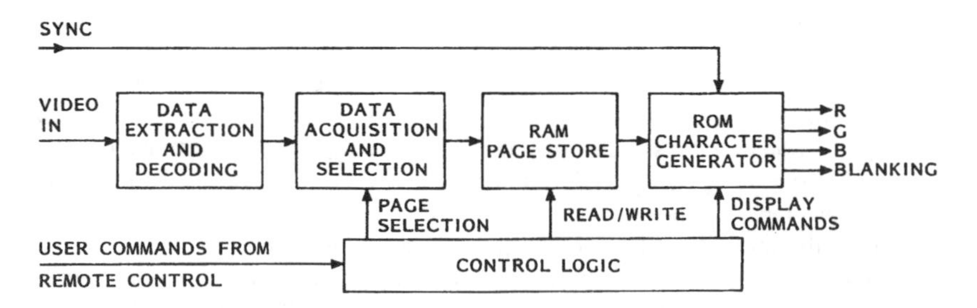

Figure 9.3 *The basic elements of a text decoding system*

bus described in the previous chapter. After serial-to-parallel conversion it's stacked in registers ready to act on the transmitted text data and the character-generator ROM.

Practical decoder

A block diagram of a single-IC text decoder chip developed by Philips Components is shown in Figure 9.4. The analogue video signal from the TV's video demodulator enters at bottom left where its black level is clamped and the sync pulses stripped out to synchronise the clock which drives the internal timing chain. Now the video signal is converted into digital form (more details in Chapter 12) and passed to a data slicer, whose function is to produce chains of clean sharp pulses for the acquisition and decoding block, in which the required page is selected. Page selection data is passed into this block from the I^2C interface, operating from the two-wire control bus described in Chapter 8. This IC has a single-page memory of 1.1k × 7 bit; other types have two-page memories or an interface block for connection to an outboard memory.

The single-page data is passed in parallel form to the character generator ROM which contains a look-up table to convert the 7-bit control data into serial pulses, locked to scan rate, which make up the numerals, characters and symbols on screen. They leave the IC at top right along with three other control signals for the RGB processing IC and one for the field timebase section: the latter de-interlaces the vertical scanning on pure text pages to eliminate interline flicker.

Memory extension

We've seen that there is little scope for increasing the number of teletext lines transmitted during the blanking interval, and that the Fastext system is only capable of reducing access time when pages are called up in a logical sequence. For fast access to a large and random selection of text pages it's necessary to provide a large data memory capable of storing scores or hundreds of sets of page data so that page selection, generally taking less than a fifth of a second, becomes a process of memory search and read rather than a wait for coincidence between requested and transmitted page numbers.

These large data memories are called Background Memory systems, and consist of an interface/data router and up to 4 Mbit of DRAM (dynamic RAM), capable of storing a maximum of 512 pages. The same decoding and display techniques as outlined in Figures 9.3 and 9.4 are used, and the data is systematically stacked in the memories as it is received. When a page request is made by the viewer, the DRAM is rapidly scanned in the full-field mode: time is saved by restricting the search to the memory area assigned to the relevant magazine and page header information. To ensure that the most up-to-date page is presented the memory is scanned in reverse order, that is from the most recently stored page to the oldest one. When the required page is located the memory is searched forward to read the data out to the one-page display memory within the text decoder IC.

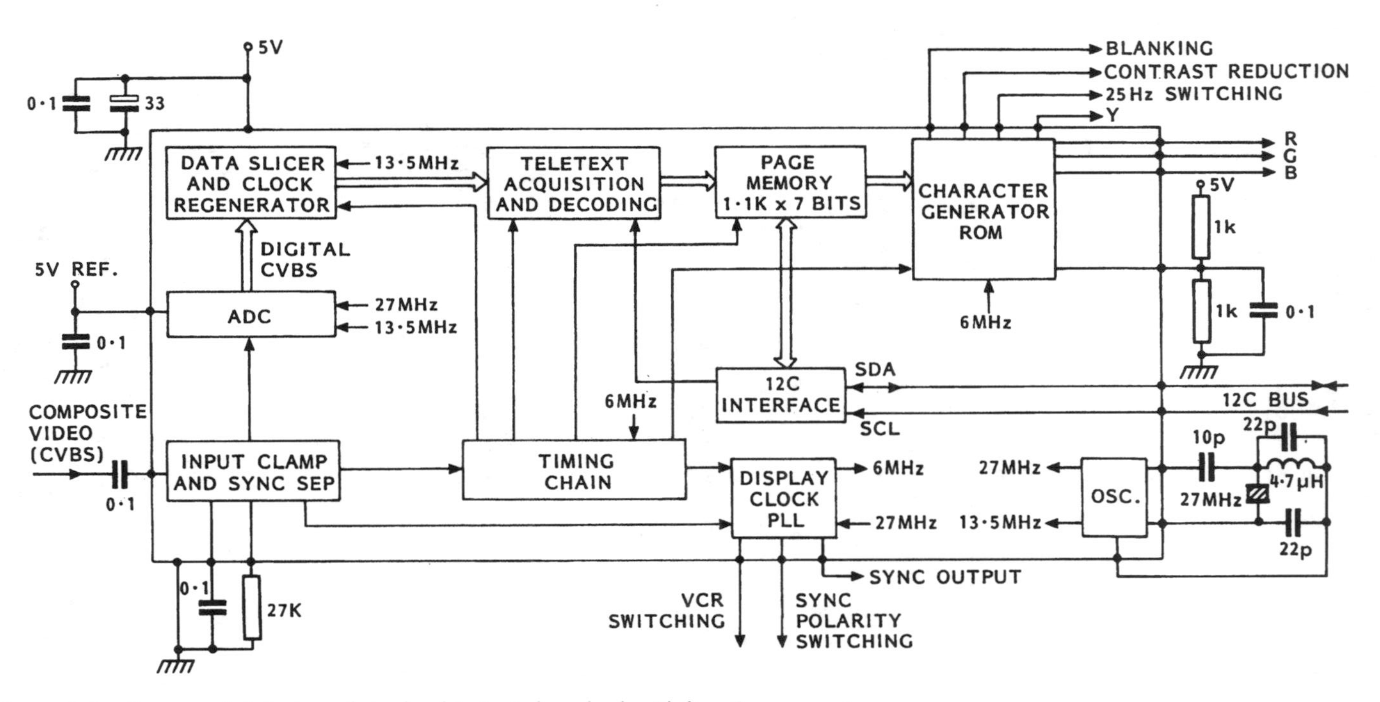

Figure 9.4 *Commercial design of single-chip text decoder by Philips Components*

Multilingual teletext

The advent of satellite transmissions which cover many countries simultaneously, and the acceptance of the UK-designed teletext system as a world standard (used in over 35 countries) has led to the development of character generators capable of holding up to 192 characters, the extras being selected by the control bits in the header row shown in Figure 9.2 and by packet 26 which specifies the supplementary character that overrides the basic one at any required point in a row and column. Similar means are used to subtitle programmes in many languages simultaneously, the desired translation being selected by the viewer for insertion in the picture. For text pages where two languages may be required on the screen at the same time, e.g. educational purposes, there are ICs with dual-ROM character generators capable of producing two alphabets. Switching between them during a single row of the display is achieved by using a control character to shift the readout as required, in similar manner to the selection of alphanumeric or graphic characters.

On-screen display

With suitable software in the TV's control system the teletext ROM can be used to generate on-screen menus and captions for such purposes as programme/channel number readout, analogue control setting displays and viewer-programmed status titles.

Teletext decoder integration

In some late designs of TV receiver all the functions of the text decoder IC shown in Figure 9.4 are incorporated in the system-control microprocessor, which thus sprouts RGB output pins, and very little else at all. On-screen captions and graphics can also be generated here, and this concept simplifies and cheapens a text-equipped set.

10 *PAL-Plus, MAC and enhanced TV*

With the coming of a new broadcast media such as cable and particularly satellite, whose bandwidth is less constricted than terrestrial broadcasts in the VHF and UHF spectra, much development work took place on new encoding systems to provide displays free of the impairments described at the end of Chapter 4. Some offer the incidental advantages of potentially better definition and elimination of flicker and *aliasing* effects in the luminance picture.

Some of them are still in use, especially MAC on satellite transmissions to Scandinavia, but in the medium term all these analogue standards and transmissions will be swept away by digital systems, which have many advantages over them.

Extended PAL

A relatively simple – and wholly analogue – enhancement system, proposed by the BBC was the *extended-PAL* system, in which the inability of a conventional TV receiver to fully display the fine luminance detail (in the presence of colour) is acknowledged by transmitting a conventional PAL signal, but with the bandwidth of the luminance component restricted to about 3.5 MHz. This gives rise to little deterioration of the picture on a conventional set using PAL techniques and a shadowmask tube, where the presence of a luminance *notch-filter* and the shadow-mask (which acts to filter out fine detail, especially in the smaller screen sizes) virtually delete the finer detail anyway. Chrominance is transmitted in the conventional way, based on the subcarrier of 4.43 MHz, so that this scheme is a compatible one (see Figure 10.1), but without significant spectrum-sharing by the Y and C signals.

The high frequencies of the luminance signal, corresponding to fine detail, are sent separately, spaced from their rightful position by exactly the colour subcarrier frequency. Thus a 5 MHz luminance component appears at 5 + 4.43 = 9.43 MHz. In suitably-equipped receivers the regenerated subcarrier can be used to drive a second luminance detector whose output, after filtering, can be added to the basic 'restricted luminance' signal to reinsert fine detail and achieve full definition

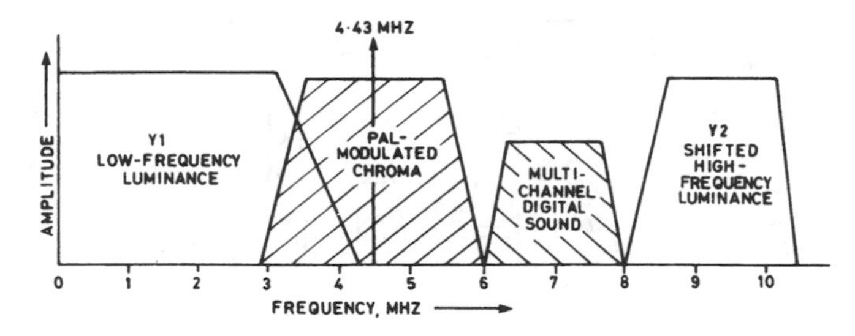

Figure 10.1 *The spectrum of an extended-PAL system. High-frequency luminance components (section Y2) have been shifted up in frequency to largely eliminate Y–C band-sharing*

pictures in colour without spurious effects. The gap in the extended-PAL spectrum between about 6 and 8 MHz is free for up to six digitally-modulated sound signals riding in a subcarrier (Figure 10.1 again).

This solution offers the advantages of compatibility and a low-cost receiving and decoding system for the enhanced colour pictures. It is, however, totally tied to the PAL system, and would have committed us to PAL and chroma subcarrier techniques for the foreseeable future.

Except in the sound channel, extended-PAL and similar systems take little account of the great advances which have been made in television technology in the decades since the subcarrier system was introduced into colour broadcasting, particularly in the area of colour picture encoding. With the economies of scale afforded by mass-production, digital and 'hybrid' ICs for use in television signal-processing can be made viable for domestic use in decoding systems which are more complex, but potentially capable of better performance.

PAL-Plus

A more sophisticated development of the PAL system, still maintaining the vital compatibility with existing equipment, is the PAL-Plus system. It offers the advantages of widescreen (16:9 aspect ratio) pictures, high-quality digital sound, and freedom from the cross-colour and cross-luminance effects (of band-sharing of Y and C components) inherent in conventional reception of subcarrier-based transmissions.

PAL-Plus pictures are seen in 'letterbox' format, with black margins above and below the picture, on ordinary receivers, because the analogue picture transmission contains 432 active lines. The others carry a digital 'helper' signal, used only in PAL-Plus receivers, which is used to build up a widescreen picture with 576 active lines. The PAL-Plus decoder has eight main building blocks as shown in Figure 10.2.

Analogue composite video (CVBS) signal from the TV's vision demodulator enters a conventional PAL decoder, the same in principle as that described in Chapter 6. The derived Y, U and V signals are A-D converted (see Chapter 12) into

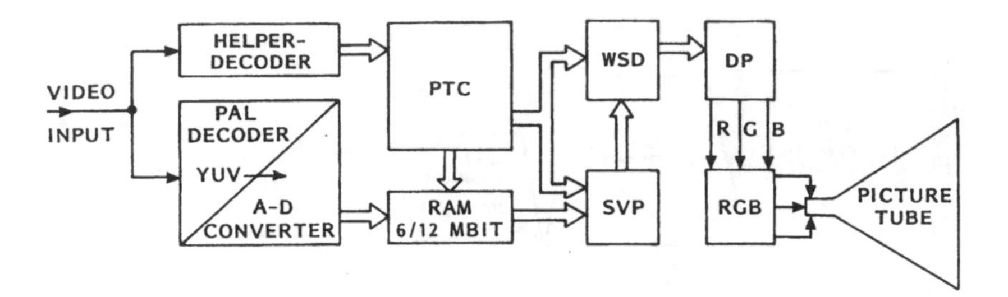

Figure 10.2 *Block diagram of the picture-processing sections of a PAL-PLUS TV receiver*

digital form for processing, and passed into a RAM capable of holding one complete picture frame: this calls for 6 Mbit capacity, or 12 Mbit in 'de-luxe' receivers with best-possible suppression of cross-colour effects. The RAM is under the control of a PTC (programmable timer controller), itself governed and instructed by a control data stream coming from the helper-decoder, whose function is to extract the control information contained in the analogue video signal's 'picture-dead' lines. The data emerging from the RAM consists of eight parallel streams of digital information, sourced sequentially from different addresses within the memory store at high speed. It enters a sequential video processor (SVP), a real-time processor which converts the 432 active lines to 576-line display data by reference to the control data furnished by the digital helper signal.

Data from the SVP is passed to the next block, consisting of a widescreen signalling decoder (WSD), an eight-bit processor which uses a control signal carried on line 23 of the vertical blanking interval. Amongst other functions it ascertains whether a 4:3 or 16:9 aspect ratio picture is being broadcast and initiates automatic scan- and signal-switching. The processor also contains a freeze-frame facility, invoked by the viewer via the remote-control handset, which cuts off the flow of data to the RAM, whose contents are continually re-read at field rate, another technique which will be described in Chapter 11. The final block in the PAL-Plus decoder is the display processor (DP), basically a D-A converter to change the eight-bit digital data back to RGB analogue form to suit (after amplification by wideband high-voltage amplifiers of the sort described at the end of Chapter 6) the picture tube. The widescreen tube itself is conventional in every way. A TV set fitted with a 16:9 tube is illustrated in Figure 10.3.

Time-division multiplex

Where several signals need to be carried over a single transmission system, various possibilities are open to us. We have already explored the interleaving mode, which is only viable where the separate signals for dispatch have a common 'base frequency'. An alternative system is frequency-division multiplex, where the various signal channels are carried on different basic or carrier frequencies. Probably the simplest everyday example of this is the conventional TV aerial downlead which contains four or more TV programmes and accompanying

Figure 10.3 *A widescreen TV receiver made by JVC*

sound tracks, all carried on different frequencies (channels) and easily sorted out by tuned acceptor circuits in the TV's front end.

A third alternative is the concept of *time-division multiplex*. Here the various streams of information are sent sequentially, each in a well-defined 'time slot' in the transmission. If we are to have a continuous signal stream available at the receiving end, this implies some form of storage system there to sustain the information flow during the periods when other information is present in the channel. A further requirement is that a synchronising signal be present at regular intervals in the composite signal to time the 'strobe' style detection and separation process.

Time-compression of analogue signals

Suppose we wish to send a five-minute spoken message over the telephone, perhaps to a far-distant country for which telephone time-charges are high. We may choose to compress our message into a time slot of $2\frac{1}{2}$ minutes by means of recording it on tape and replaying into the 'phone at double speed. If our correspondent (suitably forewarned!) were to hook the telephone to a tape recorder running at twice normal speed he could capture the message for subsequent replay at normal speed. In this way the *real-time* transmission phase is reduced by half, though at the expense of bandwidth. Inevitably the twice-speed message contains twice the frequency spectrum of the uncompressed message, and the media (in our case the telephone link) must be able to carry this. Time-division multiplex and time compression are two ingredients of the MAC (multiplexed analogue components) transmission system.

Multiplexed Analogue Components (MAC)

If we take a sample of a video signal about 700 times in each line period, and provide the means of storing the samples, we can 'hold' one TV line at a time in a memory, which can be analogue (charge-coupled device, CCD) or digital in nature, and embodied in a purposed-designed IC. We have to write in the samples at television scanning rate, but we can read them out at any rate we choose; see Figure 10.4. If we read them out at a faster rate we shall effectively have compressed the signal in time at the expense of a faster signal-rate, rather like our telephone and tape-recorder trick described earlier. The active picture period of a 625-line transmission is 52 μs (as shown in Chapter 1) so we shall need this period to write one TV line of luminance into the memory. Let's suppose we read it out in 35 μs, representing a time-compression of about 30 per cent. This will increase the data-rate proportionally, but leave a gap in the time-domain (assuming the standard TV-line 'window' of 64 μs) of 29 μs, in which we can do some time-multiplex magic to accommodate chroma and sound signals. Now for the chroma. Here we have about 18 μs available per line for transmission, and again we write the TV chrominance signal (say V, representing R – Y) into a one-line store. The read-rate here is faster than for luminance to squeeze 52 μs worth of V information into a real-time slot of 18 μs, time-compressing the V-signal by about 65 per cent, again with a proportional increase in data-rate.

The now time-compressed luminance and chroma signals are assembled in the form shown in Figure 10.5 which clearly shows the time-division-multiplex composition of the MAC signal. As only one colour-difference signal can be accommodated per data-line, V and U signals are sent alternatively on a sequential basis, rather like the SECAM system of colour encoding, where V and U signals are dispatched in the same way, though using a totally different modulation system, and operating in real time alongside the Y signal. Both SECAM and MAC require a one-line delay at the receiving end, not for cancellation-matrix purposes like a PAL decoder, but rather to yield a stored chroma line while its fellow is being received.

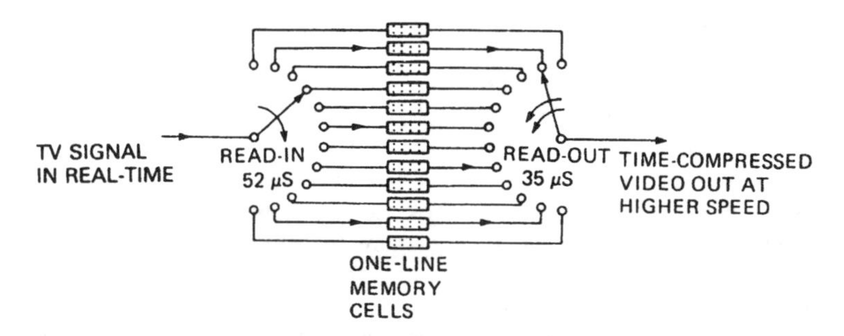

Figure 10.4 *Reading in and out of a TV line store at different rates. Each memory cell forms a short term store for a single pixel, and for a single TV line around 600–800 such stores would be required*

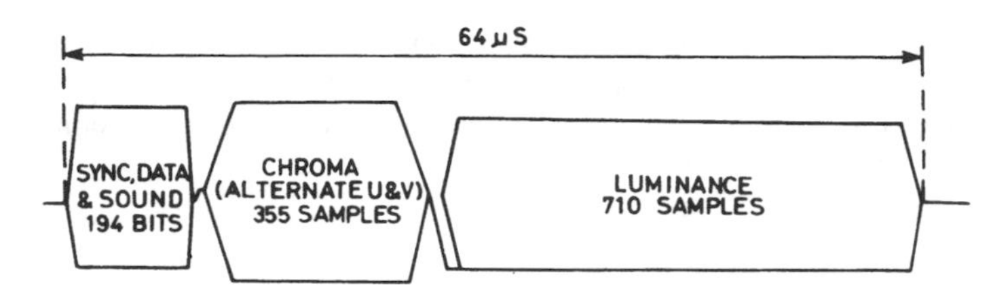

Figure 10.5 *The signal format for MAC. Full details are given in the text*

MAC sound and sync

Our 64 µs data line is now occupied by about 53 µs of picture signal information in 'compressed' form, first 18 µs of V or U, then 35 µs of Y. This leaves about 10 µs for sound and synchronisation signals. While this may not sound very much, the tremendously-high bit-rate inherent in the system affords almost 200 bits in this period, which, remember, need only sustain the sound for a 64 µs period. Thus we have room in the *line sync word* for an initial eight bits (to provide system synchronisation for sound and vision demodulation), then sound and data (teletext) signals, all in digital form. The MAC-C system has potential for eight simultaneous data channels which may be used to carry either sound tracks or teletext pages.

MAC decoding

It is not possible to go into the MAC decoding process in much detail in this book, but as with any form of decoding, the process must be the 'reciprocal' of that at the sending end. Thus we write the luminance and chrominance signals into one-line stores at the fast transmission rate and simultaneously read out of them at normal 625/50 television rate to render continuous luminance signal and line-sequential V and U signals. By means of a 64 µs delay-line store and an electronic double-pole switch a constant flow of colour-difference signals is secured, each of R – Y and B – Y being updated at 128 µs (two-line period) intervals.

MAC interfacing

At the output of the MAC decoder we have sound and vision signals in baseband form. The latter may be RGB or Y, V, U signals, and ideally they should be fed direct to the picture-tube drive circuitry of the TV set (or monitor) in use to realise the full advantage of this mode of encoding. The MAC decoder can form an add-on unit to the TV set, or be an integral part of the down-converter for satellite reception, or the terminal unit of a wideband cable system. Where the TV receiver has sockets only for UHF or baseband CVBS inputs it is necessary to recode the chroma in PAL (or SECAM where appropriate in Europe) to enable the set to display a picture, which will necessarily then suffer all the shortcomings of cross-colour etc. related above! For TV sets with just a UHF input a suitable RF modulator as used in VCRs, TV games etc., will be required as an output device in the conversion equipment, whereupon a spare channel in the receiver will be reserved for cable and satellite transmissions, and tuned to the modulator's output. More details will be given in Chapter 22.

Subcarrier versus MAC signals

The three established encoding systems, NTSC, PAL and SECAM, were designed with the characteristics and particularly the noise performance of terrestrial VHF and UHF AM broadcast systems in mind, and with the need for compatibility with monochrome equipments at the top of the priority list. Many of the constraints which existed then have disappeared, and the main stumbling block – that of channel bandwidth – has considerably eased as carrier frequencies (and now bit-reduction ratios, see next chapter) have gone up and up. With the tremendous coverage area of satellite transmissions, a different form of compatibility now has to be considered, that of the different encoding systems – and languages – of the countries within the catchment area of a single transmitter in space, or indeed the vast distance over which TV signals can be propagated along fibre-optic light-guides. The language barrier is catered for by the availability of multiple sound channels, and the idea of component coding, in which each colour-signal is preserved intact throughout the transmission chain, gives great flexibility in the types and geographical spread of receivers in use, once the terminal equipment is freely and cheaply available.

High-definition and high-fidelity TV

The path to high-definition TV consists of several discrete steps, starting with improvement and enhancement of the existing display on current standards, and ending with the realisation of a TV picture with a wide-screen format and resolution capability equal to – or better than! – the best 35 mm cine film reproduction, and sound to match. As we shall see, the technical problems associated with high-definition TV are huge, but not insurmountable. Where home and broadcast TV are concerned, the old bogey of compatibility rears its head again, and many of the proposals being made are based on the idea that any new system will be amenable to existing receivers; an incompatible format would call for dual-standard receivers and possibly programme duplication, which is wasteful of spectrum space, no matter which media (or band) is used for transmission.

The digital MAC and PAL-Plus systems just described were the first steps towards enhancement of the TV picture. Although based on the traditional 625/50 system, the elimination of the worst visible defects of subcarrier colour systems such as cross-colour offers a great subjective improvement in reproduction. By incorporating two digital field-stores in the receiver, further improvements are possible. By interpolating between corresponding lines in the two interlaced fields and drawing more lines in the display, it is possible to achieve a subjective definition equivalent to a 900-line picture. The presence of a frame-store also facilitates reading out picture information at a faster rate than the 50 fields/second used in normal transmission to eliminate the flicker effect which is noticeable in large areas of high brightness. More of this in Chapter 12.

More than 625 lines

We have seen many times in this book that more picture detail calls for more bandwidth, and this is true whether the signal is transmitted in analogue or digital mode. For a worthwhile gain in picture definition we have to go beyond 1000 lines, and systems have been suggested using between 1100 and 1500 lines. A doubling of the European 625 standard to 1249 lines has often been advocated in the interests of compatibility and Philips have done much research along these lines. To draw 1249 lines in the same field period of 20 ms the scanning spot has to move twice as fast as before which means that to draw the same features the video signal must also move twice as fast, implying a doubling of bandwidth. Having done this, however, we are left with a picture which is capable of resolving much detail provided such detail is stacked vertically to take advantage of the higher line standard.

There is little point in doubling the vertical definition without a proportional increase in horizontal definition to match. To achieve this we need to double the vision signal bandwidth again, and so we arrive at a figure of 20–25 MHz for the luminance channel alone, or for each channel if the system is an RGB one. Such bandwidths are not easily accommodated, except with bit-reduced digital broadcast techniques, an idea of which will be given in the next chapter.

Aspect ratio

The aspect ratio of current TV displays was for half a century 4:3, largely because the first picture tubes had circular screens and the TV standards were based on these. For greater realism, it is generally accepted that an aspect ratio approaching 2:1 is a reasonable compromise between the ideal (which presumably is a cylindrical or spherical screen encompassing the viewer) and the practicalities of generating the display. Certainly the wide screen goes better with stereo and surround sound, and Hi-Fi TV incorporates both these features.

High-definition TV programme sources

While 'electronic' signal sources will provide data displays and animated cartoons (along the lines of the text systems described in the previous chapter), most people are more interested in 'real' pictures, and these must come from cameras and ciné film. Solid-state image sensors have now been developed to the point where they can be used for picture pick-up in a high-definition TV system with the advantage of excellent RGB registration. Early experiments and demonstrations of high-definition necessarily used very-high definition camera *tubes*, based on the vidicon and Plumbicon principle, but usually of larger diameter, perhaps 30 mm.

The hardware for televising cine film is called a *telecine* and it uses not camera tubes but photomultipliers for image-sensing, in conjunction with an FSS (flying-spot-scan) tube to illuminate, 'swinging torchbeam' style, the film frames as they pass the gate. At the very high data rate involved in high-definition TV neither photomultipliers nor FSS tubes are really fast enough, and experiments have involved CCD line-sensors in conjunction with a prism for pick-up (Philips) and a laser beam deflected by a rotating polygonal mirror as light source (NHK, Japan). It is fortunate indeed that good 35 mm (and especially 70 mm) ciné film contains sufficient detail to do justice to these television systems!

Ironically, once high-definition TV is fully established in the professional closed circuit (as opposed to the broadcast) field, it may well signal the end of ciné film for professional purposes. Provided the necessary parameters are met, electronic film-shooting has many advantages over a conventional film camera in terms of versatility in subsequent processing; creation of special effects; correction of lighting and other deficiencies in the original scene; lower production costs in scenery and backgrounds, which can be subsequently 'added' by electronic mixing; and the ability to instantly review a sequence on a VCR monitor before sets are dismantled and personnel dispersed.

High-definition displays

The shortcomings of the shadowmask tube have already been discussed in relation to conventional television, and we saw that the shadowmask/triad combination sets a strict limit on the resolution available from this type of display. A well-set-up and magnetically-focused monochrome tube can display 1200 lines without too much trouble, but to achieve this with a shadowmask system is very difficult. There is another factor, and that is screen area. Even if we could achieve an 1149-line picture on a colour tube of conventional size (50–70 cm), it would subjectively show little resolution improvement over the present format, unless the viewer were within a few centimetres of the screen. For home use, a display area about 1.4 m × 0.7 m seems optimum for high-resolution TV, and conventionally-scanned tubes are not very amenable, in a physical sense, to that size or shape!

To display our high-definition pictures, then, with sufficient brightness and resolution it is currently necessary to use projection techniques. For very large-screen applications one of the light-valve systems described earlier can be used, and the Eidophor system is demonstrably capable of handling pictures with more than 1000 lines. For smaller screens in the 1 m^2 region, direct-view tubes are now challenging projection systems in which the requirements of registering all three images within half a pixel all over the screen make tremendous demands on the convergence system.

Figure 10.6 shows an experimental hi-fi TV projector by Philips. It can define one million pixels, and uses three high-resolution 13 cm tubes running at 50 kV

Figure 10.6 *A projector for hi-fi TV by Philips. The lens system for the R, G and B tubes is clearly seen in this photograph. Other features are described in the text*

EHT. It is intended for use with a concave screen having a light-gain of ×5. Dynamic convergence is handled by a microprocessor-based control system feeding separate correction coils on each tube.

High-definition TV summary

As our brief account has shown, the generation, distribution and display of high-definition TV pushes every aspect of TV design way beyond the limits imposed by the present TV system, in which the capabilities of all the links in the chain are fairly evenly matched. To bring them all up to standard for higher-resolution television pictures will cost a great deal of money, and require a great deal more research. Although all the technology for advanced TV already exists, and very impressive results have been demonstrated, the implementation of *domestic* systems (especially broadcast-derived ones) awaits many developments, mostly in the region of the economics of manufacturing practical fine-detail display devices.

The bandwidth problem has been solved by the development of digital TV broadcasting techniques incorporating data compression as we shall see in the next chapter. They are very complex, but are nevertheless suitable for home TV use because of the economies afforded by mass-production of the VLSI (very large scale integration) ICs involved. All the time that the conventional thermionic picture tube provides the display, however, there will be a need for the timebases, power supplies and shadowmask systems described in previous chapters.

High-definition TV has had a chequered history. One of the earliest systems was pioneered by Japanese national broadcaster NHK in 1980, using a 1125-line picture transmitted from a satellite. It was costly and greedy of bandwidth, and NHK's subsequent MUSE (multiple sub-Nyquist sampling encoding) system, while reducing the bandwidth requirement, was even more expensive in terms of the hardware of the day. An experimental HDTV system was transmission-tested in the USA in 1998 by national broadcaster CBS, not excessively expensive in terms of signal-processing hardware this time, but needing a very costly display screen to do justice to its capabilities, and at the time of writing an HDTV receiver is about twenty times more expensive than a conventional one. This brings us to the question of consumer demand, and the British public has demonstrated very clearly that most people are not interested in advanced picture systems, especially when they carry a large cost premium for home use. The new digital TV technology is quite capable of handling a high-definition picture (indeed it's an option in the MPEG system we'll meet in the next chapter) but the demand must be there before a broadcast system can be put into place. It is likely that HDTV broadcasting will be first on the ground in the USA or Japan, whose people (especially the Japanese) seem more interested than those of the UK in the technical quality of their pictures. Another factor which mitigates against the early introduction of HDTV in the UK is that DTV broadcasting is almost wholly in the hands of commercial companies whose prime interest is in selling subscription TV and collecting advertising revenue rather than advancing the technical standards of television. So much is the hardware a means to an end that (again at the time of writing) £300 digital set-top boxes are being given away in exchange for a viewing subscription and a phone-line connection for 'TV commerce'. In the trade-off between picture quality/definition and channel choice, both the public and the broadcasters have come out strongly in favour of the latter, to the extent that several channels are given over to the same movie at staggered starting times, and extra bandwidth is devoted to (for instance) providing a choice of viewpoints on a football match. In this scenario it seems that everyday HDTV is a long way off, even in the new era of digital transmissions, and may only be heralded by a breakthrough in TV display technology.

3D TV

Perhaps the ultimate step in television technology is the presentation of a three-dimensional display with depth and perspective as well as full colour and high definition. One day this may be realised by means of *holographic* techniques in which interference patterns from laser light sources are able to create an apparent three-dimensional image without the need for the viewer to wear special spectacles or other optical aids. 3D systems using more conventional display sources have been demonstrated, however, and have even found their way onto domestic

TV screens in the form of demonstration transmissions, for which special bi-colour viewing glasses are required.

One system uses two conventional TV sets, fed with video signals from two cameras viewing the same scene from different angles. The sets are mounted in 'L' formation with the screen planes at right-angles; each screen projects an image onto a sheet of reflective plate glass set at an angle of 45° between the two screens. Viewed through special spectacles the effect is very striking, with the picture appearing to project forwards into the space in front of the display surface.

An alternative approach, based on the stereoscopic principles known since Victorian times and exploited in some cinema films from time to time, involves a colour-separation process in which red images are separated from green (or cyan) and presented side-by-side on the display screen to render a picture which looks like a badly-converged colour display. The horizontal separation of the two coloured images determines the apparent depth and spatial position of the object thus outlined, and will vary to describe 'Z axis' (i.e. on a line between the viewer and the display surface) movement. This effect can be exploited in a basically black-and-white or colour picture.

A means of artificially inserting a 3D effect into ordinary colour broadcast (or videotape) pictures was introduced some years ago by European setmakers Saba and Nordmende. When invoked, the 3D facility horizontally separates the red and green images by means of an electrical time delay, and a quasi-3D effect is seen when viewing the result through the red/green spectacles supplied. There are several alternative approaches to 3D TV short of the full holographic treatment, and if and when a common and compatible system emerges it may well involve the use of polarised light.

A chronology of the main events in UK TV development is given below. The UK was first in the field with television itself; with teletext and Nicam stereo sound; and with digital TV broadcasting. America led the way with colour TV broadcasting.

1925	JL Baird demonstrates 'mechanical' TV
1929	Low-definition (30-line) TV broadcasts via radio transmitter
1936	First regular electronic TV broadcasts in the world by BBC
1939–46	TV shut down due to the war
1955	Commercial TV (ITV) begins in London area
1958	Video recording first used in Britain
1962	First transatlantic satellite TV transmission – via Telstar
1964	BBC2 begins, using UHF transmissions and 625-line pictures
1967	Colour TV starts, initially on BBC2
1969	Colour on all TV channels, live pictures from the Moon
1975	Teletext transmissions start
1982	Channel 4 opens
1988	Nicam stereo sound for television begins
1989	Satellite broadcasting direct to homes – Sky Television
1990	MAC satellite transmissions from BSB
1996	First pay-per-view TV programme – B Sky B
1997	Channel 5 launches
1998	Digital TV starts, from satellite and terrestrial transmitters
1999	Interactive TV begins

11 *TV sound systems*

In Chapters 5 and 8 we briefly touched on sound systems for TV. Here the subject will be examined more closely, with particular regard to the digital (Nicam for terrestrial and MAC for satellite) and multi-carrier (analogue satellite) systems currently used with analogue TV broadcasts. First let's explore in a little more detail the conventional FM sound system which has served us for many years and continues to do so.

FM mono system

The monaural sound is transmitted on its own frequency-modulated RF carrier with ±50 kHz maximum deviation, at a level 10 dB below that of the vision carrier. In European systems B/G the spacing is 5.5 MHz above the vision carrier. For the UK System I scheme it's 6 MHz above the vision carrier frequency, and we'll use that for descriptive purposes.

The sound carrier beats with the local oscillator in the tuner to produce an IF output at 33.5 MHz. This low-level constant-amplitude signal is passed via the vision IF amplifier to the vision demodulator where it beats with the vision carrier to produce a 6 MHz signal, still frequency-modulated, which can then be selected, amplified and demodulated. This *intercarrier* system has several advantages: tuning errors and drift have no effect on the carrier frequency, which is governed solely by the very accurately-maintained vision–sound spacing at the transmitter; the sound carrier benefits from the gain provided by the vision IF amplifier; and the sound processing circuit is a simple one.

Receiver circuit

Figure 11.1 shows a typical intercarrier sound system for a basic TV set. The wanted 6 MHz carrier is selected from the demodulated vision carrier by a ceramic filter, resonant at 6 MHz and having a bandwidth of about 200 kHz. The separated sound carrier passes into the IC at pin 1 for passage through several stages of amplification with limiting, which latter clips off any amplitude modulation caused by the vision signal: the AM rejection is about 55 dB. This clipping

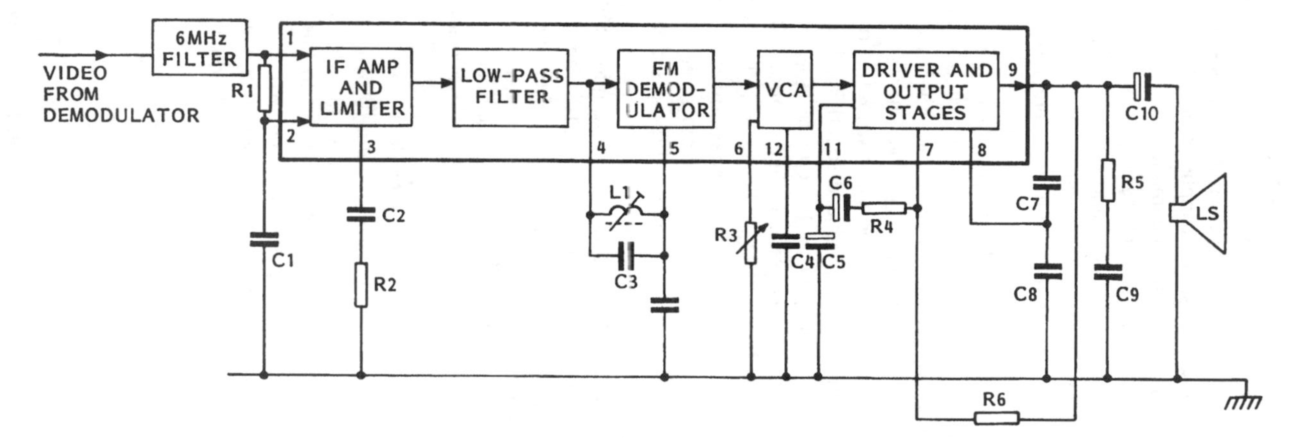

Figure 11.1 *FM receiver and audio circuit for terrestrial monaural sound*

action produces a squarewave output, rich in harmonics of the 6 MHz carrier frequency. They are suppressed in a low-pass filter which restores the carrier waveshape to approximately a sinewave for application to the FM demodulator of the quadrature synchronous type. It works on the sample-and-hold principle which we've met before in these pages. The sampling gate is opened for an instant during each carrier cycle at a time when an unmodulated carrier wave is passing through the zero point, so the sample taken corresponds to zero volts. As the phase/timing of the carrier signal advances and retreats in sympathy with the modulating sound signal, the sampling action produces an output proportional to the frequency deviation, an exact replica (after de-emphasis) of the original sound.

A reference carrier is required to produce the demodulator gate pulses. It's generated by the high-Q tuned circuit L1/C3 whose flywheel effect averages the carrier frequency, thus providing a constant-phase, 90° reference feed. In current practice this tuned circuit takes the form of a ceramic filter.

Next comes a voltage-controlled amplifier (VCA) whose gain depends on the DC voltage present at IC pin 6: the reproduced sound level is proportional to the applied voltage, which is produced by the control system as described in Chapter 8. In many modern sound processors ICs the control voltage is internally derived from data on the serial control bus as illustrated in Figure 8.12.

The demodulated audio signal is de-emphasised next, by an RC network whose C element is external to the IC at pin 12. The audio signal is now ready for application to the driver and output stages which (in the simple system shown here) is inside the same IC package. The output stage generally consists of a push–pull pair of transistors operating in Class B mode, the DC mid-point voltage being isolated from the loudspeaker by coupling capacitor C10. The value of the resistors connected to pin 7 of the IC determines, as part of a negative feedback loop, the AC gain of the output stage, the capacitors connected to pin 8 setting the amplifier's upper frequency limit. R5 and C9 form a Boucherot cell which suppresses any tendency for HF oscillation due to the inductive load of the loudspeaker and its wiring.

A small IC of the type illustrated here can provide an audio output power of about 3 W, depending on the supply voltage and the heatsink capacity (if any) provided. This type of IC, or a small power amplifier chip fed from a separate intercarrier sound amplifier/demodulator IC, easily caters for the needs of a portable TV. Although for the purpose of description here we've examined a dedicated sound channel IC, in most sets the intercarrier sound and audio preamplifier stages are incorporated in a more complex IC package which performs many other functions.

Stereo sound

Most large-screen TV sets are stereo types, fitted with two (or two sets of) loudspeakers, and relatively high power amplifiers to drive them. Balance, volume and tone controls are provided by the microprocessor control system, often via the two-wire control bus described in Chapter 8. Some sets have provision for processing a monaural sound signal to give it a 'spacial' or pseudo-stereo effect by differentially manipulating the phase or amplitude of the signals in the left and right amplifier channels. Circuits have also been devised to emphasise the difference between the L and R audio signals from a stereo source to artificially widen the perceived sound field: the loudspeakers of a stereo TV set are necessarily close together, and this 'stereo-wide' system partially compensates for this.

Stereo TV sound can come from several sources: Nicam and MAC broadcast transmissions; stereo VCRs; video discs; camcorders; and the FM stereo sound carriers of satellite broadcasts, which we shall now examine.

Satellite FM sound

With most conventional satellite TV transmissions using FM picture modulation (more of that in Chapter 13) there are FM carriers for the sound, very similar to those used for terrestrial mono TV sound. The main difference is that there are more of them! Figure 11.2 shows a typical satellite TV channel spectrum, with five sound carriers on the HF side of the baseband video signal, at 6.5, 7.02, 7.20, 7.38 and 7.56 MHz.

Thus the FM signal processing system is exactly the same as that already described for terrestrial TV sound transmissions save for the operating frequency. In, for example, Astra satellite transmissions two carriers are used to convey the left and right channels of a stereo broadcast. Thus in an analogue receiver there are two intercarrier amplifiers, filters and demodulators, tuned to 7.02 and 7.20 MHz for R and L signals respectively.

Satellite sound carriers

The auxiliary sound carriers have only ±50 kHz maximum deviation, though the audio bandwidth is 20 Hz to 15 kHz, resulting in a low modulation index. To prevent this from impairing the S/N ratio a noise-reduction system is used – with Astra transmissions the *Wegener Panda 1* type is employed. The term Panda is derived from 'processed narrow-deviation audio', a form of adaptive pre-emphasis. The dynamic range of the audio signal is compressed before transmission so that, in relative terms, high-level signals are attenuated and low-level ones boosted, the amount of compression also being frequency-dependent. The inverse of this process is applied at the receiver so that the dynamic range of the signal is restored. In the process the noise component is suppressed.

The system has much in common with the companding principle used in hi-fi VCRs (Chapter 19 and Figure 19.7) and with Dolby noise-reduction techniques. It's implemented by IC-based VCAs whose control voltages are derived from the signal itself via filters. Without noise-reduction the signal-to-noise ratio of a narrowband satellite TV sound channel is about 50 dB: the Wegener Panda 1 system provides an improvement of about 18 dB so that the subjective S/N ratio approaches 70 dB, which is very good for an analogue transmission. Table 11.1 shows the uses to which the auxiliary carriers are put, for stereo and multi-lingual sound in a typical satellite broadcast channel.

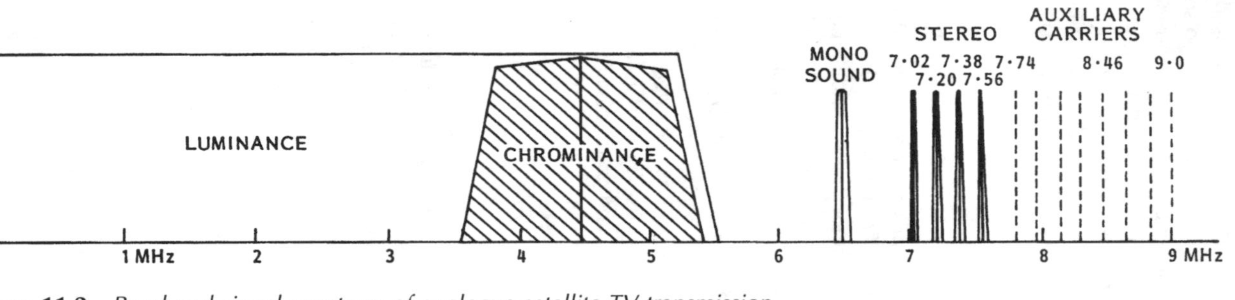

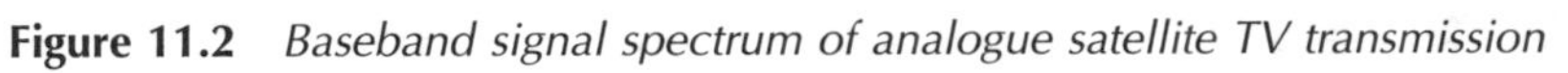

Figure 11.2 *Baseband signal spectrum of analogue satellite TV transmission*

Table 11.1 *Typical assignments of audio carriers in a satellite TV transmission*

Mode	7.02 MHz	7.2 MHz	7.38 MHz	7.56 MHz
1	language 1L	language 1R	language 2L	language 2R
2	language 1L	language 1R	language 2 mono	language 3 mono
3	language 1 mono	language 2 mono	language 3 mono	language 4 mono

Carrier conversion and selection

It is impractical to provide separate filters and demodulators for all the spot-frequency sound carriers shown in Figure 11.2, especially as a maximum of two (e.g. L/R stereo) are in use by the viewer at any one time. Instead, a superhet configuration is used to convert the wanted carriers to fixed frequencies which can be dealt with by a single pair of fixed-tuned filters and demodulators. This technique has the advantage of versatility: it works with any pair of carrier frequencies, including the piggy-back radio stations (Figure 11.2) supported by many transponders on carrier frequencies between 7.74 and 9 MHz, while in sophisticated systems the required sound channel can be user-programmed per transponder and stored in memory.

There are various ways of implementing the sound superhet technique. A common one is illustrated in block-diagram form in Figure 11.3, where two fixed-frequency oscillators run at 17.72 and 18.08 MHz. The output from one or the other is selected and fed to one of the gates of the dual-gate FET (field effect transistor) mixer, where it beats with the incoming signal to produce IFs at 10.52 MHz (R channel) and 10.7 MHz (L channel). With the 17.72 MHz oscillator switched in, the R sound comes from the 7.2 MHz carrier (17.72−7.2 = 10.52 MHz) while the L sound comes from the 7.02 MHz carrier (17.72−7.02 = 10.7 MHz). When the control system selects the output from the 18.8 MHz oscillator, the R and L sound signals come from the 7.56 MHz and 7.38 MHz carriers respectively.

Instead of having fixed oscillators a variable or programmable oscillator can be used: by varying its frequency it's possible to tune – like a radio receiver – through the band of sound carriers associated with each satellite broadcasting channel. Figure 11.4 shows how this may be arranged. The variable-frequency local oscillator is controlled by an IC containing a programmable divider and a 4 MHz crystal-controlled reference oscillator. These form a frequency-synthesis tuning system which is governed by the receiver's microprocessor control IC. The oscillator's output is fed to gate 1 of the mixer FET, while its other gate receives the whole spectrum of sound carriers from bandpass filter FL1, whose passband extends from 5 to 9 MHz.

The mixer's output contains all the sound carriers in the form of beat products in the range 8–13 MHz. Those centred on 10.52 and 10.7 MHz are selected by ceramic filters FL2/FL3 and FL4/FL5 respectively and applied to pins 11 and 14 of the dual FM demodulator IC2. By programming IC1 any pair of sound carriers

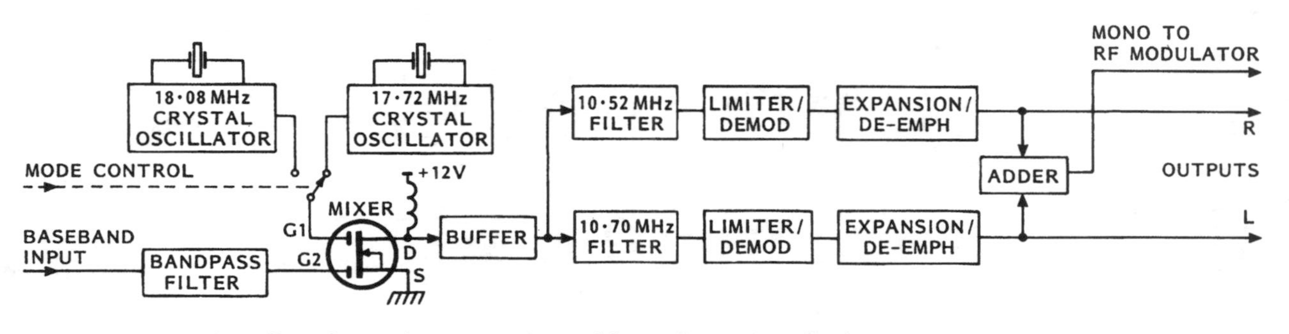

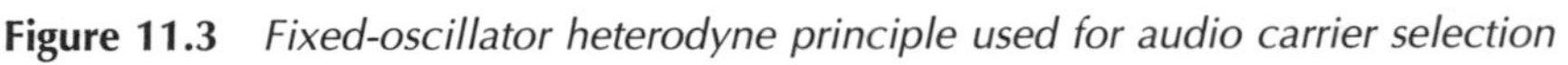

Figure 11.3 *Fixed-oscillator heterodyne principle used for audio carrier selection*

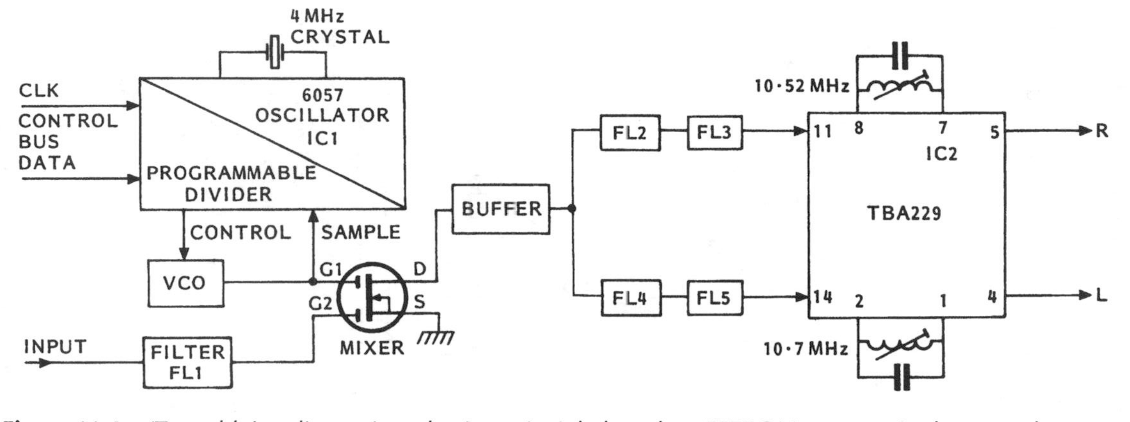

Figure 11.4 *'Tuneable' audio carrier selection principle based on EEPROM memory in the control system*

180 kHz apart (see Table 11.1) can be brought into line with the filters and demodulators. In a typical design the sound is tunable from 5 MHz to 9 MHz in 10 kHz steps, and any required point can be stored in non-volatile memory in a tuning/control system similar to that described in Chapter 8. The L and R outputs pass from the dual-demodulator IC to de-emphasis circuits which can be switched between simple linear (50 µs) or Wegener-Panda expander operation, the latter involving a purpose-designed and licensed-use IC. It's interesting to note that in this mode of audio operation the sound carriers undergo a quadruple-superhet process: their frequencies are changed at the dish-mounted LNB (low-noise block), at the indoor tuner (both to be described in the next chapter) at the vision demodulator and at the sound carrier frequency changer.

Digital TV sound systems

Nicam (terrestrial) and MAC (satellite) sound systems are very similar in their modes of operation, and the overview which follows, although based on Nicam, is equally applicable to the audio component of a MAC broadcast except in respect of the modulation and encoding of the digital data.

The Nicam system conveys audio data in phase-modulation of a low-power carrier placed just above the conventional FM sound carrier of a TV broadcast. It has capacity for two totally independent sound channels, which are generally the L and R components of a stereo signal, but can be used for bi-lingual sound tracks or data for (e.g.) downloading to a computer.

Figure 11.5 outlines the processes involved in Nicam transmission. The studio L/R sound feeds are first pre-emphasised and passed to an analogue-to-digital (A-D) converter with a clocking rate of 32 kHz, which means that samples are taken at intervals of 31.25 µs. To prevent aliasing the upper frequency limit of the L and R audio input signals is restricted by filters to 15 kHz. Each of the channels is simultaneously sampled to 14 bits to give 16 238 possible levels, after which the bit count is reduced to ten to conserve transmission bandwidth.

The 14 to 10-bit conversion process is a complex one, designed to retain most of the advantages of 14-bit processing – it's illustrated in Figure 11.6. From top to bottom of the diagram every possible bit combination for a 14-bit word is shown: where the bit value is irrelevant it is shown as an X. The shaded areas indicate the bits removed for transmission. The most significant bit (MSB) passes through regardless, but the next bits are deleted if they are the same as the MSB, up to a maximum of four. If this process leaves more than ten bits, sufficient bits are trimmed from the LSB (least significant bit) end to reduce the length of the word to ten. Thus is the digital signal compressed, and as long as the decoder (at the receiving end) knows what the compressor is doing from moment to moment it can expand the data back to 14-bit form.

A careful look at Figure 11.6 shows that 10-bit resolution is given to large signals (coding range 1), 11-bit resolution applies in range 2, 12-bit in range 3 and so on up to the 14-bit range 5 which represents the quietest and most vulnerable sounds. This 'progressive coding' system is very effective in terms of bandwidth economy, while losing little in comparison to a linear coding system.

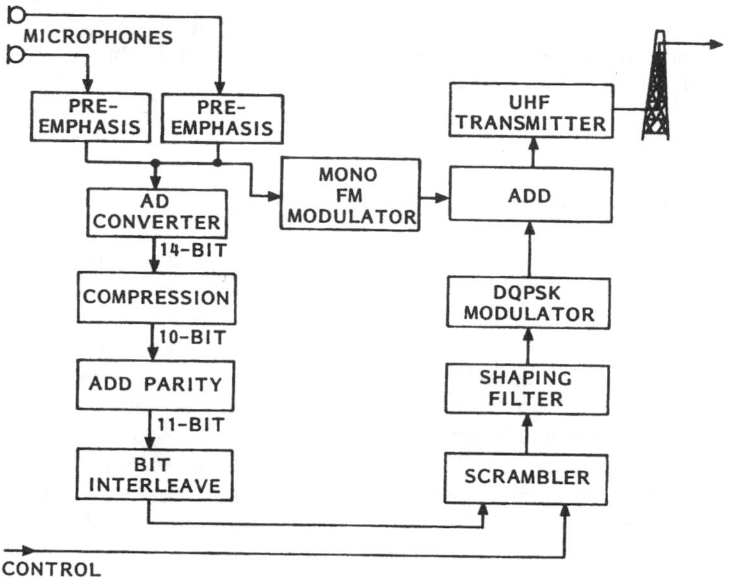

Figure 11.5 *Block diagram of the processes involved in the transmission of a Nicam stereo sound signal*

	SCALE FACTOR	CODING RANGE	PROTECTION RANGE
0 1 X X X X X X X X X X X X	111	1	1
0 0 1 X X X X X X X X X X X	110	2	2
0 0 0 1 X X X X X X X X X X	101	3	3
0 0 0 0 1 X X X X X X X X X	011	4	4
0 0 0 0 0 1 X X X X X X X X	100	5	5
0 0 0 0 0 0 X X X X X X X X	010	5	6
0 0 0 0 0 0 0 X X X X X X X	000 OR 001	5	7
0 0 0 0 0 0 0 X X X X X X X		5	7
1 1 1 1 1 1 1 X X X X X X X		5	7
1 1 1 1 1 1 1 X X X X X X X		5	7
1 1 1 1 1 1 0 X X X X X X X	010	5	6
1 1 1 1 1 0 X X X X X X X X	100	5	5
1 1 1 1 0 X X X X X X X X X	011	4	4
1 1 1 0 X X X X X X X X X X	101	3	3
1 1 0 X X X X X X X X X X X	110	2	2
1 0 X X X X X X X X X X X X	111	1	1

X = IRRELEVANT BIT   (shaded) = BITS DELETED FOR COMPRESSION

Figure 11.6 *Nicam coding/compression table showing how the 14-bit words are reduced to 10 bits for transmission. Bits in the shaded areas are not transmitted*

Ordinarily a 10-bit data word offers an S/N ratio of about 60 dB in the reproduced signal. With the companding system described here the S/N ratio approaches the 84 dB theoretically available from a 14-bit system.

Protection and scale factor

A *parity* bit is added at the end of each 10-bit word to provide a check on its validity at the decoder. Here even parity is used so that the first six bits of the received word, when added to the parity bit, give a '0' result. The parity bits are modified according to a look-up table held in ROM at each end of the chain to signal the scaling factor in a three-bit word shown right of centre in Figure 11.6. If the missing bits are on the left-hand side of the diagram the scale factor data enables them to be recreated at the decoder; if, however, they are on the right of the block they are lost forever. The words of which these latter form a part are not crucial ones in terms of noise, though, because they represent loud sounds or transitions between two widely-different levels, where the noise is masked.

For scale-factor signalling purposes the 11-bit (10 plus one parity bit) words are grouped together in blocks of 32, each block occupying 1 millisecond. Since only one scale-factor message is sent for 32 consecutive words, some of them do not get optimum expansion accuracy. In fact the scale-factor conveys data on the magnitude of the largest sample in each block. While this inaccuracy increases the quantisation noise in the sound signal, it only happens where the 'busyness' of the signal itself disguises or masks it. The scale-factor data is sent frequently enough to track the fastest perceptible loudness transitions, and that is the key to subjective noise reduction.

The parity system gives good protection against corruption of the data words for single-bit errors, but impulsive interference can take out complete data words, making necessary further protection. It's achieved by a simple interleaving system in which the data is written into memory and then read out non-sequentially according to an address sequencer held in ROM. The data readout order ensures that bits which were originally adjacent become separated by at least fifteen other bits. An error burst in the received data can corrupt several consecutive bits, but when the words are reassembled at the receiver (by a memory-plus ROM system reciprocating that at the encoder) the errors are distributed among several of them so that the damage to each is usually minor, and capable of repair by parity correction and/or error concealment as necessary, techniques which are universally used with PCM sound systems. The digital signal now has enough protection to enable it to pass unscathed through all but the worst propagation conditions.

Sound/data block

Along with the audio data it's necessary to send 'housekeeping' data to control the decoding process. Figure 11.7 shows the composition of a single stereo broadcast

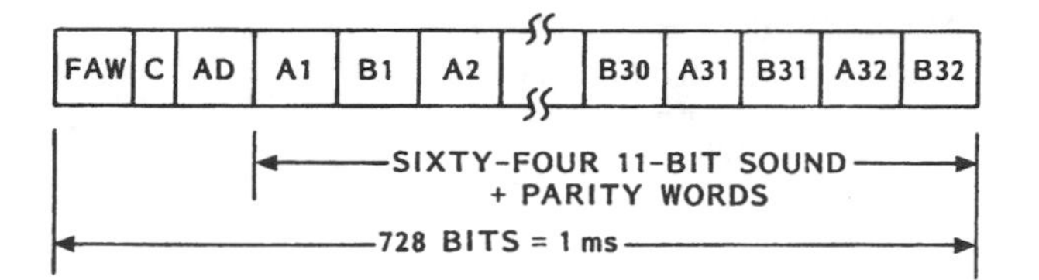

Figure 11.7 *Nicam data frame as transmitted for a stereo sound signal*

data frame, each of which spans 1 millisecond and contains 728 bits. The frames are sent end-to-end with no gaps between them.

The block starts with a frame alignment word (FAW) which synchronises the decoder at the receiver. It's the same in every frame, and consists of the eight-bit sequence 01001110. Next come five control bits C to define the type (stereo, bi-lingual, data) of signal being sent and thus to control the decoder's operating mode and the switching and routing of its output signals. The next data block AD is reserved for 'additional data' and can be used for various purposes.

The initial section of the frame has used 24 bits. The remaining 704 convey the stereo or dual-channel sound data in 64 11-bit words, the A channel (stereo left) and B channel (stereo right) samples being sent alternately throughout the period, 32 of each. The two sound channels (plus the control/data preamble) are thus transmitted as a single serial data stream, and the bit rate of each is approximately doubled as a result. This is a form of time-division multiplex (TDM) and the required time compression is achieved by writing each set of data, A and B, simultaneously into memory then reading them out alternately at double speed. TDM is used in many data storage and transmission systems to match signal density to channel bandwidth. The order of A and B samples shown in Figure 11.7 represents the sequence before bit interleaving, described above, takes place. Only the 704 bits of sound data are interleaved.

Data scrambling

The Nicam data stream contains certain fixed patterns. If it is used as it stands to modulate a carrier, the sidebands produced will contain fixed patterns likewise, and that is undesirable from the point of view of interference to other users of the broadcast band, so the sideband energy is evened out by 'scrambling' the data. It's not possible to scramble the data bits in a completely random way because they could not be recovered at the receiver. A simple circuit which generates binary digits in what appears to be a random way but whose output is predictable and repeatable is called a pseudo-random sequence generator (PRSG). Identical PRSGs are used at both ends of the chain, and run in synchronism throughout the data frame, starting at the end of the FAW. Their outputs are added modulo-two (in an exclusive-OR gate) to the data. Thus the data are scrambled for transmission, giving the effect of a completely random bit-stream, then descrambled at the decoder.

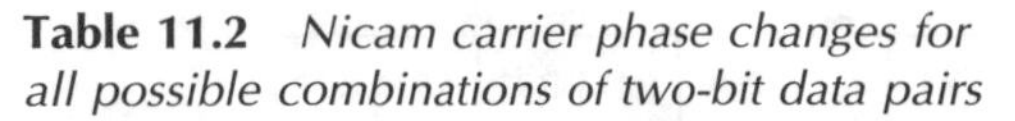

Table 11.2 *Nicam carrier phase changes for all possible combinations of two-bit data pairs*

Input data	Carrier phase
00	0°
01	−90°
10	−270°
11	−180°

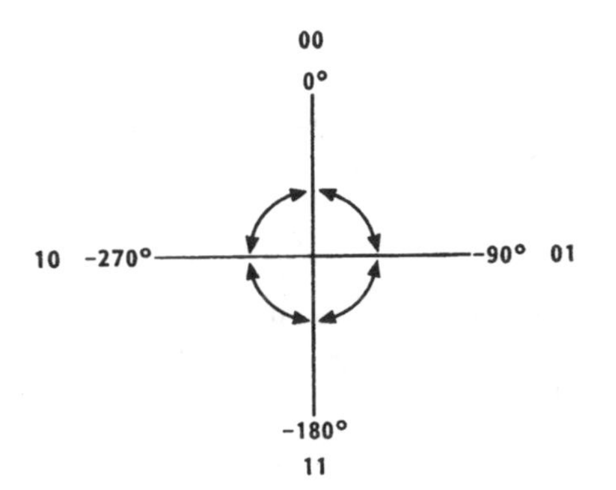

Figure 11.8 *Carrier phase shifts for the four bit combinations of Table 11.2*

Modulation and transmission

A relatively simple carrier modulation system is used for Nicam; it's called digital quadrature phase shift keying, DQPSK for short. The DQPSK carrier is constant in frequency, but has four possible phases. At the transmitter a serial to two-bit parallel converter changes the serial data into a series of two-bit pairs which can take one of four forms: 00, 01, 10 and 11. Each of these moves the carrier phase by a different amount, as shown in Table 11.2, from its previous state. The effect is shown in the phasor diagram of Figure 11.8. Carrier phase changes thus take place at two-bit intervals with a maximum shift of 180° or half a cycle of carrier frequency.

In practice the phase does not jump from one state to another, but moves smoothly due to a low-pass filter in the data path to the modulator.

Figure 11.9 shows the spectrum of a terrestrial TV broadcast with mono FM and Nicam carriers at the HF end of the channel slot. The Nicam carrier, at +6.552 MHz is broadcast at the low level of –20 dB with respect to the vision carrier, representing a *power* ratio (peak vision carrier to Nicam carrier) of 100:1.

Nicam reception

The phase-modulated Nicam carrier is tuned at the TV receiver or VCR as part of the sound/vision 'bundle'. It beats with the local oscillator inside the tuner to

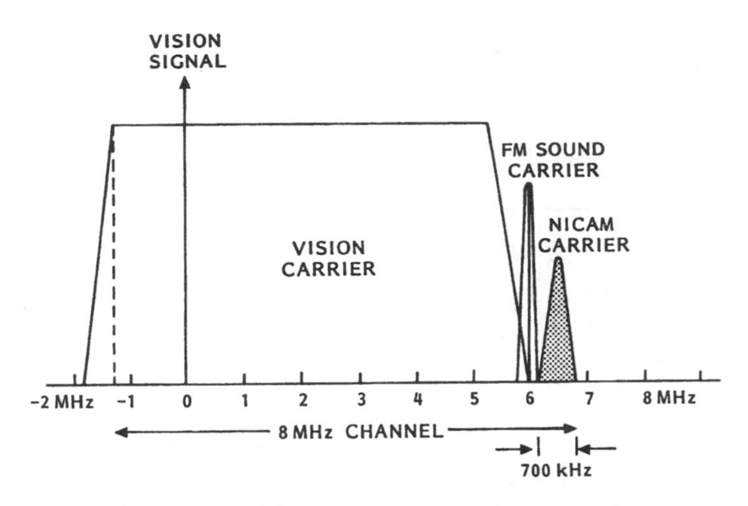

Figure 11.9 *The position of the Nicam carrier in the demodulated signal spectrum*

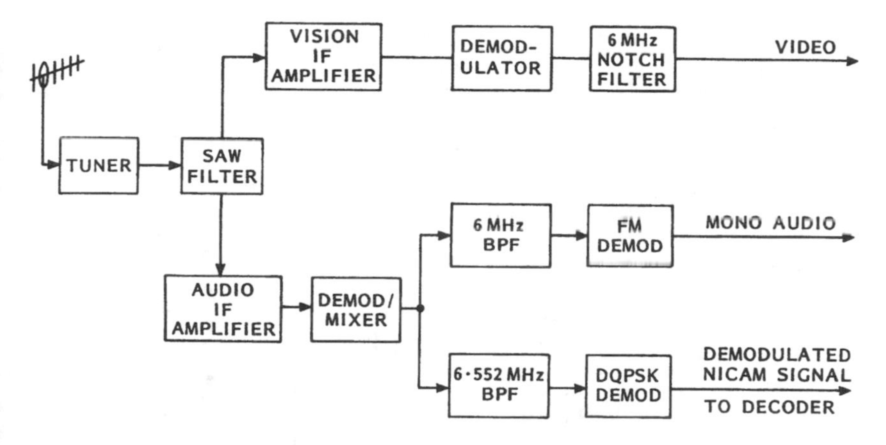

Figure 11.10 *Derivation of the Nicam carrier signal in a 'parallel' IF system. Some receiver designs use a single-output SAW filter and extract the Nicam carrier from the vision demodulator via a 6.552 MHz filter*

produce its own IF signal, still phase-modulated, which is allowed to beat with the vision carrier (just like the 6 MHz mono sound carrier) to generate a difference signal at 6.552 MHz for the UK system I transmissions. A diagram of a typical set-up for Nicam reception and carrier-handling is given in Figure 11.10.

Demodulation and decoding

The Nicam decoder's operation, in grossly simplified form, is conveyed in Figure 11.11. The four-phase modulated carrier at 6.552 MHz is intercepted by a filter

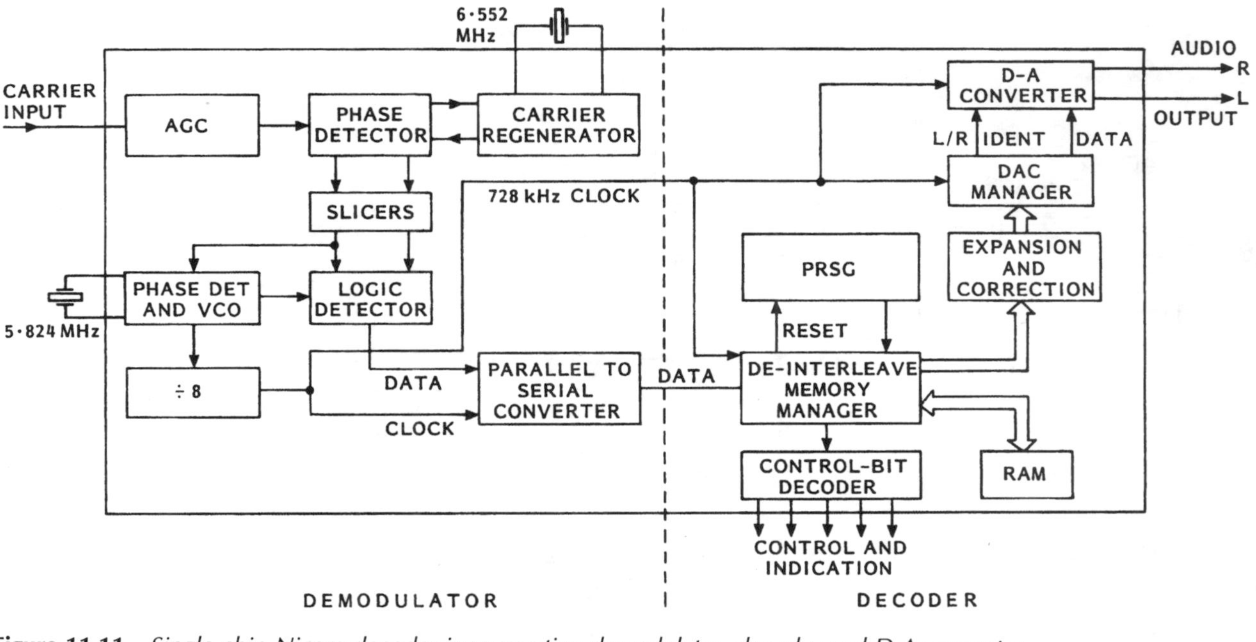

Figure 11.11 *Single-chip Nicam decoder incorporating demodulator, decoder and D-A convertor*

and passed into the Nicam IC, where it first encounters an AGC control stage to iron out level fluctuations. Next comes a phase detector, timed by a synchronised crystal oscillator running at the carrier frequency of 6.552 MHz. The phase detector's outputs pass into a pair of adaptive slicers, which optimise the points at which the signal is sampled in the following stage, the QPSK detector, gated by switching pulses from a data-rate clock running at 5.824 MHz, again governed by a quartz crystal.

Now the signal has been demodulated into bit-pairs, and the next stage is to convert them to a serial datastream: the parallel to serial (P/S) converter is clocked by the 5.284 MHz oscillator mentioned above, which is synchronised to incoming data by its own phase-locked loop.

The demodulated data now undergo decoding, which is the inverse of the processes described earlier; the operation is illustrated on the right-hand side of Figure 11.11. The decoding is synchronised by the FAW and governed by the control bits. The local PRSG is reset at the end of the FAW sequence and added to the data to descramble it. De-interleaving is carried out with the help of the on-board RAM (typically 3 of 64 × 11-bit) whose readout, in the order dictated by the built-in 'codebook' ROM, re-creates the original 11-bit data words for passage to the 14-bit expander stage, whose operation depends on the scale factor shown in Figure 11.6 and the parity bits which convey it. The parity bits are also used here in the processes of reconstruction, by repair or discard-and-interpolate, of any damaged or corrupt data. Data for L and R channels are produced alternately and in rapid succession.

The final process inside the Nicam IC is D-A conversion, in which the data carried by each 14-bit word are translated into a voltage level, generally by charging a capacitor at constant current for a period dependent on the data contained in the word. Separate sampling capacitors are provided for L and R channels, synchronously selected by the L/R ident switching signal derived from the system clock. Emerging from the IC, then, are two sets of step-waveforms representing an analogy of the original L and R audio waveforms.

Post decoder

The 'raw' L and R signals are filtered in low-pass circuits ($ft = 15$ kHz) to smooth them and eliminate the discrete steps they contain. Finally de-emphasis is applied in a frequency-selective feedback circuit to restore tonal balance and suppress noise. Now back in 'hi-fi' form, the stereo signals undergo a switching/routing/selection process before application to their amplifiers and loudspeakers, or to futher processing for recording on tape.

DTV sound

In digital TV systems the (multi-channel) audio signals are sent in digital form as part of the programme stream multiplex. We shall examine it in the next chapter.

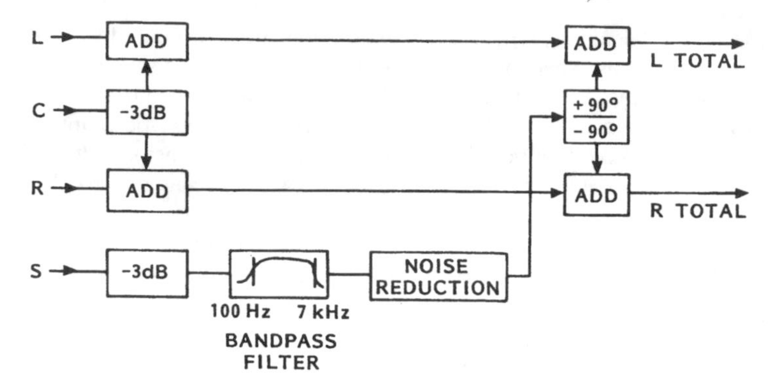

Figure 11.12 *Principle of operation of a Dolby surround encoder*

Surround sound

Home stereo sound sources, including commercially-produced video cassettes, Nicam and analogue-FM transmission systems, are capable of carrying *surround-sound*, encoded on a plan devised by Dolby Labs. The system, originally designed for use in cinemas, is now available to the 'home cinema' enthusiast in commercially-available decoders, available in black-box form or incorporated into TV sets, VCRs and satellite receivers.

The surround sound 'package' is carried solely by the two (L and R) stereo channels in the broadcast or storage medium, but does depend on them having well matched phase and amplitude responses. Figure 11.12 shows how the Dolby surround encoder works. It accepts four separate input signals, left, centre, right, and surround (L, C, R, S) and creates from them two outputs: left-total and right-total – Lt and Rt. The L and R inputs pass straight to the Lt and Rt outputs without modification. The C input is split equally between the Lt and Rt channels with a 3 dB level reduction to maintain the correct proportion of acoustic power in the mix. The S input (carrying the 'surround' channel effects) is also divided equally between Lt and Rt channels, but it first undergoes three processing steps. Initially it's bandwidth-limited in a filter which only passes frequencies in the range 100 Hz to 7 kHz. Next a form of noise reduction is applied in a modified Dolby-B NR encoder. Finally the C-signal undergoes plus and minus 90° phase shifts to create a 180° phase differential between the signal components added to Lt and Rt.

The surround-encoded signal now passes over the broadcast medium or into the record system. When reproduced by ordinary stereo equipment there is no loss of compatibility; the full separation between L and R signals is maintained. Neither is there any theoretical loss of separation between the centre and surround signals: since the surround signal is recovered by taking the difference between Lt and Rt, the (identical) centre channel signals in Lt and Rt cancel out in the surround output. Similarly, since the centre channel is derived from the sum of Lt and Rt, the equal and opposite surround channel components cancel out in the centre output. Maintenance of low crosstalk between centre and surround signals requires close matching between the two transmission/storage channels. If a mismatch between them causes the centre channel components in (e.g.) Lt to differ from those in Rt, centre information appears in the surround channel as unwanted crosstalk.

Pro-Logic decoder

There are two forms of surround decoder, the passive and the Pro-Logic types. Of these the latter is the most popular and effective; an outline of its operation is given in Figure 11.13. The Lt and Rt signals pass relatively unmodified through the decoder to form the main L and R outputs. Lt and Rt also pass through a Pro-Logic adaptive matrix, shown at the bottom of Figure 11.13. Here they are separately bandpass filtered to 100 Hz – 13 kHz before being fed to four full-wave rectifiers, two directly, one via an L/R adder and one via an L/R subtractor. The DC control voltages thus derived are processed in logarithmic-difference amplifiers whose outputs indicate the *dominance vector* in the soundfield, regardless of absolute signal levels in the channels. It is used to apply directional separation enhancement to a degree governed by the dominance in the transmitted signal, and works not only in the four directions represented by L, R, C and S, but at any point in the 360° surround-sound field in which the viewer sits.

The two control voltages are bipolar, and represent, respectively, dominance along the left/right axis and along the centre/surround axis. Each control voltage is continuously monitored by a threshold switch which – once the relative dominance exceeds its trigger level – switches the circuit to a fast mode of operation, quickening its response to change in level. A polarity splitter resolves the two bipolar dominance signals into four unipolar control voltages E_L, E_R, E_C and E_S which represent the soundtrack dominance in electrical terms embodying psychoacoustic properties. They are applied to the signal-cancelling VCA (voltage-controlled attenuator) stages. With two input signals (Lt and Rt) and four control voltages, the eight VCAs can generate eight variable sub-terms; with the Lt and Rt inputs, ten individual terms are available. To construct a decoded output signal, portions of each of these ten terms are selectively added or subtracted with a predetermined weighting factor in the combining networks. Selection of the appropriate magnitudes and polarities for the 40 summed components gives directional enhancement and non-dominant signal redistribution to the reproduced soundfield while maintaining constant accoustic power for the signal components.

Returning now to the top section of the decoder block diagram in Figure 11.13, the surround signal passes through an anti-alias filter, rolling off at about 7 kHz. Next it encounters a variable time delay, in practice a RAM store, whose period (typically 20 milliseconds) can be set by the user to match the propagation time of the sound from the rear (surround) loudspeakers to the listening position. The idea is not to create a 'spacious' effect by means of an echo, but to improve the sense of clarity and direction of *front* channel sounds. It is based on the *Haas* or precedence effect. Emerging from the delay stage the S signal is again filtered to 7 kHz before passage to the noise-reduction decoder, which applies a reciprocal effect to that operating at the sending end. Finally the S signal, along with the

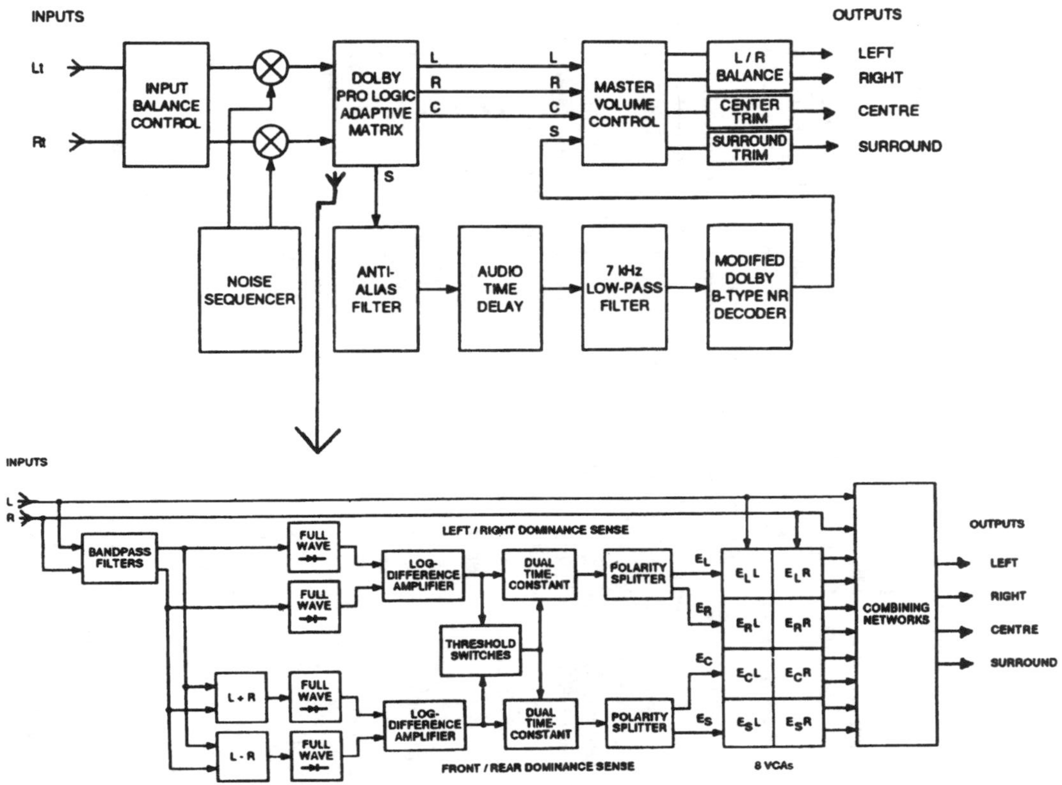

Figure 11.13 *Pro-Logic surround decoder. The upper part of the diagram shows the decoder as a whole, while the lower section enlarges on the adaptive matrix*

others, passes through a VCA master volume control on its way to the power amplifiers which drive the loudspeakers.

In practice two loudspeakers are generally used at the rear of the listening room to provide the surround sound, while the centre speaker is placed very close to the viewing screen, primarily to reproduce dialogue, which needs to be tied closely to the picture, direction-wise. Low frequency sound has no perceptible directional characteristic, and can be reproduced by a single woofer unit placed almost anywhere in the room.

The noise sequencer shown in Figure 11.13 is used for setting up and matching to room-size and accoustics in an inbuilt computer-controlled installation program. The entire Pro-Logic decoder is embodied in a single IC, typically with an outboard RAM to provide the delay for the surround channel. Several operating SFP (soundfield processing) modes are available, examples of which are Church, Cinema, Stadium, Disco and Concert Hall. They alter echo and reverberation characteristics, and can be stored in EEPROM (user's software) and called up at will by the remote control keys. They are conveyed to the surround decoder chip by a two-wire serial control bus like that described in the 'control systems' section of Chapter 8. Current Pro-Logic decoders are implemented entirely in the digital realm, with A-D converters at the Lt and Rt inputs, and separate D-A converters for each of the four outputs shown in Figure 11.13.

Dolby AC3

Also known as Dolby Digital, AC3 offers five separate audio channels: left, right, centre, L rear and R rear, each independently preserved throughout the transmission and reception chain. The availability of an additional sub-bass channel leads to the system being termed 5.1. Originally developed for cinema sound, it is used for some DTV systems and with the DVD players which we shall meet in Chapter 26. A wholly digital processing system (typical data-rate 384 kb/sec), AC3 lends itself well to current transmission and storage media, and has the advantage over Dolby Surround of offering full frequency response (3 Hz–20 kHz) in all five main channels, as well as complete separation of the sound in the rear speakers so that surround effects can provide left/right as well as front/back directivity. AC3 decoders are built into some DVD players and amplifiers, usually alongside Pro-Logic surround decoders.

THX

THX and its digital variant THX 5.1 work with Dolby Surround and AC3 respectively to adapt and optimise them for home use, as opposed to the cinema environment for which they were designed. THX requires more functions in the surround decoder and offers a 'differential' effect in the rear/surround channels of a Dolby Pro-Logic set-up; a better acoustic match between the sound from front and rear channels; a better response (X-curve correction) in 'living-room' environments as opposed to large theatres; and an opportunity to *bi-amp* the speaker system, which involves separate power amplifiers for the different drive units within the loudspeaker enclosure; here the frequency-crossover filters work at low level in the amplifier feed network rather than at high level in the speaker cabinet.

Home cinema

Many enthusiasts seek to recreate the effect of a cinema at home, in some cases right down to the popcorn and posters! The main ingredients are a large wide-screen TV; a movie-subscription DTV tuner; a surround-sound decoder, amplifier and loudspeaker system; a VCR; and a DVD player, whose ideal link to the TV set is an RGB one. The audio coupling between player or receiver and the decoder can be made by a fibreglass optical or RF co-axial link for lossless transfer.

12 *Digital TV*

Long before digital vision became a reality, digital sound was well established. First introduced as a domestic technology in the form of Compact Disc in 1982 by Philips, it had been used by broadcasters for point-to-point transmission in the Sound-in Syncs system for some years before that. Digital vision techniques have been harder to implement because of the sheer volume of fast-changing data involved, calling for large memory stores and very large-scale integration ICs, all of which have only become available, in a form compatible with home TV systems, relatively recently.

The digital video signal

The concept of quantising analogue waveforms was briefly described in Chapter 1, and following on from that we must now look in more detail at the way in which a TV picture signal is encoded in binary form. Figure 12.1 represents the video waveform over a period of 1 μs which, on a typical domestic TV with a 56 cm diagonal screen, would occupy about 9 mm of *one* scanning line. The picture pattern on this section of the scanning line can be seen to start at mid-grey, brighten up, then pause at 'bright grey'; before dropping to a lower level. At the end of our 1 μs period the brightness is increasing again towards peak white. On the left-hand scale of the diagram is shown the quantising levels between 0 and 256, while along the bottom scale time intervals are shown up to 1 μs. At regular intervals of about 74 ns (ns, nanosecond, one thousand-millionth of a second) the video waveform is sampled to establish its level at that instant. In Figure 12.1, then our samples at 74 ns intervals are 143, 160, 223, 241, 234, 200, 196, 203, 175, 105, 103, 140 and 170, making 13 samples. These numbers are produced in a fast A-D (analogue-to-digital) converter, not in the form of our decimal numbers, but in binary code; an eight-bit (bit, binary digit) word is used, with the necessary range of decimal 0–255. Binary coding is the simplest way of representing decimal numbers in 'pulse' form; reading from the *right* the 1s represent the numbers of 1s, 2s, 4s, 8s etc. Our first three samples, then, are 10001111, 10100000 and 11011111, and so on throughout our 1 μs 'window'. The digits come out in serial form, necessarily fast! All eight digits must be produced in the 74 ns period, so that for real-time digitisation we must expect around

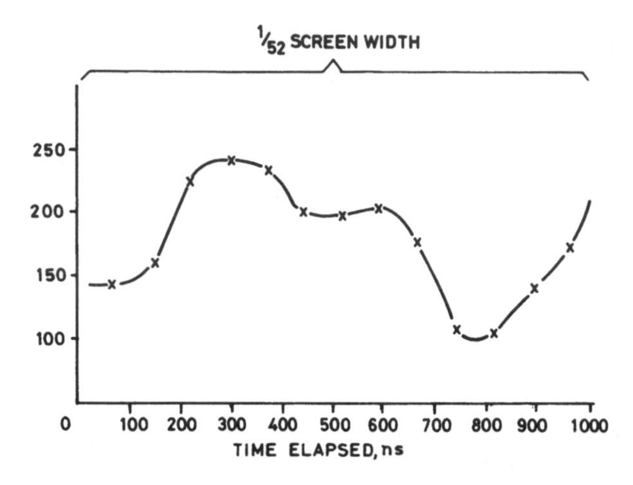

Figure 12.1 *Quantising a video signal: the waveform represents a small portion of the video signal of one TV line. Quantising levels are represented by the crosses*

100 bits per microsecond, representing a bit-rate of 100 Mbits/second, though the signal is now in binary form, and has only two states, 1 and 0. Various techniques can be used to reduce this bit-rate without impairment, as we shall see.

There are two approaches to quantising video signals. Either the complete video signal (encoded, with subcarrier, to PAL, NTSC or SECAM standard) can be quantised at a high sampling rate to render a composite digital signal, or (better) each of the Y, V and U signals can be separately quantised to give three digital pulse streams for combination before transmission. In this system, called *component coding,* the sampling rate for each of the colour-difference signals need only be about half that for the luminance signal due to their lesser bandwidth, and correspondingly lower *Nyquist* frequency. Because colour-difference signals, as we have seen, can have positive or negative values, zero is taken as level 128 from which the colour-difference signal can go down towards zero for negative signals and up towards 255 for positive signals. The total of 256 levels available imply an eight-bit word as before. The advantage of component coding is the compatibility of the system, in which each component, Y, V and U is effectively kept separate throughout the transmission and reception chain. Upon receipt, they can be fed separately to the RGB matrix and display channels of a colour receiver, or encoded in whatever form is required for the country or area intended.

Back, now, to our quantised video signal of Figure 12.1. At the end of the chain the digital pulse stream is presented to a D-A (digital-to-analogue) converter, whose output can describe 256 different levels, corresponding to the numbers conveyed in the bit-stream. This is the recreated facsimile of the original analogue signal, and at the sampling rate employed, it is a very good analogy indeed, as shown in Figure 12.2. Over our 9 mm section of a single scanning line it is subjectively indistinguishable from the original signal, except for the complete absence of noise and other spurious effects! It is not hard to see that noise (and virtually all the noise – or snow – we see on our TV screens is picked up along the way) would not be reproduced on the digitally-transmitted picture; provided the 1s and 0s of the pulse stream were distinguishable from each other in the receiver's decoder, no noise at all would be visible in the displayed picture.

A-D conversion

The A-D converter used for this relatively high speed application is known as a 'flash' type, shown in Figure 12.3. It produces an eight-bit output, using 256 separate comparators. Each of them has two inputs, one of which is connected to its own tapping point in a chain of 256 equal-value resistors, while the others

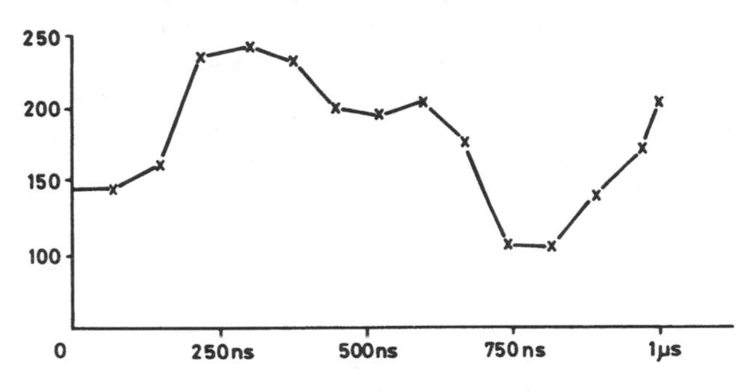

Figure 12.2 *The video signal of Figure 12.1 reproduced after passing through a digital coder. It is a very good facsimile of the original signal*

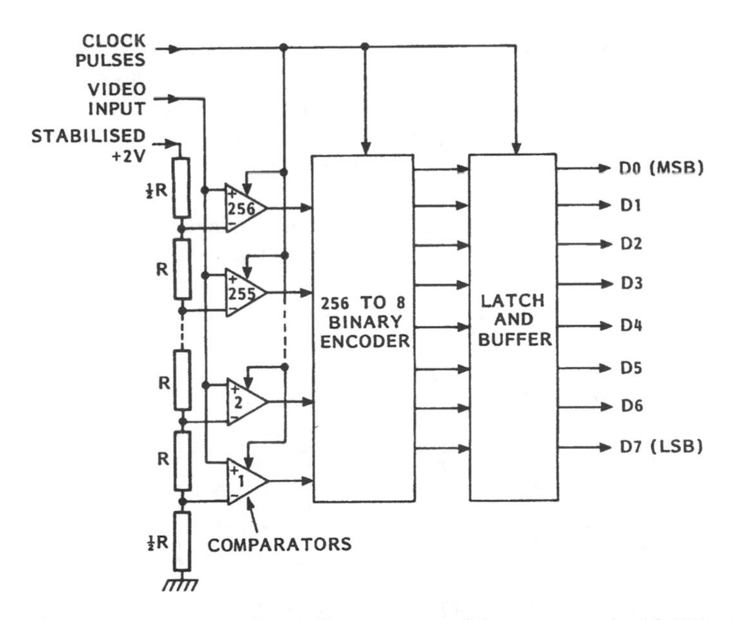

Figure 12.3 *Fast-acting A-D convertor of the type used with TV picture signals*

are all connected together and to the video input signal. A stable reference voltage, typically 2 V, is maintained across the resistor chain so that each resistor drops one 256th of the total voltage. The video input is gain-controlled to exactly 2 V peak-to-peak while its black level or sync tip is clamped at a fixed level such that the peak white video level just turns on the topmost comparator in the chain. As the video signal swings from black through grey to white it turns on progressively more comparators in 256 steps of 7.8 mV (0.0078 V).

When a clock pulse arrives the outputs (0 or 1) of the 256 comparators are loaded into a 256-to-8 binary encoder which converts the quantised samples into eight-bit bytes for passage to a latched buffer. On each clock pulse a fresh sample is taken and a new eight-bit byte produced. The frequency of the sampling clock varies with the application and the required picture quality, and may range from 13.5 MHz for the sort of picture definition we are currently used to in broadcast reception, to 30 MHz or so for high-definition systems. To avoid 'aliasing' and consequent spurious effects, a Nyquist filter limits the incoming video frequency to less than half the sampling clock rate. At 13.5 MHz we get 676 samples per 52 μs active line period: increasing the clock frequency or omitting the line blanking interval permits a greater number of samples to be taken. In a component coding system the luminance (Y) signal would be processed in the way just described, while the U and V components are separately dealt with in their own A-D converters whose principle of operation is identical. Since the bandwidth of the colour-difference signals is only about a quarter that of the Y component they require less frequent sampling. In practice they are often sampled at the same rate as the luminance signal, after which three in four samples are discarded by clocking the data out of the shift register at a quarter of the basic rate, i.e. 3.375 MHz for a Y sampling clock frequency of 13.5 MHz.

Thus for every eight-bit luminance byte there are two bits of U data and two bits of V data, complete U and V bytes being produced in the time taken for four luminance bytes. If the bit streams are interleaved in time (time-division multiplex, see previous chapter) a complete YUV 'bundle' consists of twelve bits as shown in Figure 12.4. The bit interleaving is carried out by *formatting logic* between the A-D converters and a 12-bit register.

Now that we have got the vision signal into digital form, having only two levels, it is amenable to storage in a digital memory IC, and to transmission over links whose only requirement is the ability to pass a high-speed string of 0 and 1 pulses with sufficient discrimination between them to permit recognition of 'on' or 'off' at

D0	D1	D2	D3	D4	D5	D6	D7	D8	D9	D10	D11	BITS	
Y0	Y1	Y2	Y3	Y4	Y5	Y6	Y7	U0	U1	V0	V1	WORD 1	13·5MHz
Y0	Y1	Y2	Y3	Y4	Y5	Y6	Y7	U2	U3	V2	V3	WORD 2	3·375MHz
Y0	Y1	Y2	Y3	Y4	Y5	Y6	Y7	U4	U5	V4	V5	WORD 3	
Y0	Y1	Y2	Y3	Y4	Y5	Y6	Y7	U6	U7	V6	V7	WORD 4	

Figure 12.4 *Interleaved luminance and chrominance data make 12-bit words to fully describe the characteristics of each picture element*

the receiving end. Thus high noise levels, for instance, can be easily tolerated, with no picture disturbance at all. The signal now consists merely of a 'blueprint' or 'typing instructions' for the picture, which is built up anew in the receiver, specifically at the D-A converter to be described later in this chapter. First we shall examine the storage of a digital signal in a large RAM, and take as our example the display of a flicker-free picture in a domestic TV receiver.

100 Hz pictures

We have seen how the PAL-Plus system uses digital technology to improve reception of analogue TV broadcasts. Two impairments of conventional PAL/625 pictures are interline flicker, where sharp horizontal edges of objects in the picture and fine horizontal lines have a 25 Hz repetition rate; and large-area flicker, which is particularly noticeable with large screens and bright pictures. It arises from the inability of the human eye and brain to fully integrate successive pictures with a 'flash rate' of 50 Hz as currently used in Europe. In the USA, Canada and Japan where a 60 Hz field rate is standard, the flicker effect is lessened; at a flash rate of about 90 Hz it disappears under all picture brightness, viewing angle and ambient light conditions. The use of a large RAM IC storage system removes the need for the picture to be scanned in real-time synchronism with the transmission and opens the way to the use of a 100 Hz display for flicker-free viewing.

The basic idea is shown in Figure 12.5, and is simple in concept. The video signal is converted from analogue to digital form and written into the memory at the 50 Hz broadcast rate. It's read out of memory at the faster rate of 100 Hz, converted from digital to analogue form and displayed on screen at double-speed line and field scan rates.

Memory ICs

The field store memory typically consists of a 3 Mbit fast DRAM (dynamic random access memory) with sequential addressing. There are several alternative ways of storing the data: eight-bit Y data can be stacked in its own 2 Mbit (e.g. 2 of 4 bits × 256K) section while the four-bit UV data is held separately in another 1 Mbit IC; or three 2 Mbit stores can be provided, one each for Y, U and V data. While this latter system requires an expensive 6 Mbit capacity it provides a YUV sampling ratio of 4:4:4 and full bandwidth in all three channels, which is useful for RGB operation.

With the 4:1:1 ratio scheme the RBG inputs have first to be filtered then matrixed to obtain YUV form and then stored for reconversion to RGB format after D-A conversion in the 100 Hz part of the circuit. This is not such a problem as it seems: no encoding, modulation or interleaving is required, and with an RGB signal most of the information in the three channels is common to them all, and thus redundant, as we saw in Chapter 3.

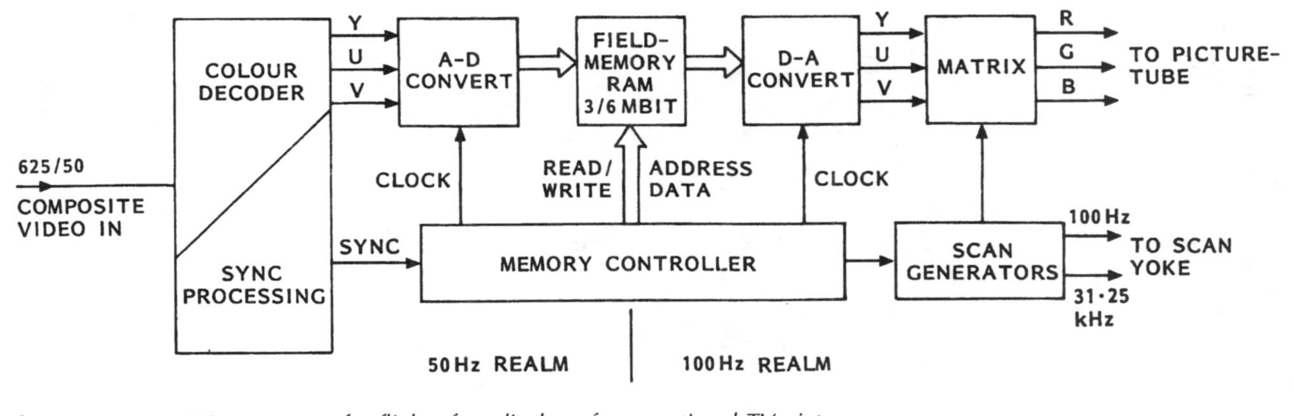

Figure 12.5 *100 Hz processor for flicker-free display of conventional TV pictures*

Digital noise reduction (DNR)

The presence of a picture field store in a TV receiver or monitor opens the way to a noise reduction system. There are many reasons why noise intrudes on an analogue video signal, amongst them weak signal reception and tape noise in VCR playback. Noise is random in nature, with no correlation between successive video fields, while those of a still video picture have total correlation. Thus if several video fields are integrated the noise cancels to virtually zero while the picture-signal components add.

Electronic noise reduction can be effected in a field-store TV set by progressive integration of the present and previous fields. Figure 12.6 shows the basic idea in block diagram form. The block marked K is a *recursive filter* whose characteristic determines the noise-reduction factor: the higher the K factor the greater the noise reduction. In practice the recursive filter is governed by a motion detector which continually adjusts the K factor, reducing it for moving parts of the image (to prevent blurring due to integration) and increasing it for stationary picture areas. The effect of this is that the longer an object in the picture remains still, the better and more noise-free it appears.

Memory readout

A memory controller IC governs the read and write processes and generates the sync pulses for the 100 Hz/31.25 kHz timebases. Figure 12.7 shows in block diagram form a system designed for domestic TV sets which uses two sets of field memories, DNR, 4:4:4 sampling ratio and field and line flicker reduction. The sampling and read/write rates can be selected by the setmaker or in software to suit different scanning rates and standards. At the higher rates the picture definition is greater than that required by the CCIR standard, but all rates call for very short DRAM access times. If the time-constant of the voltage-controlled oscillator (VCO) controlling the memory readout is long, the system takes on the characteristic of a timebase corrector (TBC) ironing out timing jitter in the video signal. This will, for example, provide better off-tape picture stability. Plainly it's not possible to combine this with effective DNR because the pixels in successive fields are no longer time-coincident.

Some other features of the system shown in Figure 12.7 will now be described. The delay in the Y data stream is part of the bit interleaving/data formatting system which produces the 12-bit YUV words of Figure 12.4. IC1 carries out noise- and cross-colour reduction by recursive filtering. Downstream from it IC2 implements

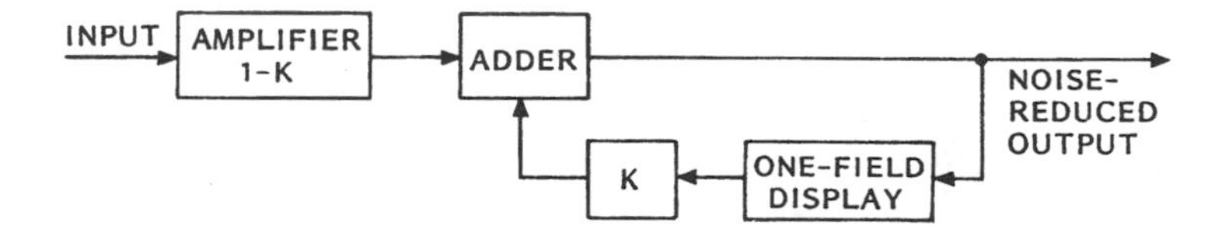

Figure 12.6 *Noise reduction by successive integration in the digital realm*

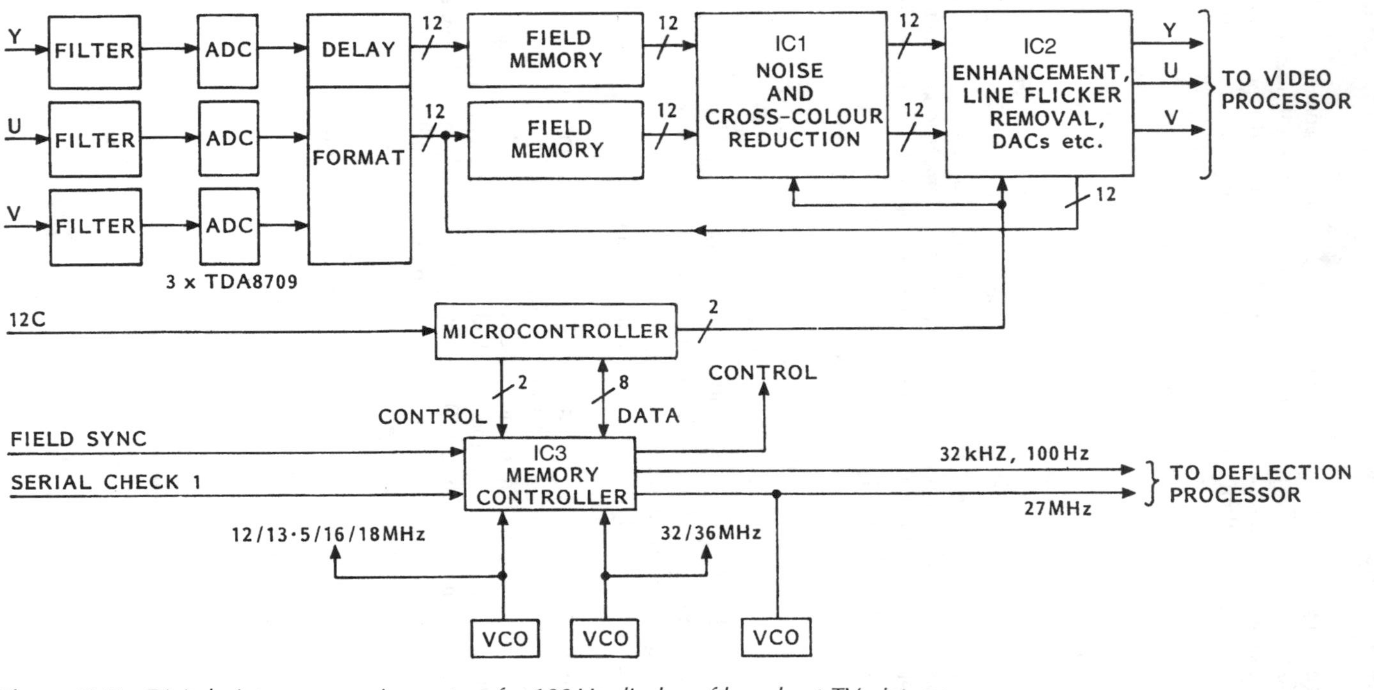

Figure 12.7 *Digital picture processing system for 100 Hz display of broadcast TV pictures*

interline flicker removal, U and V signal reformatting, digital colour transient enhancement (which subjectively sharpens up vertical edges of coloured picture objects) and Y signal peaking. It also contains the necessary high-speed D-A converters to bring the signal back to analogue form as shown on the right of Figure 12.5. Two additional features of this IC are called Zoom 1 and Zoom 2. The former provides vertical picture expansion so that 4:3 aspect ratio letterbox pictures can be correctly displayed on a 16:9 aspect ratio picture tube. Zoom 2 offers magnification by two of the displayed picture, both horizontally and vertically, as a viewer option: in effect each pixel is quadrupled in size in the 'magnified' picture. All the memory-control and post-memory processing functions can be software controlled via a serial bus which is also governed by the user's infra-red remote control handset.

Special effects

In addition to the noise-reduction and picture zoom features described above, various other possibilities exist in a system incorporating a digital field store. An obvious one is a perfect freeze-frame facility: the memory writing process is halted while the 100 Hz readout continues as long as the viewer requires. Other possibilities are *pixelation*, in which the D-A converter clock rate (see later) can be progressively slowed – in practice progressively halved – by the viewer to give the picture an increasingly mosaic-like appearance; and *solarisation*, where the two, three or four least significant bits are removed from each byte. This reduces the number of analogue signal levels, giving the reproduced picture a surreal effect.

Where the field-store memory bank is part of a video effects console rather than a TV set, further possibilities are exploited. It's possible to squeeze, zoom, reverse and 'wave' the picture by changing the memory reading rate and direction; to superimpose two pictures whose original scanning generators are not locked in sync with each other; to provide *chroma-key*, in which another picture is inserted over (e.g.) all blue areas of the main image; and so on, depending on how much computing power is available in the system- and memory-controllers.

D-A conversion

When all the tricks and manipulations have been carried out on the digital data containing the picture (and this may include compression, transmission and reception, see later) it's time to reconvert the data into analogue form to suit the requirements of the display device, be it a picture tube or a solid-state panel.

Where necessary the digital data stream is de-interleaved to separate it into its Y, U and V components for application to three separate D-A converters. Each of them works on the principle shown in Figure 12.8. The first step is serial-to-parallel conversion: the eight-bit serial data words are stepped through a register by the bit-clock pulses at the memory read rate, typically 27 MHz for 100 Hz TV display or 13.5 MHz for domestic 50 Hz applications. The eight-bit words are transferred to a slave register at the byte clock pulse rate, thus making parallel

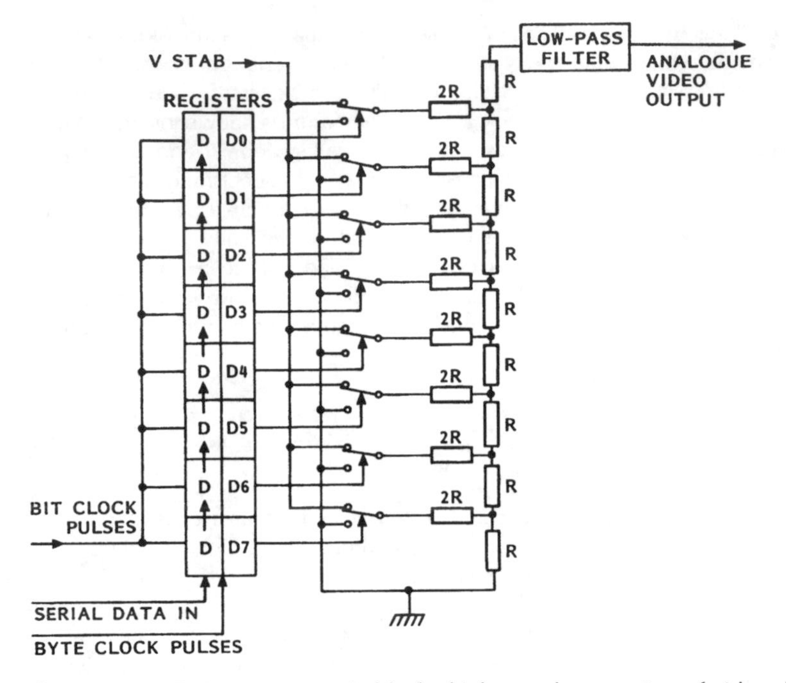

Figure 12.8 *D-A converter suitable for high-speed processing of video data and signals*

data words available to the D-A converter proper. Each bit in the word being converted controls a two-way switch connected to an R/2R ladder network with eight 'rungs'. The ladder has a closely-stabilised voltage applied across it. The effect of each bit in the word depends on the position of its switch down the resistive ladder. Thus the most significant bit (MSB) D0 at the top of the chain commands 50 per cent of the output voltage swing, the next significant bit D1 25 per cent and so on down to the least significant bit (LSB) D7 which influences the output voltage by a mere 1/256th or 0.4 per cent.

A 256-step video signal cannot, when viewed on screen, be distinguished from a non-processed analogue-derived one by the human eye. The other advantage of eight-bit processing (it's possible to get away with six- or seven-bit quantisation) is that the 'noise' effect introduced by the digital processing is very low at −54 dB, virtually imperceptible in normal viewing. In the 100 Hz picture system described above a low noise figure is essential for a high quality picture.

D-A conversion is followed by a sharp cut-off analogue filter (usually an LC type) to smooth out the quantisation steps and suppress the remnants of the clock waveform. The Y filter cuts off sharply at a frequency rather less than half that of the memory-read clock, say 12 MHz for a 27 MHz readout rate, and 6 MHz with a 13.5 MHz rate. Since the frequency of the U and V signals is so much lower than the D-A clock rate their filter requirements are much less demanding. To capture all the available picture detail the bandwidth of the analogue circuits downstream of the Y converter, and in the RGB channels, must reach up to the transition frequency of the Y filter.

Digital picture broadcasting overview

The stumbling-block, for many years, in digital television transmission has been the very fast bit-rate involved in fully describing all the detail in every pixel at the rate of 50 fields per second. The bit-rate for pictures of the quality we are presently accustomed to is 216 Mbit/second, corresponding to a sampling rate of 13.5 MHz and a 12-bit word expressing the luminance and chrominance values in 8 + 2 + 2-bit form as described earlier in this chapter. High-definition TV gives rise to an even faster bit rate: for a 1250-line, 50-field, 16:9 aspect ratio picture data is generated at the rate of 1.2 Gbit (1200 Mbit)/second. While these rates can be handled in studio and closed-circuit applications (digital operation has been standard in some control and editing suites for many years) they are quite incompatible with conventional broadcast and cable distribution systems, whose bandwidth is limited and costly. To reduce the bandwidth and storage capacity required for video datastreams, two techniques are used: bit-rate reduction, and what might be called economical modulation systems.

Bit-rate reduction principles

An alternative term is *digital compression,* whose object is to reduce the bit-count as far as practical while facilitating the re-creation of the best possible picture at the receiver. It is made possible by the fact that most television scenes contain much redundant information. This redundancy takes two forms: *statistical,* in which picture and pixel samples are closely correlated with adjacent samples, both in the same frame and in previous and successive frames; and *pyschovisual,* in which advantage is taken of the tolerance of the human eye and brain to certain types of distortion in the image, dependent on its context and content. Bit-rate reduction removes the statistical redundancy and distributes any remaining distortion in time and space on screen so that its visibility is minimised. Some scenes have great redundancy: a still image, or stationary part of a picture, needs sending only once and can then be repeatedly read out from a store at the receiving end. Imagine a scene consisting of a crowded beach beneath a blue sky. The sky section of the image has almost total redundancy, so that data capacity 'saved' in transmitting the sky image can be 'spent' in reproducing the detail and movement of the lower part of the picture. Plainly the most difficult images to reproduce by this compression system are ones that are highly-detailed and rapidly changing: examples are jump-cuts between 'busy' pictures; and camera zooms amid sharply-defined images.

The bit-rate chosen for transmission depends on many factors. To convey a picture subjectively equal to that of a standard (low-band) VCR replay image, a rate of about 2 Mbit/second is adequate. At progressively higher rates

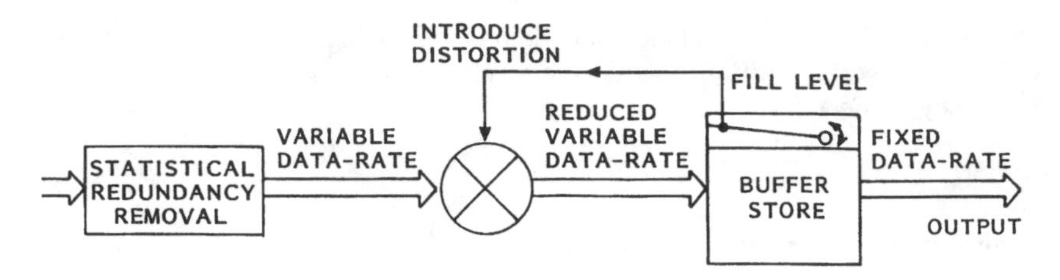

Figure 12.9 *Principle of bit-rate reduction as used in digital TV systems*

the reproduced picture improves. Figure 12.9 outlines the basic bit-reduction technique. The incoming fixed-rate high-density data passes first through a stage which removes the statistical redundancy to render a datastream whose bit-rate varies according to the characteristic of the video signal: as we have seen, some pictures (and parts of them) have more redundancy than others. To match this varying data rate to the fixed rate transmission channel a buffer store is provided, typically with a digital storage capacity of two frames. The buffer store acts as a regulator with a constant rate of output data. Its content is continuously monitored, and when it nears overflow a 'slow-down' signal is fed back to what might be called a distortion processor whose purpose is to reduce the buffer's input rate and thus maintain equilibrium. The manner in which the distortion is introduced is very critical: it's based on an algorithm carried out in a discrete cosine transform (DCT) processor, a microprocessor-based block capable of performing foward and reverse transform operations in real time. It processes the data in 'chunks' corresponding to 8 × 8 pixel blocks to produce coefficients representing the amplitudes of the various frequencies present, and the DC level of the whole block.

By scanning the data matrix in a controlled manner, called the *matrix transform*, long runs of zero and near-zero values are generated, reducing the amount of significant data to be sent. The remaining coefficients are quantised to reduce the number of non-zero values in a technique called *adaptive quantisation*. The number of bits in each sample is further reduced by using a 'shorthand' coding system based on the work of Huffman, in which the most frequently occurring symbols are allocated the shortest possible codes. Finally *run length* codes are used, in which, for example, the sequence 3,3,3,3, can be coded 3,4 (3, four times) to minimise the number of symbols transmitted. By these means compression ratios of 50:1 and more can be achieved without significant subjective deterioration of the reprocessed picture. At the decoder the process is reversed, with the help of a two-frame digital picture memory store and suitable processing circuitry operating from a look-up table containing the 'codebook'. The digital data recreated in the decoder is D-A converted and reproduced on screen as a picture with excellent quality and virtually imperceptible noise.

Data compression

We have seen that a 'raw' digital colour TV signal to the European 625/50 standard involves a bit-rate of 216 Mbit/second, and although its transmission would offer consummate picture quality, it would gobble up so much bandwidth in its passage as to be impossible in the crowded broadcast bands. A great deal of the information in the pictures and the succession of pictures which form the moving images is effectively redundant, and can be discarded with very little loss of quality or detail on the viewing screen. The data compression system is based on a set of standards set by the MPEG-2 protocol. The first step (after the A-D conversion process described earlier in this chapter) is to group the coded samples into blocks each consisting of 8×8 samples of luminance plus 4×4 samples of each of Cr (red colour data) and Cb (blue colour data) as illustrated in Figure 12.10a. These blocks are grouped in fours to form *macroblocks*, representing 16 lines by 16 pixels as shown in Figure 12.10b. These are the basic digital building blocks which sometimes become visible on the viewing screen when things go wrong! It's at this macroblock level that the data compression processes are carried out. Time-successive macroblocks, end to end, form a data slice to describe a segment of the picture as shown in Figure 12.11. This slicing arrangement – in reality a string of data words – facilitates the error-detection process, in that a slice containing errors can be discarded. The motion-prediction, in which

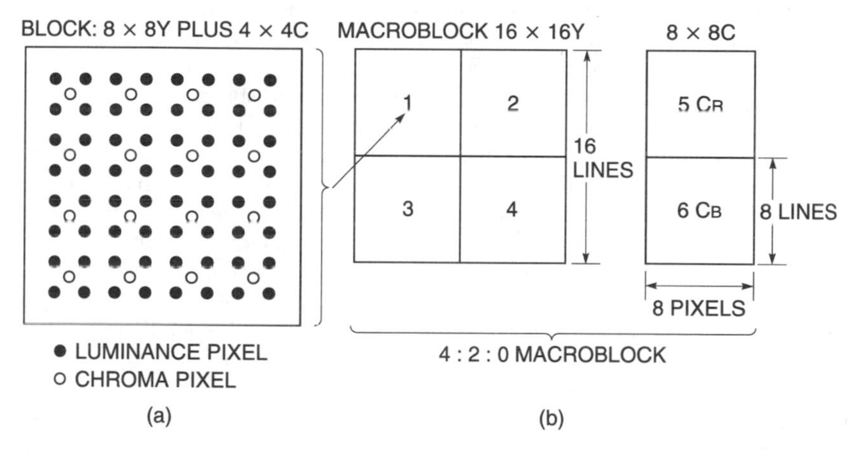

Figure 12.10 *Assembling picture elements into blocks and macroblocks for digital processing*

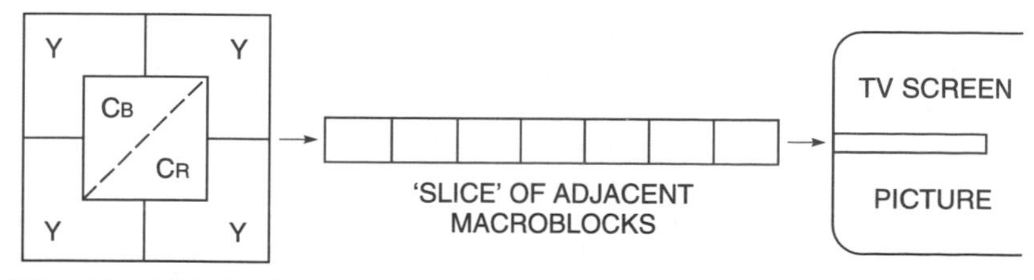

Figure 12.11 *Macroblocks, grouped together in sequence, form a picture 'slice'*

macroblocks are used to predict motion in the picture, is based on what has gone before.

The first stage in data compression is to compare successive frames, retaining and coding only the differences between them to provide an 'update' for the display picture. This *interframe* comparison eliminates temporal (in time) redundancy in the image, and the data thus generated varies from none at all on a still picture to a great deal, in fact an overload, on a jump-cut from one picture to another. if we could see the 'temporal update' image on a studio newscast there would be a small area of activity, relatively slow-moving, in the region of the reader's head and face. On other scenes there would be 'slivers' of information representing movement of picture objects, and at each complete change of scene the whole screen would momentarily come to life for a single frame which with practical coding systems would be very short on detail, quickly filled in on subsequent frames.

Having discarded a great deal of temporaly-redundant data, each individual picture frame is now examined in terms of its brightness and colour content. Again taking the case of a newsreader, and setting him against a plain blue background, a single set of Y, Cb and Cr values suffices to describe the background, and can be sent once with instructions to 'print' the result over the whole area of the backdrop. This is an extreme example, but in most real pictures there are areas whose brightness and hue is constant. Skies and seas, grass and particularly man-made and -sprayed subjects have areas of identical colour and shade, and can thus be shortly dismissed in terms of picture data. The most difficult situation for this *intraframe* compression is where there is a very slight change of hue in a large coloured area, perhaps represented by hazy clouds in a blue sky: with low bit-rate systems this could lead to objectional 'blocky patches' on the picture, with sharply-defined corners and edges, containing a slightly different hue to the surrounding area. If we could see on screen the effect of intraframe compression it would be rather like an outline drawing of the picture objects, with edges and transitions giving rise to bursts of energy/data, and the areas between, generally large plain ones which have similar pixel values, generating little or no data.

Exploitation of spatial and temporal redundancy greatly reduces the amount of data which needs to be sent, but several more techniques are available to further reduce the bit rate. A relatively simple one is *motion compensation*, for situations where complete picture chunks (e.g. when the camera is panning) change position without altering their form or shape. Here entire blocks of 'existing' pixels can be directed to a new screen position by the transmission of vector codes. Allied to this is *motion prediction*, in which the direction, speed and rate of change of the motion of a picture object (football?) can be used to compute its likely trajectory, involving relatively little data to correctly position the relevant blocks of pixels. The prediction process is based on luminance values only; the accompanying chroma data is taken as read, as it were.

DTV transmission takes place in groups of frames, twelve in succession to form a picture group, see Figure 12.12. The first of them, an I (Intraframe) type, acts as a reference for those which follow: it is not highly compressed, having had only spatial redundancy removed. It is reconstructed and stored in memory to help build up the following eleven frames. Following the I frame come two B (Bidirectional-predicted) frames, composed by comparing the contents of future

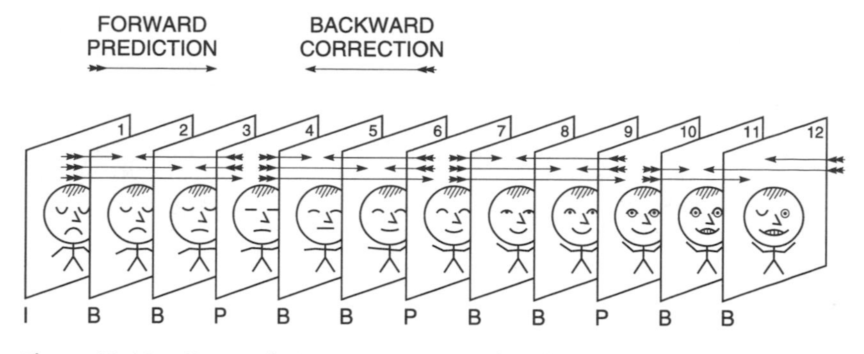

Figure 12.12 *Group of pictures consisting of twelve frames. The arrows show how all but the first frame have picture-data contributions from preceding and following ones for B frames, and earlier ones for P frames*

(P-type) and past (I-type) frames. The fourth frame, P (Predicted) carries more compression than I-frames; its content is derived from forward prediction of the change in content of the preceding I or P frame. Next come two B frames, drawing their content from previous and future P frames, followed by P and B frames, reconstituted in similar fashion, up to frame 12. At this point a new I frame is sent to update the whole caboosh, and correct the errors which have built up over the almost half-second (12 frames, 24 scanning fields in a conventional display) period of memory manipulation, prediction and interpolation. Thus an acceptable picture is sent and reconstructed with only one in twelve frames actually containing anything of real substance. Of course a 'future' frame is made available for processing by delaying 'current' fields in memory until it becomes available.

DCT

Discrete cosine transformation is a complex mathematical process based on the work of Fourier. Working on the 8 × 8 pixel blocks shown in Figure 12.10, the pixel values are converted from the time to the frequency domain. The conversion process, fundamental to the MPEG system, renders coefficients of the frequency components, visualised as a matrix of numbers in an 8 × 8 block, wherein the most significant is that in the top left hand corner – it conveys the DC level of the entire block. As we move away from the top left corner (Figure 12.13) of the block in any direction, the information very rapidly thins out unless the picture is a very busy and detailed one. The values shown are 'difference' ones, represented by the plus and minus numbers, and are rounded up or down to reduce the number of symbols to be sent.

The next step is to regroup the matrix values into serial form, carried out by 'scanning' the block in a zigzag fashion starting at the top left-hand corner to read off the DC value and the low-frequency components first. The resulting bitstream contains (for most pictures) long strings of zeros towards the end of each block, facilitating the run-length and variable-length coding techniques which we'll examine shortly. The end of the block is signalled by a special code. Except in I frames, even the data in the top left of the block need not be sent in its entirety: a

284	3	−4	−1	0	0	0	0
−4	−2	−1	0	0	0	0	0
−2	2	0	0	0	0	0	0
1	0	0	0	0	0	0	0
0	0	0	0	0	0	0	0
0	0	0	0	0	0	0	0
0	0	0	0	0	0	0	0
0	0	0	0	0	0	0	0

Figure 12.13 *DCT block showing frequency coefficients and difference values after time-to-frequency conversion*

short code to indicate its difference from adjacent frames can be used. This is called differential coding.

Quantisation

The quantisation process is not linear – it takes advantage of the noise- and detail-discerning ability of the human eye. Thus a weighting system is used, with small quantisation steps for low-frequency information and large ones for the higher-frequency components. This weighting factor is modified when necessary by feedback from the output buffer store which acts to maintain a constant data rate in the output bitstream: strict limits on this are imposed by the transmission system. To high a data-rate would lead to a very visually-objectionable 'jump' in the picture caused by a skipped frame; to low a data-rate may lead to a repeated frame and thus a momentary freezing of the image. Too fast or long a data-burst from the coding section results in a filling-up of the buffer-store and, before it overloads, a fed-back message to the quantising stage to coarsen the sampling steps. The on-screen result varies from a soft or fuzzy image to – in the worst case – a 'blocky' one, in which pixel groups become momentarily visible. The feedback loop is shown in Figure 12.9.

Code abbreviation

Further economics in bit-rate are achieved by sending short descriptive codes rather than strings of symbols which have any sort of pattern. Thus *run-length coding* (not used for the top left block in Figure 12.13) abbreviates 4, 4, 4, 4, 4 to 5, 4, four lots of five; and *variable-length coding* assigns a short code to the most common symbol groups, rather like Morse code. In the receiver this is applied to a ROM-based look-up table which regenerates the required symbol group.

Data rate

Figure 12.14 shows the steps in the MPEG process which we have seen. The pulse-train output has a constant data rate which depends on the spectrum space available in a broadcast channel or the storage capacity of a recording system, be it optical or magnetic. In a practical digital broadcast multiplex conveying (e.g.) six programmes, a system of *statistical multiplexing* (STATMUX) is used, in which the total available bandwidth of the transmission channel is shared as required between the programmes, each making demands on the bitstream proportional to the 'busyness' of its picture from moment to moment. Here there is a link from what might be called an overall flowmeter to the quantising stage in Figures 12.9 and 12.14 and those of the other programme coders involved. By this means a football game, motor race or action movie can 'borrow' from a quiz show, panel game or snooker match on the same multiplex. We shall see later how the video, audio and other components of several programmes are combined into a single datastream and modulated for transmission. The makeup of a MPEG picture-data packet is shown in Figure 12.15.

MPEG sound encoding

We saw in Chapter 11 how the first TV digital sound systems, those for Nicam and MAC, are implemented and how they economise on bit-rate by using the

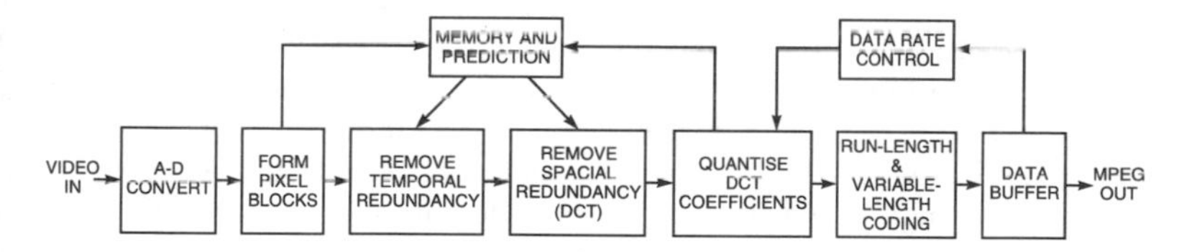

Figure 12.14 *The entire MPEG coding process for TV pictures in much-simplified block form*

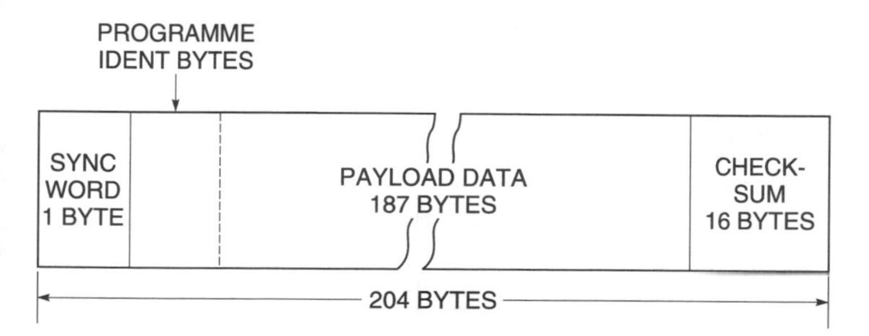

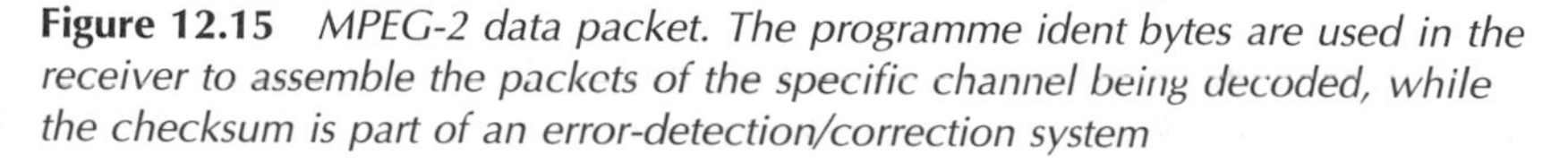

Figure 12.15 *MPEG-2 data packet. The programme ident bytes are used in the receiver to assemble the packets of the specific channel being decoded, while the checksum is part of an error-detection/correction system*

noise-masking principle. In MPEG DTV transmissions the same technique is used, but in a more sophisticated way. The masking effect is due to a basic characteristic of the human ear. A loud sound masks a quiet one when the dominant sound is lower in frequency and the two are fairly close in frequency. This enables the coder to discard some information, that which would not be heard anyway. The system is called MPEG layer II audio coding (MUSICAM, Masking pattern-adapted Universal Sub-band Integrated Coding and Multiplexing). Here the audio signal spectrum is divided by a digital wave filter into 32 separate, equal width sub-bands for the purpose of coding: see Figure 12.16. The contents of each of these sub-bands are sampled at a rate depending on the quality required and the room available in the broadcast path or storage medium: 32 kHz, 44.1 kHz and 48 kHz rates are possible, and an indication of the chosen rate is sent with the audio data in code form. Alongside the filter the audio signal is fed to a Fourier transform block for analysis in terms of its frequency coefficients. Allied to this is a psycho-acoustic model representing the characteristics of the human ear. It instantaneously and continuously calculates noise-hearing threshold values to control the scale factor of the audio quantising block and govern the scale factors used in the individual sub-bands. The data is encoded in three groups of 12 samples in each sub-band, making a total of 1152 samples per audio channel; scale factors are common to two or three of these groups when expedient, only changing when audible distortion threatens. Stereo (capable of carrying Dolby surround signals) and MPEG 5.1 Surround are supported by the MPEG2 audio system. The encoding system gives the best possible compromise between sound quality and bit rate economy, further assisted by 'almost-transparent' techniques of backward adaptive quantisation and adaptive differential PCM. Typical bit rates for audio vary between 64 and 256 kbits/sec. Figure 12.17 shows how a MPEG audio packet is constructed: the first (header) bits convey identification and synchronisation data; CRC (Cyclic Redundancy Check) provides error detection and (limited) correction facility; bit allocation data describes to the decoder the resolution of the samples; and the scaling bits indicate the scaling factor, not exactly, but in a 'good-enough approximation', as described in connection with Nicam sound in the previous chapter. The sound packets go on to be time-interleaved with video and other data packets as we shall see next.

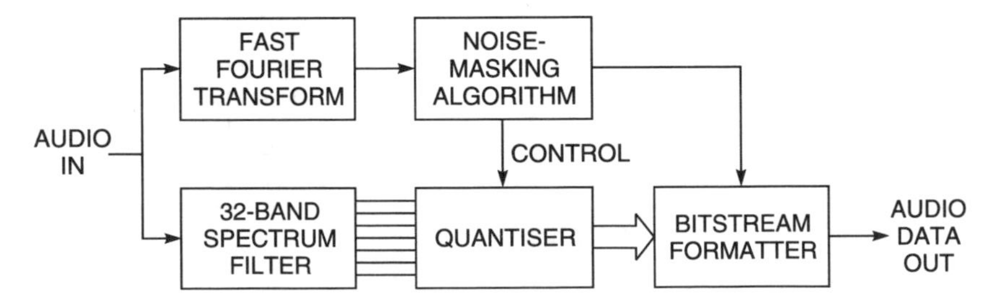

Figure 12.16 *Audio coding for MPEG transmission, based on frequency-selective noise-masking and quantisation*

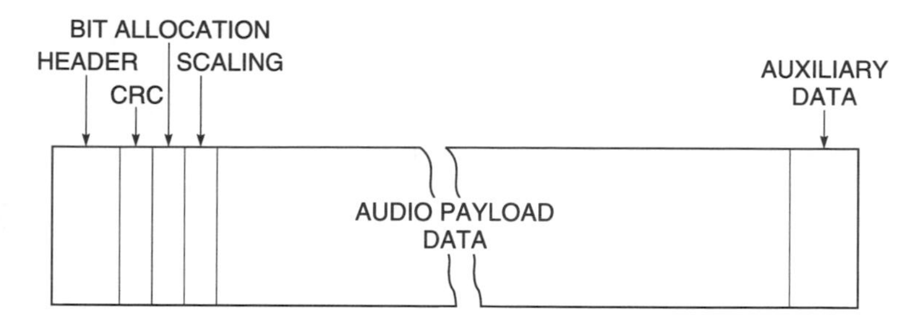

Figure 12.17 *MPEG audio packet, with ident, control, error-correction and decoding data preceding the sound information*

Housekeeping data and PES

We have seen in Figures 12.15 and 12.17 how the video and audio datastreams are neatly parcelled up into data packets, each with its own electronic label, for onward transmission on the 'conveyor belt' represented by the transmission system. For each programme there is a third data-parcel known as the service data packet: it carries programme ident information, conditional access codes, key flags for descrambling, user- and control data and possibly address pointers for network routing etc. This and the other two datastreams are assembled sequentially to form the PES, Programme Elementary Stream in a *packet multiplexer*, see Figure 12.18.

Transport stream

PES data from several programmes, typically four or five, come together in the transport stream multiplexer which assembles them end to end in a continual series of 188-byte data packets, each containing a 4-byte header and 184 bytes of video, audio or housekeeping data. The header carries first a sync word for decoder clock synchronisation then ident-, type- and priority data. This transport stream multiplex is now almost ready for transmission.

Forward error correction

In a perfect transmission medium there would be no need for error correction technology, which (for the first time in this chapter!) actually *adds* to the transmitted bit rate. In practical systems datastreams become distorted and corrupted in their passage, necessitating a comprehensive system of error-checking and – in most cases – error correction at the receiver. All digital systems involving transmission or magnetic/optical storage use error protection techniques, tailored for the characteristics of the system and its vulnerability to reproduction errors. For the broadcast systems we are examining here, FEC (Forward Error Correction) is

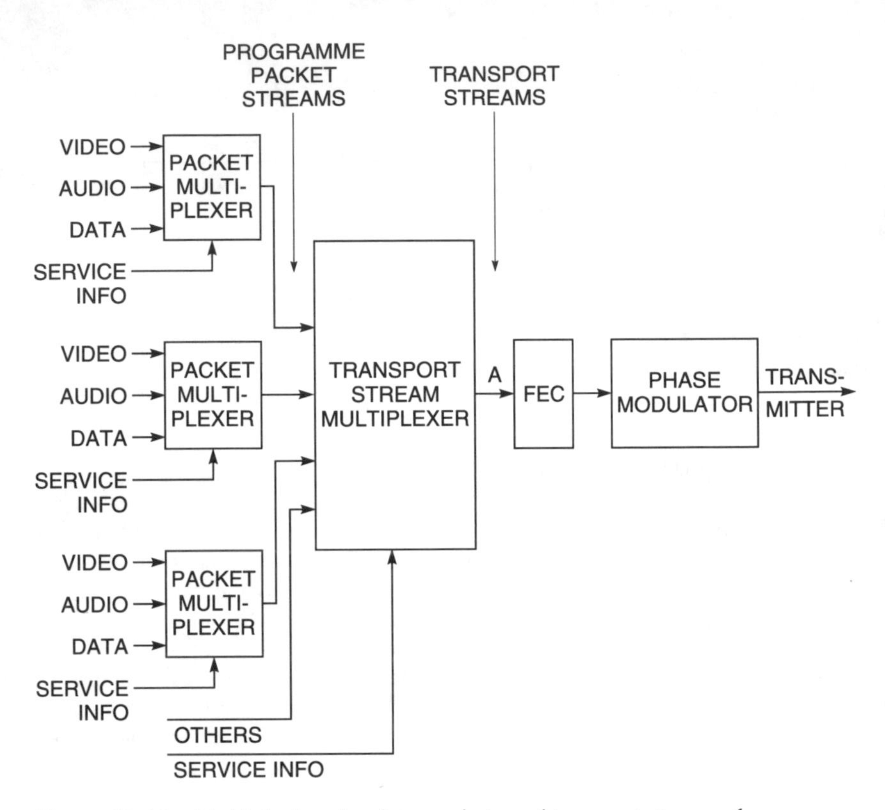

Figure 12.18 *Multiplexing the data packets and transport streams for transmission*

applied to facilitate detection and repair of errors in the distant receiver. The first process is 'randomising' of the data, using a pseudo-random sequence generator of the sort we met in connection with Nicam on page 155: it removes any DC 'bias' in the pulse train; provides sufficient binary transitions to ensure reliable clock synchronisation in the decoder; and disperses the signal energy evenly over the available spectrum. The next process, *outer coding,* employs the well-known (e.g. from CD players) Reed-Solomon system of error coding using parity symbols. Here 16 parity bytes are tagged onto each packet, lengthening it to 204 bytes. This cannot, however, cope with a 'burst error' where heavy corruption (due, perhaps, to impulsive interference) takes out a large number of adjacent bytes in a frame. To combat this, *convolutional interleaving* of the data is carried out. After de-interleaving at the receiver the errors are dispersed and thus amenable to repair by parity-based correction. The *Forney* interleave formula used, involving 12 progressively longer pulse delays, time-switched on a byte-by-byte basis, is reciprocated at the decoder. Finally, for satellite and over-air terrestrial DTV broadcasting, *inner coding* is employed.

Inner coding, known as convolutional or Viterbi coding, involves the modulation and simultaneous transmission of two separate data trains, X and Y, each of which contains the whole of the transport stream, though their timing (and thus 'phase') is not the same. The transmission of two complete datastreams thus, while providing 100% redundancy and very reliable reception, is wasteful of spectrum space, so a compromise is provided by serialising the X and Y bitstreams and leaving out some of the bits in a ratio which depends on the reliability of the transmission path. This *puncturing ratio* may typically be two-thirds or three-quarters. Thus armoured, the bitstream is modulated for transmission.

Modulation

The three transmission media satellite, terrestrial and cable have very different requirements in terms of their modulation systems to best match their capabilities and weaknesses. We shall look at each in turn. Simple modulation techniques like amplitude- or frequency-shift keying are very inefficient in terms of spectrum usage, and do not fit in at all with the Scrooge-like philosophy applied to the coding of DTV signals!

QPSK

Satellite transmissions are characterised by relatively high noise level, the availability of a comparatively large bandwidth, no multipath reception problems, and little interference. Relatively low power is available for transmission and directional high-gain aerials are used at both ends of the link. This scenario lends itself best to DPSK (Differential Phase Shift Keying) in which fast-changing symbols can be transmitted at constant amplitude. This is the same type of modulation as used for the terrestrial Nicam signal we saw in the previous chapter, though here the rate of change of carrier phase is much greater. Figure 12.19a shows the four possible phase angles of the carrier and the bit-pairs they convey. In a relatively noisy signal the presence of only four phase conditions and no amplitude constraints (save that of the noise floor) makes the signal easy to demodulate. Normally a phase-detection process requires the presence of a timing reference signal as we saw in Chapters 3 and 6 for analogue colour subcarriers. In the QPSK system used for DTV carrier modulation, however, it is possible to derive a carrier reference from the signal itself so long as each successive phase change is relative to the previous one, a technique called *differential coding*. Of all the types of DTV demodulator, that for QPSK is easiest to implement in the receiver.

QAM

Quadrature amplitude modulation is a development of QPSK, primarily used in cable distribution systems. Here (unlike terrestrial broadcasting) the entire signal path is under the control of the cable operator, who takes care to maintain a 'comfortable' passage for his signal. In general a cable system has a bandwidth of about 8 MHz, a relatively low noise floor (and hence good S/N ratio), some possibility of signal reflection due to impedance mismatch (this gives the same effect as multipath propagation) and the risk of low-level electrical interference

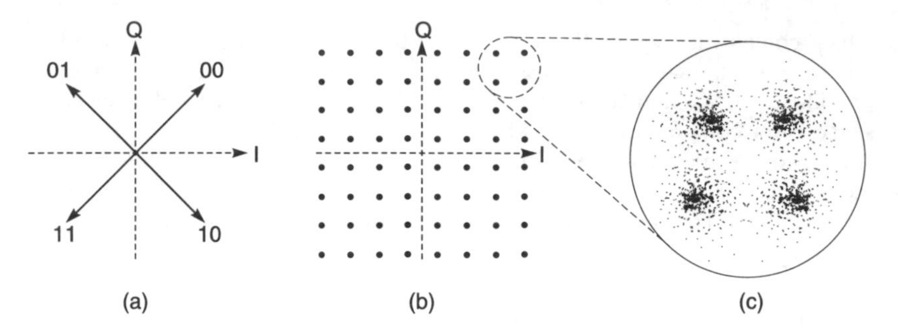

Figure 12.19 *Modulation systems: (a) DPSK; (b) QAM, showing 64 symbol-codes; (c) the effect of noise on QAM reception, increasing the risk of bit-errors*

where the conductors are copper rather than glass fibre. The ideal modulation system, then, is a slower-moving one than used for satellite, with the many more levels and phases that this medium can afford. Here 64-QAM modulation of a single carrier wave is used. Figure 12.19b shows that it has 64 possible combinations of phase and amplitude, each conveying a 6-bit symbol. Thus the information is much more tightly packed than with 4-state QPSK, but with much less tolerance to noise, which tends to spread out in time and space the 'targeting' of the carrier components as shown in Figure 12.19c. The lower the system noise level the more certain becomes correct symbol detection.

COFDM

Terrestrial DTV broadcasting involves the most difficult and variable path conditions so far as bit-corruption is concerned. Interference comes from electrical impulses, co-channel transmissions and other sources; multipath reception and frequency-selective fading are common, especially in built-up areas; signal strength can vary enormously; and aerial systems, being entirely in the hands of householders in most cases, cannot be relied upon in the same way as satellite dishes or cable feeds. Indeed many people have learned to expect to receive terrestrial TV on a set-top aerial, and see no reason why they should not continue thus with digital broadcasts!

The DTV modulation system used here then is COFDM, Coded Orthogonal Frequency Division Multiplex, whose 'coded' prefix refers to the forward error correction artifice already described; the main virtue of COFDM is its tolerance of multipath propagation. In the UK the 2 K variant of COFDM is used, with provision for 2048 separate carriers in the 8 MHz-wide channel available. 1705 are actually used for transmission of a transport stream (multiplex) typically carrying five TV programmes, see Figure 12.20a. Each carrier is allocated a space of about 4 kHz and is 64-QAM modulated with relatively few symbols such that the loss of some of them has no effect on reception once error-correction has been carried out. The orthogonal spacing of the carriers, Figure 12.20b, prevents mutual interference and gives a 'sweet-spot' for detection of each: the adjacent carriers'

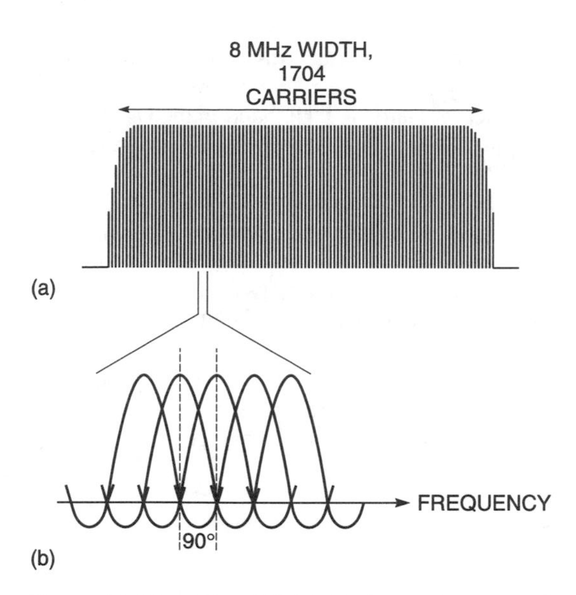

Figure 12.20 *2 K COFDM. (a) the carriers completely fill the 8 MHz (UK) channel allocation, carrying several programmes simultaneously in a* multiplex. *At (b) is shown the interleaving of individual carrier frequencies*

frequency spectrum has a null at each centre point. The large number of carriers permits the duration of each COFDM symbol to be longer than that of the original transport bitstream, and this permits, at the receiver, a delay in signal evaluation until all the 'echoes' have arrived. In fact the echoes can effectively *reinforce* the main signal. The echo-rejection process is aided by the provision of a guard interval between transmitted symbols, during which echoes from the previous symbol will be disregarded, and inter-symbol interference minimised.

It is not possible (or necessary) in practice to generate such a large number of modulated carriers by separate oscillators. In fact they are derived by an inverse fast Fourier transform (IFFT) which in effect permits the carrier signal to be generated in the frequency domain and then transformed into the time domain for transmission. At the receiving end an inverse process is used for demodulation. Of all the DTV modulation systems currently used, COFDM demands the most expensive and complex demodulation system at the receiver.

COFDM in practice

The perfectly even distribution of power and energy in the broadcast channel (Figure 12.20a) compares very favourably with that shown in Figures 3.13 and 5.2 for conventional AM TV broadcasts. Because of this and the relative ruggedness and immunity to interference, noise and signal echoes the carrier power required for transmission is very low compared with that for ordinary terrestrial TV channels. Equivalent coverage can be achieved with a carrier power 20 dB (100 times) lower than for AM broadcasts. The risk of interference in co- and adjacent channels is reduced accordingly, opening the way to the use of channels which would be 'taboo' for conventional broadcasts. This characteristic can be used to achieve greater geographical coverage or to minimise transmitter power. Currently analogue and digital broadcasts share the UHF band in the UK during a transition period; eventually analogue TV broadcasting will be discontinued.

A potential advantage of COFDM is its amenity to SFN (Single Frequency Network) broadcasting, in which the same channel-slot can be used country-wide for the same programme multiplex. In an analogue system this would lead to intolerable 'ghosting' on the picture, especially at times of high barometric pressure: the original 1960s transmitter siting and frequency plans for UHF bands 4 and 5 were designed to avoid it. With 2000-carrier COFDM the simultaneous presence of signals from local and distant (probably off-beam for the aerial) transmitters does not give rise to problems in most cases, though 8K COFDM (8000 carriers)performs better still in an SFN scenario. The hardware limitations which prevented 8K COFDM being adopted for the UK in 1998 have now been overcome.

Propagation of DTV signals

To avoid changing systems, bands, sending- and receiving aerials and (for instance) hardware which is irretrievably floating about in space, DTV transmissions occupy the same frequency slots, bandwidths, satellite transponders etc. as the pre-existing analogue services they are replacing – indeed the design of the modulating systems we have examined are tailored to this compatibility. For details of terrestrial transmission and reception see Chapter 5; satellite links and cable networks are described in the next two chapters. Aerial/dish alignment and cabling/termination can be more critical for DTV, and the use of spectrum-analyser equipment gives the best indication of received signal. DTV receivers generally have in an installation menu an indication of signal 'goodness' based on a BER (Bit Error Ratio) reading from the error-correction section. It's typically presented in bar-graph form, and some sophisticated field strength meters, designed for rooftop use, also have a BER readout, much more revealing than a simple indication of mere 'brute' strength in the received signal.

The DTV receiver

Apart from the first and last sections, the tuner/demodulator and the final output stages, a DTV receiver is a wholly digital device, in effect a purpose-designed computer with a central processing system, a great deal of memory of various sorts, and a fast-and-wide operating system. Increasingly software-driven, the receiver box – in many cases – incorporates a telephone modem and a hard-disk drive for interactive TV operation and programme-recording respectively. Only in its power-supply, tuner, remote-control and interfacing sockets does it bear any resemblance to a conventional analogue receiver.

Figure 12.21 offers an outline of the processes involved in DTV reception and decoding. The tuner automatically scans the appropriate broadcast bands, locking onto transmissions and identifying them, then storing the tuning points in memory: no tuning action is required of the viewer or installer. The

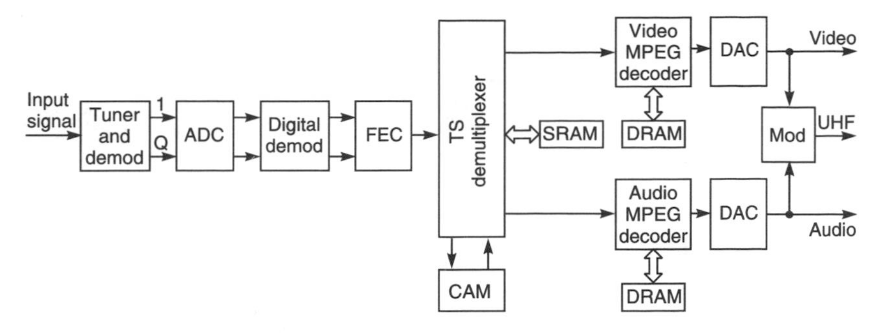

Figure 12.21 *Outline of DTV receiver system*

demodulator, of a type appropriate to the system in use, presents I (in-phase) and Q (Quadrature phase) carrier components to the A-D convertor and digital demodulator. Now the original bitstream has been reconstituted for application to the FEC 'repair' system which feeds a transport stream demultiplexer. Here the data packets which make up the required programme are selected and assembled – with the aid of short-term storage in the SRAM – for onward passage to the video and audio MPEG decoders, each with a DRAM in which, respectively, the picture frames and audio datastream are built up. Analogue video and audio signals are derived in D-A converters to leave the set-top box or inbuilt digital module as conventional waveforms containing 625/50 (Europe) picture waveforms and baseband stereo audio signals. We shall now examine the receiver blocks in more detail.

Channel decoder

The block diagram of Figure 12.22 is representative of a set-top box for DTV reception. The first four blocks constitute the *channel decoder,* whose design varies with the carrier frequency and modulation system in use. The system drawn here is designed for satellite reception with QPSK. From the descrambler block onwards the system is common to all three DTV media, satellite, terrestrial and cable.

The tuner downconverts the incoming carrier to an IF frequency while rejecting adjacent- and image frequencies as described in Chapter 8. The IF signal is demodulated into I and Q components which are next A-D converted into two 6-bit datastreams for digital demodulation and subsequent application to the FEC section, in which the Reed-Solomon check/correction, de-interleaving and Viterbi decoding are carried out. The bitstream emerging from the FEC block is a clean and error-free representation of the transport stream assembled at point A in Figure 12.18. It enters the *transport demultiplexer* chip.

Transport demultiplexer

The operation of this IC may be likened to that of a mail sorting centre, in which a multiplicity of labelled parcels are assigned destinations and assembled in a specific order for dispatch to two main destinations: the video and audio MPEG

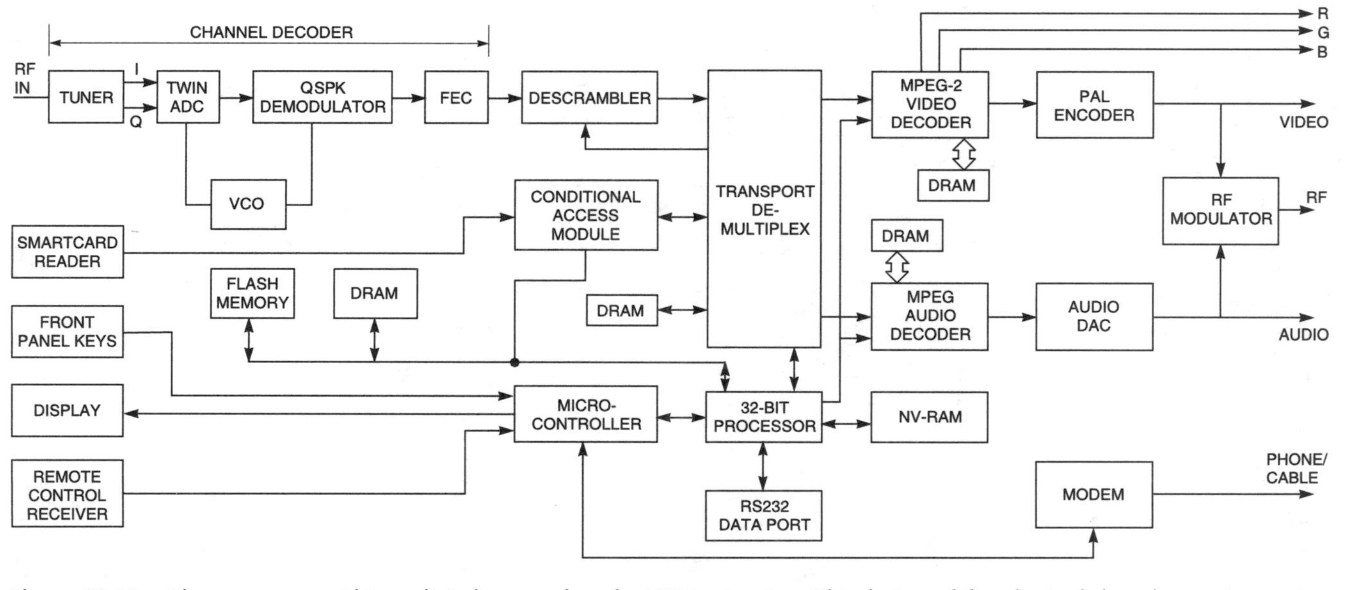

Figure 12.22 *The processes within a digital set-top box for DTV reception. The design of the channel decoder section varies between satellite, terrestrial and cable types*

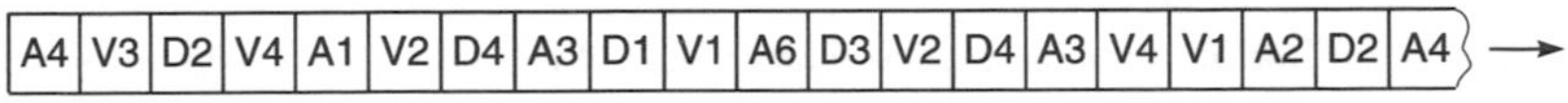

A4	V3	D2	V4	A1	V2	D4	A3	D1	V1	A6	D3	V2	D4	A3	V4	V1	A2	D2	A4

→

Figure 12.23 *Interleaving of data packets in the transport stream. Here six programmes are present*

decoders. As Figure 12.18 showed, the data packets for the various programmes in the multiplex are interleaved in time: each consists of a 204-byte sequence of sound, vision or housekeeping data. The sequence may typically be as shown in Figure 12.23. If, for example, the user has selected programme four in the channel bouquet, the transport demux chip pulls out of the transport stream packet numbers 4, 16 etc. and stacks them sequentially in the DRAM (typical capacity 4 Mbit) shown on the left of the main chip to form the programme four video data. Packet numbers 1, 20 etc are also extracted and stacked elsewhere in the transport DRAM to form the audio bitstream. Although these packets enter the DRAM at high speed and irregular intervals, they are read out at a different rate (controlled by the microprocessor) to suit the vision and sound decoders. Thirdly, the de-mux chip extracts service packet numbers 7 etc, which contain identification- and instruction- bytes relative to the selected programme 4, which might be, for example, BBC News 24. From the packets which make up the data for programme four, the de-mux chip gets PID (Packet Identification Data), to govern the packet selection process, and time stamps to ensure the simultaneous release of sound and vision data: thus is good lip sync of picture and audio achieved. Clock frequency for the de-mux chip is 25 MHz.

Decryption

In the case of an encrypted transmission the data is diverted through a CA (Conditional Access) module. It works in conjunction with (a) a smart card, for which a subscription may have been paid; (b) data in the transmission, which defines scrambling and authorisation codes, and provides part of the decryption key; and (c) the control microprocessor, which validates the subscription for the service requested. The CA module slots into an aperture in the rear panel of the receiver.

Digital video decoder

The MPEG video decoder is the most complex section of the DTV receiver: it has to compose a complete and continuous moving picture from the transmitted I, P and B frames. Along the way it must perform data expansion, inverse DCT, interpolation and error correction. The complete, reconstituted frames are built up inside the DRAM shown just below the video decoder in Figure 12.22. Its capacity is typically 16 Mbit, corresponding to three frames: one each for the current I and P frames and one for the display. The decoder's video output pulse train contains all the values for Y, Cr and Cb for all the pixels in the picture. It emerges from the decoder chip on an 8-bit bus to a combined D-A convertor and (in the case of a set-top box) a PAL encoder for compatibility with existing TV receivers. Two main clock trains are used within the MPEG decoder: one at 55 MHz for memory manipulation and processing, and a 'master' at 27 MHz to time and synchronise the decoding process.

Audio decoder

The audio datastream is strobed out of the transport de-mux IC to the MPEG sound decoder, where it is decompressed and reconstituted in the associated DRAM to provide pulse trains for the D-A converters. Memory readout is delayed by up to one second while the associated video signal 'catches up' after its processing, especially long in the case of an encrypted signal; the delay is governed by the time stamps sent with the broadcast data. Also present in the broadcast data is a code to indicate the audio sampling rate 48, 44.1 or 32 kHz, so that the audio decoder can work to the same standard. In some receivers the video and audio MPEG decoders are contained within the same chip.

Practical DTV chipset

Figure 12.24 gives an idea of the distribution of functions between ICs in a terrestrial DTV set-top box or the receiver/decoder section in an integrated DTV set. This chipset is designed and manufactured by LSI Logic. Device type L64780 performs initial A-D conversion and digital demodulation for COFDM signals in 2K or 8K mode. IC L64724 caters for forward-error correction, feeding the transport stream to the third chip of the set, type L64108, TS demultiplexer, based on a 32-bit 54 MHz MIPS processor. It contains an 8 kB instruction memory, 4 kB data memory and programmable hardware-assisted section filters. The fourth IC of the ensemble, L64105, contains both video and audio decoders, and works in conjunction with a 16 Mbit SDRAM.

MPEG decoder programming

Sound and vision decoders are governed by the same 32-bit control microprocessor as the transport de-mux section. The software for this processor is stored in the flash memory chip drawn near the left of Figure 12.22. This can be updated

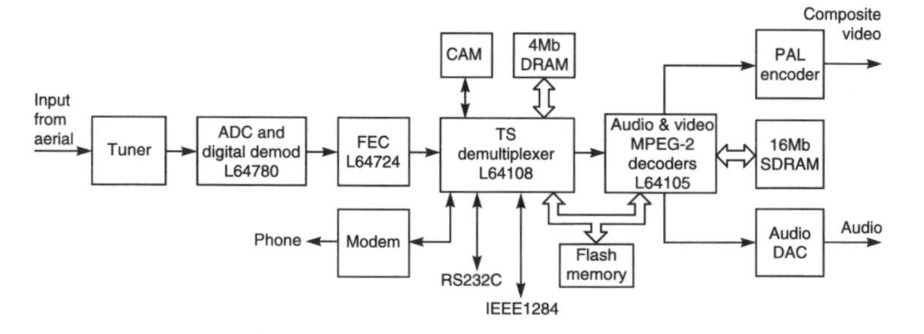

Figure 12.24 *Commercial chipset for terrestrial DTV receiver, showing IC functions*

with new programs over the air by the broadcaster as necessary: upgrading is done by first downloading the received software into the DRAM chip then transferring it into the flash IC.

Receiver control

We have seen that the microprocessor is primarily concerned with the transport demultiplexing and MPEG decoding processes. The more mundane general housekeeping functions (mostly common to TV and VCR circuits, and including remote control processing, front panel display, tuning control etc) are undertaken by a separate microcontroller, shown in our diagram at bottom centre. It works with an I^2C bus of the sort shown in Figure 8.12. At initial switch-on the microcontroller instructs the tuner to sweep-search the band to find the 'home channel' which contains an EPG (Electronic Programme Guide). At this time, too, all the processors in the receiver are reset; and data is loaded into the decoders, processors etc. from the flash and ROM memories. The RS232 port, by which the receiver can be interfaced to a PC (in the workshop for faultfinding, data-downloading and control, or in the home for interactive and 'extension' operation) is wired to the microprocessor IC, and has four main links: serial data send, serial data receive, and request lines for each direction.

Modem link

An important component of the receiver or set-top box is the modem, shown at the bottom right-hand corner of Figure 12.22. In the future it may well be connected to a satellite uplink for direct communication with the service providers, but at present the link is via a conventional telephone line to a dedicated switching centre, in similar fashion to the Internet link to and from a home PC. The modem facilitates home banking and shopping, pay-per-view services and Internet links. It also permits software updating of the receiver and other 'technical' communications such as malfunction detection and remote fault diagnosis, also TV audience measurement. Subscriptions can be dealt with by the modem link, by (e.g.) reading a plugged-in everyday debit or credit card, or by 'recharging' the viewer's credit for viewing.

The call can be initiated at either end. The modem incorporates a non-volatile memory chip to store the received data for action, or the accumulated data for sending. Also present is a programmable dialler and a modulator/demodulator to convert the data to and from a form which can be handled by a telephone line whose frequency response is designed for speech traffic extending little beyond 3 kHz. The signal carrier, a tone, can be modulated by the data in several ways: amplitude-, frequency- or phase-shift keying. For two-way (*duplex*) operation the 3 kHz-wide phone-line frequency spectrum is split into two bands each 1.2 kHz wide, in a frequency-division multiplex technique involving two modulated carriers at frequencies of about 1.5 kHz and 2.7 kHz.

A much fuller use is made of the telephone network with the ADSL system, a description of which appears in Chapter 14.

13 *Satellite TV*

A relative newcomer to the broadcasting field is the practical realisation of space transmitters. Before we attempt to describe them in any detail, we need to understand a little of how a satellite is positioned and controlled so as to appear stationary to an earth-mounted receiving installation.

As we are well aware, the force of gravity is a dominant one on earth, tending to pull loose objects towards the centre of the earth. There is another effect called centrifugal force which operates on spinning objects, this time pulling outwards, away from the centre of the spin-arc, as demonstrated by the domestic spin-drier, certain fairground rides and a conker on a string. The strength of centrifugal force depends entirely on the orbital speed of the object in question. For any given orbital speed, then, there must be a point in space where centrifugal force acting on an orbiting satellite just cancels the gravitational pull of the earth, so that these balanced forces hold the satellite at a fixed distance above the earth's surface without any need for motive power from the satellite itself. If we require the satellite to appear to stand still relative to Earth, it will need to orbit at exactly the same speed as Earth itself – once every twenty-four hours. With the orbital speed thus rigidly fixed, the distance from Earth for the satellite comes out at 35 800 km (22 200 miles) to satisfy the gravity versus centrifugal force equation. So we have two fixed parameters for our *geostationary* satellite, orbital speed (which comes out at 11 069 km/hour for one revolution per day) and the altitude of 35 800 km. It must necessarily be over the equator to prevent an apparent daily North–South oscillation. When all these parameters are satisfied, the satellite will effectively hang stationary in the sky. From time to time drift will inevitably take place of its position and attitude, and this is corrected by small on-board jet motors which operate on a *servo* principle to stabilise and maintain the satellite's attutude to earth.

Thus we can set up a transmitter in the sky, but at first sight its position above the equator is bad news for the UK, situated as it is some way into the Northern Hemisphere. Plainly, a radio beam from the satellite will come in to us at an oblique angle rather that directly downwards; in the UK the elevation angle from the horizon varies from 17° in the Shetlands to almost 28° in West Cornwall. Except in built-up areas, a clear line-of-sight to the sky can usually be secured at this angle from ground level, so that receiving aerials can be mounted at ground level in many domestic situations, perhaps in a back or side

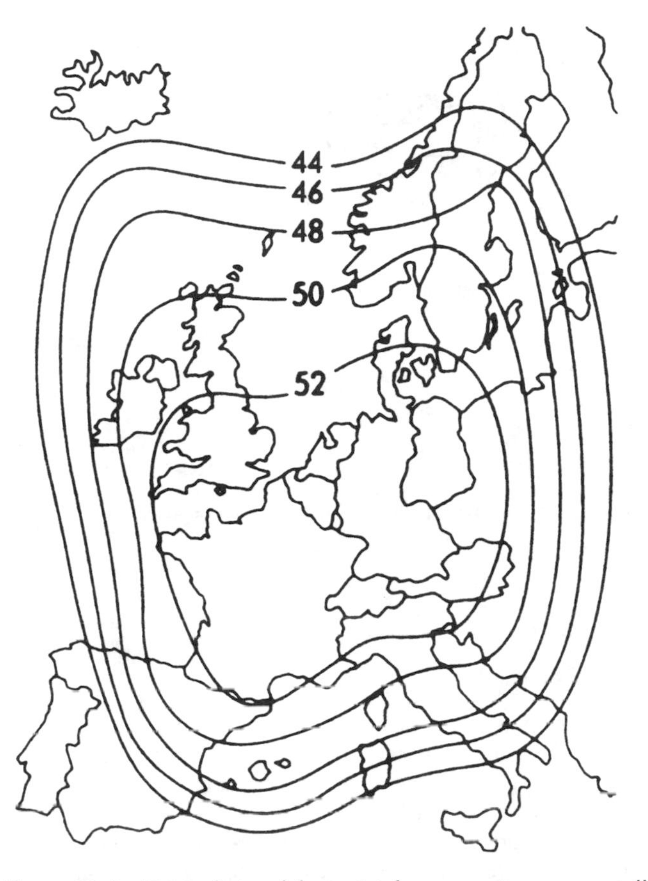

Figure 13.1 *Typical signal footprint for a pan-European satellite. This one shows that of the Astra 1A beam*

garden; only in the shadow of buildings or other obstructions will it be necessary to mount the receiving dish on a wall, roof or tower. In urban areas where these problems exist they can be solved by the use of a 'community dish' with local cable distribution to individual homes.

The slant 'radio-illumination' effect stemming from the satellite's position over the equator means that a narrow transmission beamwidth can cover the entire UK; for a *footprint* like that shown in Figure 13.1 the beam need only be 1.8° in the 'vertical' plane and 0.7° across. Because all the transmission energy can be concentrated in this very narrow beam by highly directional (almost like a gun!) transmitting aerials at the satellite, the use of radio energy is very efficient indeed, though the narrow beam calls for very great accuracy in the alignment of the satellite in space, and the receiving dish on the wall! They have to be held within ±0.1° and ±0.5° of nominal pointing position respectively.

To give an impression of the radio-torchbeam-from-space effect, and the reason for the elliptic/potato shape of the footprint, Figure 13.2 represents the relative positions of the satellite, the equator, and the UK.

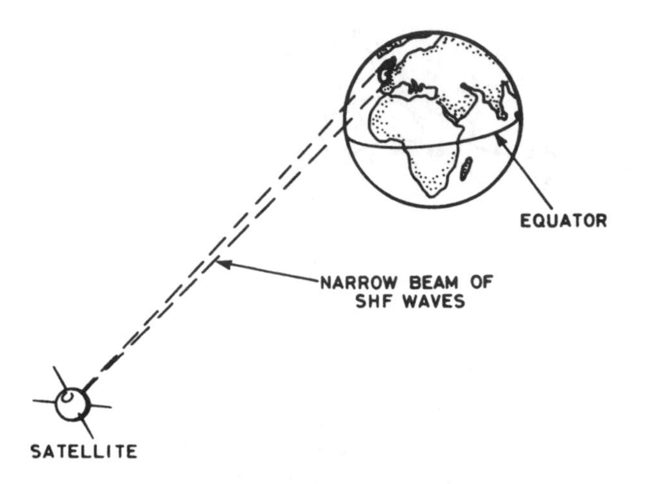

Figure 13.2 *The relationship between the Earth, equator and a geostationary satellite*

Satellite power

In Chapter 5 we discussed the effect of a directional radiating aerial on the apparent power of the transmitter, and saw that ERP was the product of true RF power multiplied by the 'gain' of the transmitting aerial. The very narrow beam from a satellite (in the UK case, $1.8° \times 0.7°$) concentrates the power beautifully, and from an RF output power of about 150 W we can realise an ERP of 3000 kW! The only type of radiating aerial capable of launching such a narrow RF beam is a parabolic dish, and this is the basis of the satellite's transmitting element. It may be between 1 and 3 m in diameter, depending on the beam angle required. The true RF power of the satellite contrasts remarkably with that required for terrestrial broadcasts – on the one hand 50 W for good coverage of the whole of the UK, on the other many many megawatts, indifferent coverage of some areas and 1000 individual transmitters to provide and maintain, in each case for one TV analogue channel.

The power to run the transmitter and the peripheral equipment in the satellite itself comes from banks of solar cells drawing energy from the sun. These are mounted on 'wings' springing from the satellite body, and are servo-controlled to keep them at right-angles to the sun's rays. This accounts for the characteristic dragonfly shape of a broadcast satellite: see Figure 13.3.

The DC power provided by the solar cells needs to be about 2 kW; much depends on the efficiency of the transmitters and the requirement of other, 'housekeeping' equipment on board. Some countries, due to their size and position relative to the equator, require greater beamwidth than others, and for larger beamwidths the RF power required increases rapidly. As an example, if the UK were directly below a satellite on the equator we would need a fivefold increase in power to achieve the same field strength over the same area.

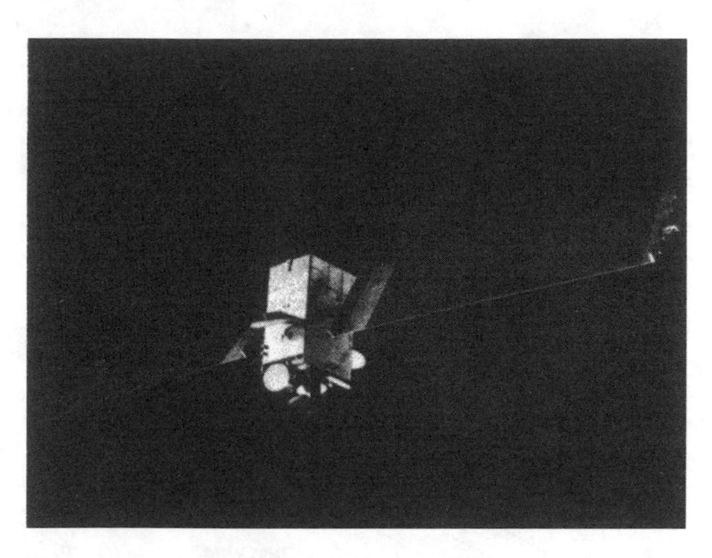

Figure 13.3 *A typical TV broadcast satellite*

The satellite as an 'active reflector'

Of course, the programmes are not originated on board the spacecraft, so it must also be fitted with a receiving aerial and the necessary transposing equipment. Coming up to it from earth will be the programme material and signal routing instructions, along with engineering control signals to cater for 'housekeeping' and positional control. This uplink also takes place in the SHF band with a narrowbeam signal between parabolic dishes. Slotted into the satellite's output signals are data streams containing information about the internal workings of the spacecraft and its motor, power and signal-processing sections for monitoring on Earth. The up- and down-links are shown in Figure 13.4.

The active life of the craft comes to an end when the fuel for the correction motors runs out, assuming that no catastrophe occurs to the other systems to disable the satellite. As in all space-equipment design, crucial components are duplicated, and a great deal of potential redundancy is designed into every part of the vehicle. At the end, the satellite can be 'kicked off' into obscurity in deep space by a final burst of the positional motors at the very last, to make room for its successor!

Satellite signal reception

Leaving aside now the 'Star Wars' atmosphere of the orbiting satellite, it is time to return to the back garden to see what will be required in the way of receiving aerials for space transmissions at 11 GHz. While in theory a dipole with reflector and directors might work, the reality is that at the wavelength of a 11 GHz carrier (about 2.5 cm) the conventional aerial would be small enough to slip into a shirt

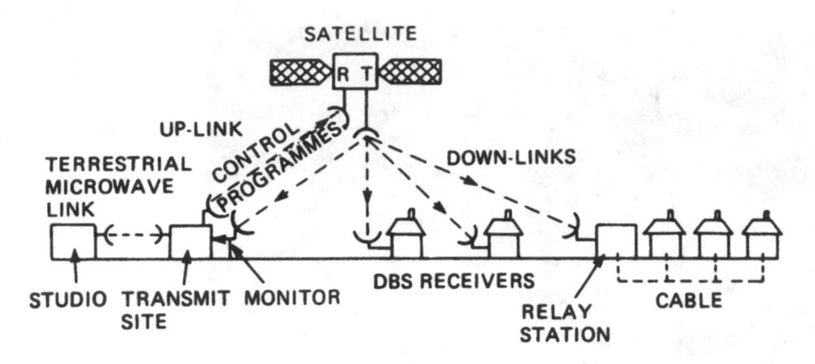

Figure 13.4 *Arrangement of up- and down-links for the DBS satellite*

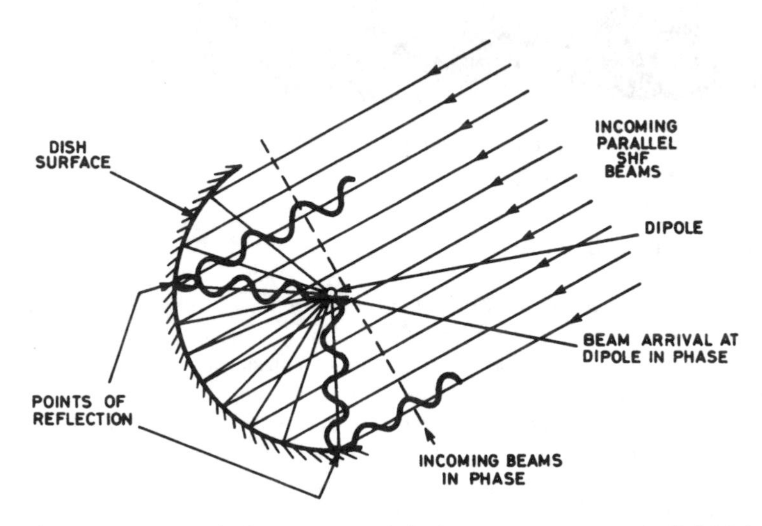

Figure 13.5 *Parabolic receiving dish showing incoming parallel RF beam and the in-phase reflection to the pick-up device*

pocket and would certainly not have enough 'meat' in it to extract a usable (or even detectable!) signal. We need to capture the beam over a relatively wide area and focus it on a dipole to achieve sufficient field strength to make a usable signal. For this a parabolic dish is called for, with some form of dipole at its focal point.

The larger the dish, the more RF energy it can intercept and concentrate on the pickup element; thus dish size has a counterpart in element-count in a conventional Yagi array. Again, the question of *phase* is relevant because all the reflected cycles of RF carrier must arrive in phase at the dipole in order to reinforce each other; see Figure 13.5. If any arrive out of phase a cancellation effect will take place with consequent loss of gain. This requires tremendous accuracy in the manufacture of the dish which must not deviate from a true parabolic shape, and must have a good surface finish. Any physical damage, warping or contamination of the dish surface will reduce the efficiency of the receiving aerial, which in good conditions should be about 60 per cent.

Detail design of SHF dishes depends on the individual manufacturer, but the system is engineered to provide a carrier-to-noise ratio of better than 14 dB for 99 per cent of the worst month weather-wise, with a dish of 60 cm diameter in the primary service area. A larger dish offers improved gain, but is more difficult to align physically.

Offset dishes

All other factors being equal, and so long as the pick-up device is at the focal point, the strength of the received signal is proportional to the *area* of the reflector. For convenience in manufacture, assembly and installation (and perhaps a better aesthetic effect) small fixed-dish installations use an *offset* dish, with the interception electronics mounted on the end of an arm protruding from under the dish as shown in the photograph of Figure 13.6. The principle here is that the dish, no longer parabolic in shape, conforms to the contours of a section of a much larger parabolic reflector, Figure 13.7, with the two advantages that the arm and head unit do not make a 'shadow', signal-wise, on the dish to reduce its efficiency; and that the disc surface is more nearly vertical so that water, snow and other matter is less likely to settle on it.

Figure 13.6 *Offset metal mesh dish for satellite reception*

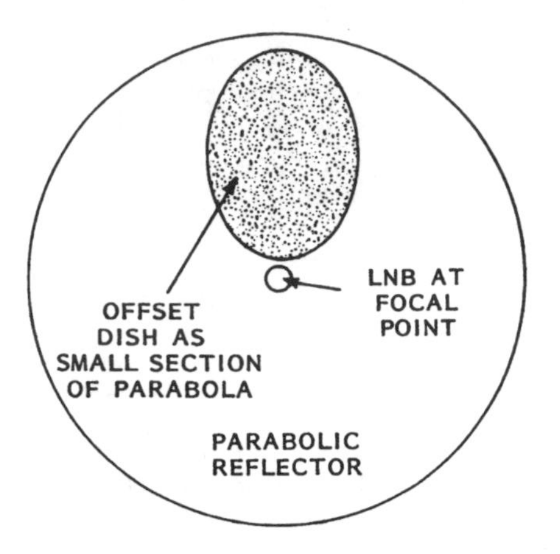

Figure 13.7 *An offset microwave dish may be regarded as a small section of a large parabola shape*

Head-end convertors

At the point of pickup, the signal is in 11 GHz FM form, and at that frequency the only possible transmission medium is a waveguide in the form of a large and rigid metal pipe. Any form of co-axial cable would introduce intolerable attenuation, so a frequency converter is mounted at the receiving dipole, in similar fashion to the masthead amplifier of some terrestrial signal-receiving installations. This converts the signal, by means of a heterodyne process, to a lower frequency around 1 GHz which is amenable to feeding indoors by means of very low-loss co-axial cable.

Service areas

Unlike conventional TV transmitters, the coverage of satellite broadcasting is not limited by constraints of line-of-sight, nor (except in a very local sense, as described earlier) by topographical features. For this reason the service area is much more homogenous and predictable for DBS (direct broadcasting by satellite), falling off gradually from the centre of the footprint. Within the primary service area (see Figure 13.1) a 0.6 metre dish should suffice, and in the secondary service area a similar standard of reception can be achieved with a larger (higher-gain) receiving dish, typically 0.8 metre in diameter. Most parts of the UK are within at least the secondary service area of many European satellite services.

Dish installation

Although dishes for use at 11 GHz are not so large as those required for UK reception of the earlier 4 GHz satellites such as the Russian Gorizont, a fair amount of weight and surface area are inherent in them. This means that their mounting must be very strong and stable, not only from a safety-hazard point of view in strong winds, but also because their very critical angular alignment can only be maintained if they are mounted very rigidly.

Once a stable fix has been achieved the dish must be aligned onto the satellite's beam. The first step is to calculate the azimuth and elevation of the satellite at the receiving site, and this requires an exact knowledge of the position of the site, obtainable from an Ordnance Survey map, or a marine satellite navigator. From this the required angles can be calculated to several places of decimals and the dish aligned with reference to a magnetic or gyro compass. A typical 11 GHz receiving dish and mount are illustrated in Figure 13.6.

Multi-satellite reception

Where reception from more than one satellite is required, there are several single-dish techniques available: UK planning laws generally forbid the mounting of more than one dish on a dwelling. The most common requirement, perhaps, is to receive signals from two satellite clusters a few degrees apart, for example Astra and Eutelsat/Hot Birds. Here a special bracket, mounting two LNBs with view angles 6° apart is mounted on the boom and aligned so that each LNB sees a dish-reflected signal from only its 'own' signal source. The dish may be primarily aimed at one of the two positions or at some intermediate point, governed by relative field strength and the likelihood of interference from other satellites nearby. With this system it is not possible to get optimum reception – in terms of signal strength and interference immunity – from *both* LNBs.

An alternative technique, giving similar results, involves a motor driving a single LNB across the face of the dish to achieve different 'look angles' into the sky. This is limited to a few degrees of arc because the efficiency of the dish/LNB combination drops rapidly off as the LNB alignment moves away from dish centre.

Extension of the concept of the dual-feed bracket leads to a specially-designed wide-angle dish fronted by a curved bracket mounting several LNBs, each aligned and optimised for a particular satellite or cluster. As with the bracket, this offers the advantage of simultaneous reception from all the selected satellites if required, though its appearance is not attractive.

The fourth (and most expensive) choice is a motorised dish, which under manual or automatic control can scan the entire satellite belt from horizon to horizon, stopping at whichever signal beam is required. A special (and very stable) mount and careful alignment are required to establish and maintain good tracking and thus optimum reception throughout the scanning range.

LNB switching and DiSEqC

We shall see later how polarisation switching of the LNB is achieved by changing the LNB supply voltage between two levels. *Universal* LNBs are capable of operating on two bands, low (10.7–11.7 GHz) and high, 11.7–12.75 GHz, the latter typically used for DTV transmissions to the UK. It's achieved by using a dual-frequency local oscillator running at 9.75 or 10.60 GHz, and is switched between them by the superimposition of a 210 mV rms 22 kHz tone on the LNB supply line. It comes on to select high-band, and is thus always present at the aerial socket of a Sky digibox.

The DiSEqC control system is capable of governing the selection, band-switching and polarisation settings of several LNBs, automatically by the receiver if required and programmed into the installation/control software. It can be one-way or bidirectional, in the latter case involving feedback for the purpose, for instance, of dish positioning control. At the bottom end of the cable is an encoder or modem which modulates the 22 kHz tone with digitally-coded commands; at the top is a decoder or modem which governs LNB functions and perhaps controls motor drive. As we saw in the last chapter with telephone lines, it seems that engineers are never satisfied unless an analogue-signal-carrying cable is burdened with other (preferably digital) traffic and functions!

Dual-feed LNBs

A standard LNB has a single output for connection to one receiver. For use with two separate receivers a *dual*-LNB is available, containing two independent sets of electronics. This permits independent setting of polarisation etc. for the two receivers while still fulfilling the one-dish requirement. Another type of dual-feed LNB has separate, *unswitched* outputs for each polarisation plane. It is intended for use in MATV distribution systems, typically in a block of flats, where each user's polarisation setting is addressed to a switch associated with the line amplifiers rather than the LNB itself.

Preamplifiers for SHF

We have seen that down-conversion is carried out at the dish by a superhet technique. Before mixing, however, a very low-noise preamplifier must be used to bring the carrier signal up to a suitable level. The noise generated in the first stage virtually determines the S/N ratio for the entire installation, and the performance of this amplifier is thus crucial. The FET (field-effect transistor) is used here, and designs using GaAs FET SHF amplifiers have been developed to offer very low noise figures: less than 0.5 dB at a gain of 17 dB. The tuned circuits for selection and filtering are etched onto a glass-fibre-based PC board, see Figure 13.13.

As SHF front-ends are developed with progressively lower inherent noise, advantage may be taken of this to reduce dish size (making alignment less critical) or alternatively to utilise the same size dish for reception of fringe (outer-footprint) reception of transmissions intended for other areas of Europe. Some readers will be aware that for no perceptible noise on a conventional analogue-derived TV picture an S/N ratio of about 42 dB is required; with FM modulation the carrier-to-noise ratio of 14 dB quoted earlier is quite adequate to provide an acceptable display picture.

The question of interference is a relevant one. Any interference to the up-link is very unlikely, and at the receiving site man-made interference is not significant at SHF, especially with the encoding and modulation systems used. Heavy rain at the receiving site will degrade the signal strength by about 4 dB, but in the primary service area this should not be detrimental to picture quality provided such factors as dish size and alignment and preamplifier gain are adequate.

Satellite TV modulation and encoding

All analogue TV satellites use FM modulation systems. There are three main types of emission: MAC variants, discussed in Chapter 10; digital broadcasting, which we examined in Chapter 12; and FM radiation of subcarrier-based analogue colour signals in Secam, NTSC or PAL, which when demodulated correspond to those of terrestrial transmissions, and require the same sort of post-demodulator processing and colour-decoding circuits as have been described in previous chapters of this book.

Analogue receiver configuration

Satellite receivers, then, can be built into VCRs and TV sets, but are more commonly produced as stand-alone units with sockets to interface with the AV ports of audio, TV and video equipment. In the case of MAC receivers, the high-quality picture transmission system permits direct connection to the RGB or YUV input sockets of a TV or monitor because the MAC decoder is built into the satellite receiver box as shown in Figure 13.8. For PAL (or Secam or NTSC, for that matter) satellite transmissions, the satellite receiver does not contain a decoder because it would offer no improvement in picture quality, merely duplicating that already provided inside the TV. The set-up here, then, is as depicted in Figure 13.9, with a tuner, demodulator and sound circuits to produce baseband video (CVBS) and audio (stereo where applicable) signals for cable-linking to TV or VCR. A UHF modulator (mono sound only) is also provided for compatibility with equipment not fitted with AV input sockets and with home-distribution systems.

Receivers designed for use with digital satellite transmissions contain, alongside the tuner, demodulator and control system, a complete MPEG decoder and RGB converter, affording RGB connection to the TV – Figure 13.10 – for best picture quality. For use with VCRs, and with TVs which have no RGB input port, they also sport a PAL (or whatever is appropriate to the local terrestrial broadcast system) encoder for compatibility. To use this, however, is to throw away many of the advantages of digital broadcasts, as described in Chapters 10 and 12, and to reintroduce the impairments due to bandsharing of Y and C picture components. The same applies to the subcarrier encoders incorporated into MAC receivers. In the rest of this chapter we shall look at satellite receivers designed to receive FM PAL transmissions, and take the popular Astra series of satellites as our transmitter example. In terms of broadcast parameters, ERP, channel-spacing and polarisation of transmissions, other analogue-broadcast satellites are similar, except that those

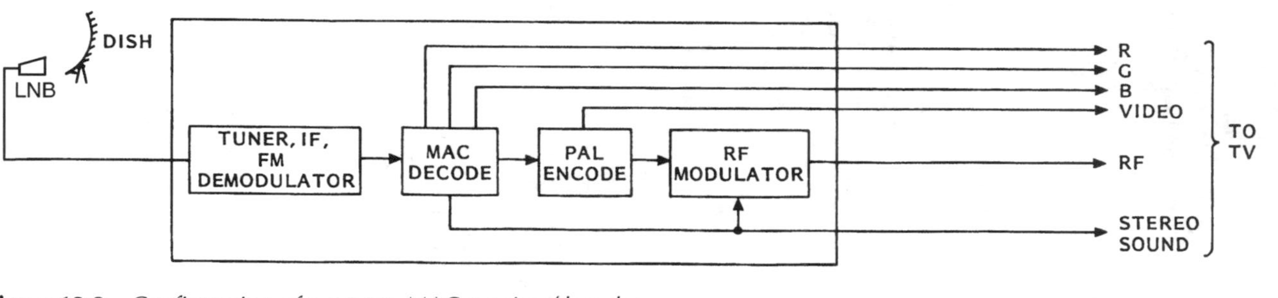

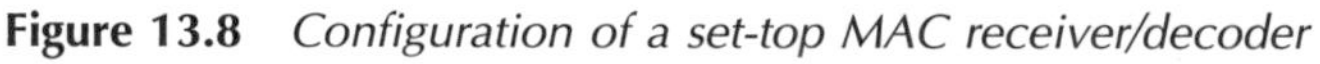

Figure 13.8 *Configuration of a set-top MAC receiver/decoder*

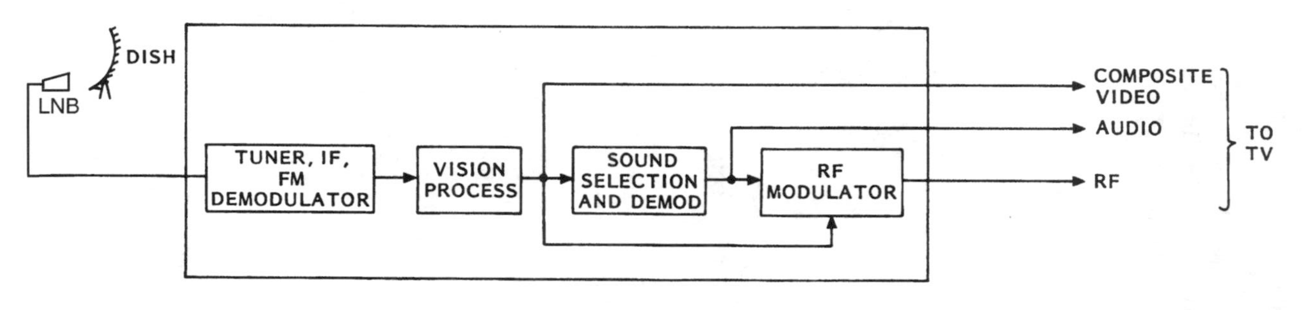

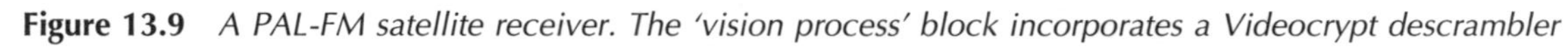

Figure 13.9 *A PAL-FM satellite receiver. The 'vision process' block incorporates a Videocrypt descrambler*

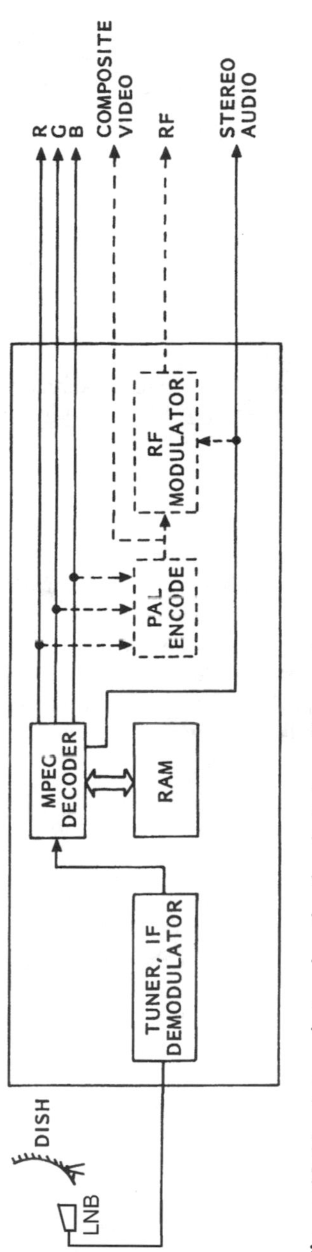

Figure 13.10 *Internal signal path of a digital satellite receiver*

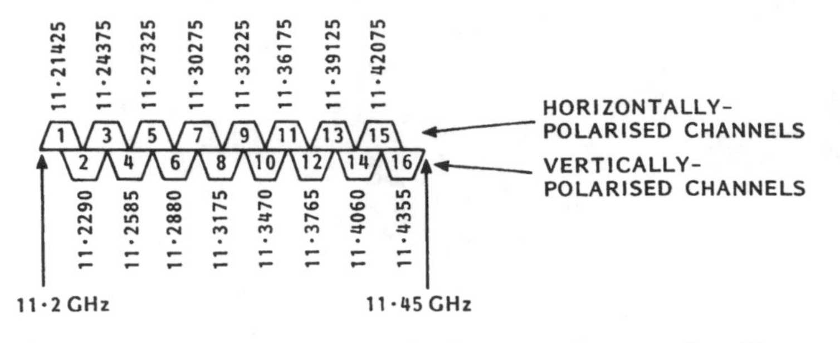

Figure 13.11 *Carrier spacing and polarisation for a set of satellite transmissions*

radiating MAC signals very often use circular polarisation as an alternative to the fixed (horizontal/vertical) mode to be described below.

Satellite transmission

Astra 1A is one of a group of co-sited entertainment satellites in geosynchronous orbit at 19.2E. It has 16 transponders, each radiating a power of about 45 W into a highly-directional dish radiator to give a coverage footprint similar to that shown in Figure 13.1. The carriers are in the FSS (fixed satellite services) band, and range from 11.2 to 11.45 GHz as depicted in Figure 13.11. Horizontal and vertical polarisation is used alternately throughout the frequency range to facilitate channel overlap without crosstalk or interference. Each channel occupies a bandwidth of 27 MHz.

LNB

The microwave signals bounced off the dish first encounter a low-noise block (LNB) mounted at the focal point. They are conveyed by a waveguide to a micro-patch, whence the vertically- and horizontally-polarised transmissions are coupled to separate printed lands on the PC board, each connected to its own GaAsFET low-noise transistor preamplifier stage, selected as required by the receiver's control system. This selection is typically achieved by switching the LNB supply voltage between 13 V and 17 V: the DC operating voltage for the entire LNB comes up the cable by which the signal passes to the indoor unit.

The operating principle of an LNB is given in Figure 13.12. The RF amplifier uses two or three FETs to provide an overall gain of about 20 dB. After bandpass filtering the SHF signal is applied to a mixer consisting of a Schottky-type diode of either silicon or GaAs construction: its inherent non-linearity provides the mixing/beat action. The local oscillator is again based on a low-noise FET. It works in conjunction with a *dielectric resonator* to produce a very stable and noise-free CW (continuous wave) signal at 9.75 GHz, on the low-frequency side of the incoming carriers.

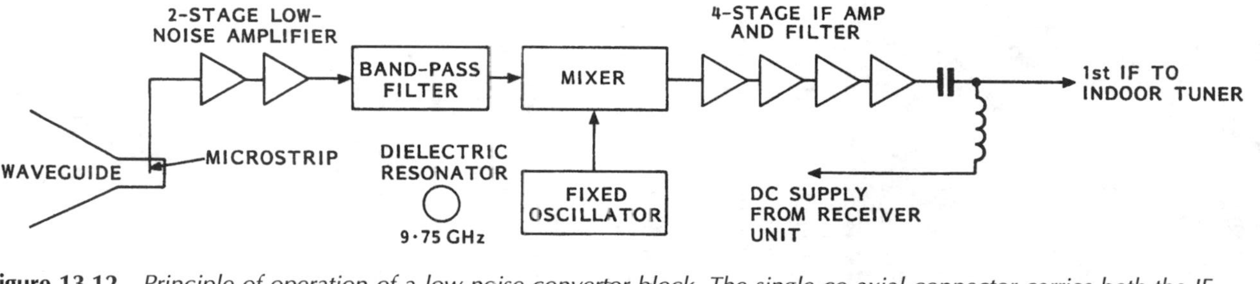

Figure 13.12 *Principle of operation of a low-noise convertor block. The single co-axial connector carries both the IF output signal and the DC operating voltage*

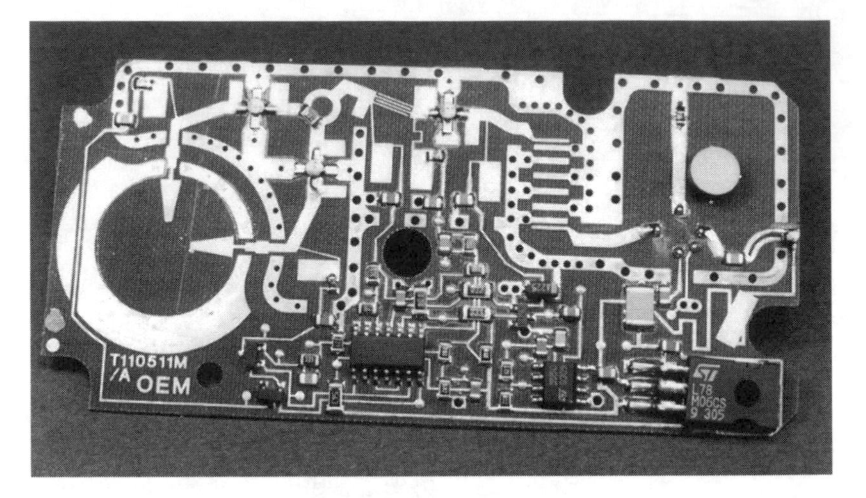

Figure 13.13 *Internal view of an LNB printed panel. At left are the micropatch interceptors for H- and V- polarised signals, at right the oscillator 'puck' and output terminal. Many of the print patterns function as bandpass filters*

Each broadcast vision carrier produces an IF corresponding to the difference between its frequency and that of the local oscillator. Since all the carriers beat simultaneously with the oscillator signal, the result is block-conversion of the whole spectrum of channels to a lower frequency band which is selected by a wideband filter at the mixer's output. The channel spacing and modulation characteristics remain the same. To take as an example Astra 1A ch. 1 at 1 121 425 MHz, the difference frequency produced when it beats against the 9750 MHz local oscillator signal is 1 464 250 MHz or 1.46425 GHz. This passes down the feed-wire as one of the sixteen IF frequencies from Astra 1A. In practice the indoor receiver can tune over the range 950–2050 MHz, which embraces the IFs of all the Astra-group transponders in the several co-sited satellites. The overall gain of the LNB is about 50 dB. A photograph of the internal layout of an LNB appears in Figure 13.13.

Indoor tuner and receiver

The indoor tuner selects the satellite TV channels. It's very similar to the UHF tuner described in Chapter 8 – see the block diagram in Figure 13.14. The main difference is that the second IF is much higher in frequency: 140 MHz, 200 MHz or most often 480 MHz. A SAW filter is shown as part of the tuner in Figure 13.14 because in satellite receivers the IF amplifier and FM demodulator are packaged in a single screened module with the tuner section. For this application the SAW filter has a bandwidth of 27 MHz to match that of the transmission. The oscillator within the indoor satellite tuner is governed by a phase-locked loop (PLL) and a reference crystal in a frequency-synthesis tuner system similar to that used in a TV

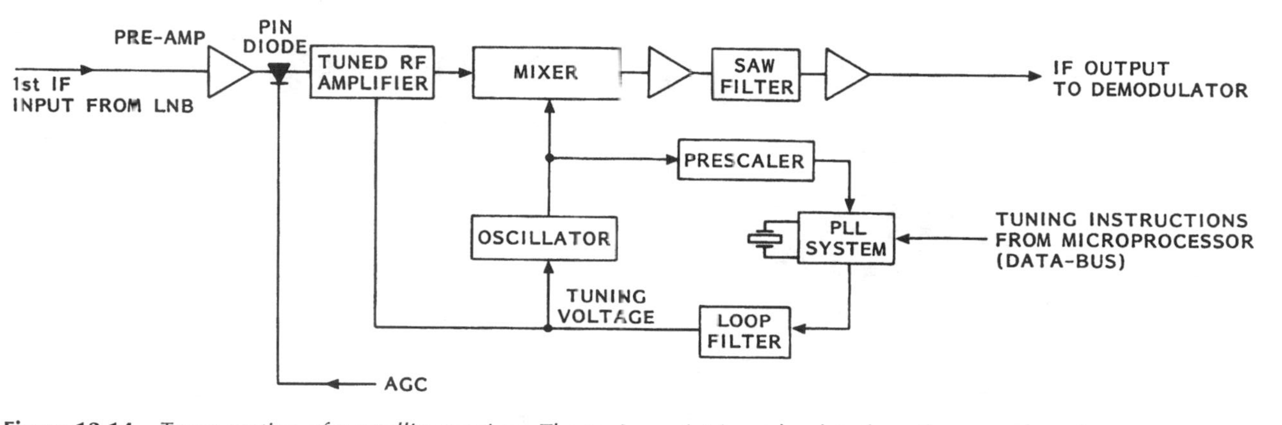

Figure 13.14 *Tuner section of a satellite receiver. The tuning point is set by data from the control section*

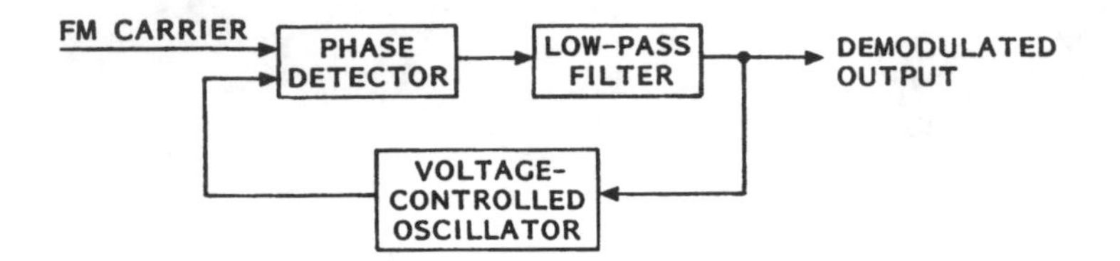

Figure 13.15 *Operating principle of PLL FM demodulator*

set or VCR. The whole operation is orchestrated by a microprocessor, and the tuning data (and many other operating reference points) are stored in a large EEPROM as described in the TV control section of Chapter 8.

Two post-tuner techniques used in satellite receivers remain to be described because they differ from conventional (AM) TV receiver practice. They are FM vision demodulation and energy-dispersal clamping. The FM demodulator is immune to amplitude modulation and impulsive interference so long as its input signal level exceeds a certain threshold. The PLL principle, illustrated in Figure 13.15, is used in a simple circuit which is invariably buried inside an IC. A linear VCO (voltage-controlled oscillator) free-runs at the second IF, say 480 MHz. Its output forms one feed to a phase detector whose second input is the IF itself, a frequency-modulated carrier. The output from the phase detector, which is an error signal representing the difference between the frequencies of its two inputs, is used to steer the VCO which thus faithfully tracks the vision carrier deviations. The result is that the error voltage, after suitable low-pass filtering, is a facsimile of the original modulating signal: the video waveform complete with the sound carriers and where applicable the teletext data.

Energy-dispersal technique

To prevent fixed patterns appearing in the sideband structure of an FM satellite transmission – and the consequent risk of interference to other services – an energy-dispersal waveform is added to the video signal before modulation. It takes the form of a triangular waveform at 25 Hz, half the field frequency. After demodulation it is removed to prevent flicker on the picture. The action is carried out by a circuit like that in Figure 13.16, where a pulse, timed to coincide with some fixed amplitude point in the video waveform (e.g. front porch, black level or sync tip) restores the charge on the right-hand plate of coupling capacitor C1 to a reference potential once per television line.

Encryption

Most satellite picture broadcasts are *encrypted* to give the broadcaster control over who watches them so that subscription systems can operate and so that geographical copyright of programme material is not infringed. The process of

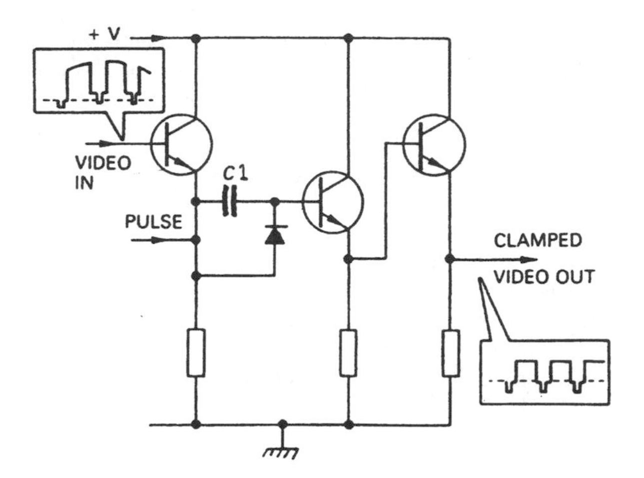

Figure 13.16 *Video clamp circuit for stabilisation of the black level in a TV signal*

encryption, in most cases, consists of changing the timing order in which TV line signals are broadcast, and sending a 'key' along with the TV signal, which carries instructions on how to reassemble them in the correct order to produce the original image. For D-MAC transmissions the encryption system used is called *Eurocrypt,* whose principle is based on individual addressing of receivers (each of which has its own unique identity number, and hence memory-address) over the air.

With the PAL/Secam transmission formats used in Europe, different encryption systems are used: the most common of these is *Videocrypt,* which depends on an individually-addressed smart-card sold by the broadcaster and inserted by the viewer into a slot on the front of the receiver. The card contains an IC-based algorithm, in effect a set of instructions held in memory. The Videocrypt concept is to cut each TV line in two, and splice the halves with two other line fragments. There are 256 possible points in each line period where the cut can be made; in a transmission the cut-points are selected in a pseudo-random sequence. The result is a scrambled and totally unrecognisable picture on screen, only sorted out into a coherent picture when the line segments have been reassembled into correct order by a decoder or descrambler. The descrambler is instructed on where the cuts have been made, and the way in which they have been shuffled, by data transmitted (in similar fashion to Teletext) on unused lines in the field blanking interval. The data itself is also encrypted, requiring processing by the algorithm contained in the smart card to define the cut points. The Videocrypt data lines convey 32-byte messages at the rate of about 24 per minute. At the smart card it is converted to a 60-bit value as an instruction to the decoder to generate the cut and shuffle points to reconstruct the picture. Also contained in the 32-byte data are control commands for the individual smart card, defined by its unique address; and a four-byte authorisation code. The former can activate or deactivate the card, and trigger on-screen messages from it, while the latter ensures that only 32-byte instructions from a legitimate source (the broadcaster or his agent) will be actioned by the smart card and descrambler. This elaborate and complex system is designed to prevent unauthorised viewing by 'hackers' and 'pirate' cards.

Satellite sound

The two systems used for broadcast satellite sound are FM and digital. In the case of digital and MAC broadcasts the audio is digitally encoded as described in Chapter 11 and slotted into the digital data stream in similar TDM (time division multiplex) fashion to that illustrated in Figure 10.5. The transmission and reception systems used for PAL/Secam satellite sound were described in Chapter 11, and for DTV transmissions in Chapter 12.

Receiver architecture

Figure 13.17 gives an idea of what happens inside the satellite set-top unit. The supply voltage for the LNB is switched – to select polarity of reception – by the control system, while its 950–2050 MHz block-converted IF output signal has one of its broadcast components selected by the tuner under the control of the master microprocessor and with reference to the data stored in the EEPROM and the crystal in the frequency-synthesis circuit. These processes are identical to what goes on inside a micro-controlled terrestrial-reception TV set. The second IF output is selected and tuned by a SAW filter, then amplified for application to the FM demodulator.

Emerging now as a baseband signal from the demodulator, the signal takes three paths. It's routed out via a socket for application, if required, to an external decoder, whose descrambled input re-enters for possible selection in a video routing switch operated by the control microprocessor. It also passes, inside the box, to a video de-emphasis network, followed by an energy-dispersal clamp as described above. Now the video signal is processed by the descrambler, to emerge (if the subscription has been paid!) as a 'studio'-state waveform for selection in the routing switch to the video (e.g. 21-pin SCART) output socket and the UHF modulator, which acts as a mini closed-circuit TV transmitter with a carrier output on a UHF channel. The FM sound carriers are filtered out from the demodulated video signal, as described in the previous chapter, and the audio signals they convey are passed to the UHF modulator and to the AV output sockets for connection to the TV/monitor, a VCR, and/or a separate stereo audio system, any of which may incorporate a surround-sound processor for a 'home-cinema' effect.

The microprocessor control system, communication highway, data bus, EEPROM programme/mode memory and (usually switch-mode) power supply system work to exactly the same principles as those described in the appropriate sections of Chapter 8 where their application to terrestrial TV receivers was examined. An internal view of a PAL satellite receiver designed for reception from the Astra group of satellites is shown in Figure 13.18.

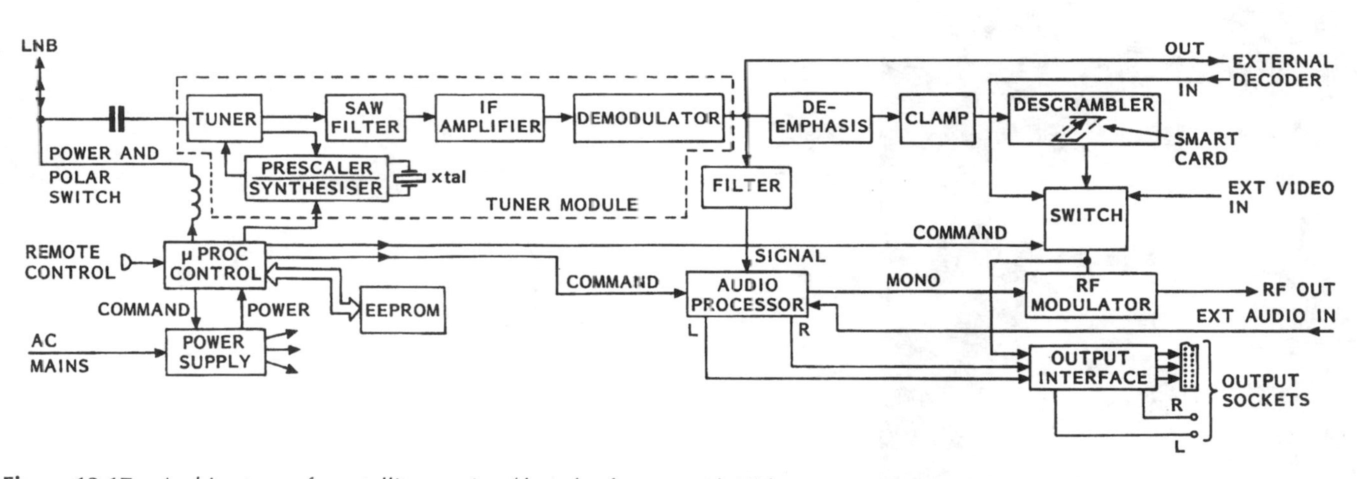

Figure 13.17 *Architecture of a satellite receiver/decoder for use with Videocrypt analogue-FM transmissions*

Figure 13.18 *Internal view of an IRD (integrated receiver-decoder) for reception of satellite transmissions. The switch-mode power converter is on the left; the smartcard slot and reader top centre; and the timer/demodulator at bottom right. (Pace)*

14 *Cable TV*

Originally the philosophy of cable television was simple: to bring television to those who could not receive it by means of conventional RF aerials for reasons of local topography, the impracticability of installing an aerial, or aesthetics as judged by the local council or planning authority. Many CATV (communal-aerial television) systems had their origins in the dark ages of 405-line television on VHF, with few and scattered transmitting sites. With the spread of the UHF terrestrial broadcasting network (transmitters have been installed for communities of fewer than 300 souls) the necessity for CATV as an alternative to listening to the radio (!) has all but disappeared; blocks of flats, hotels and similar domiciles can be served by MATV (master-aerial TV) which is a small-scale cable system, working from a master aerial and distributing signals to tens, rather than thousands, of TV sets.

In some cases, out-of-area programmes were made available on a community cable network as a 'bonus' to subscribers, although the advent of Channel Four sometimes meant dropping this facility, as many cable systems (especially wired-pair HF and some co-axial VHF networks of long standing) had a maximum capability of four vision channels. Because of government policy, cable operators could, in general, only distribute the programmes of the national broadcasters, and this (in the UK at least) tended to limit the popularity of cable systems. Further problems for old cable networks were the propagation of teletext signals through the network, the difficulty of maintaining good bandwidth and delay characteristics for colour signals, and the incompatibility of commercial home VCR machines with the special receivers (called *terminal units*) used with some cable systems.

In the more recent past, interest in cable networks has revived, with encouragement from the UK government, and many networks are now in place. An incentive to viewers is the provision of cheaper telephone calls, and another attraction of new cable franchises is that alongside 'off-air' material they are permitted to broadcast exclusive programmes, not obtainable except over the cable, and thus create a demand from the viewing public and a financial incentive for the cable operators. It is envisaged that large-scale 'cabling' of the UK will follow, with its possibilities of Pay TV, and *interactive* services in which the viewer can communicate with a central information and data exchange via the cable, in the same way Internet users are able to send and receive videotext over the telephone network. The advent of satellite services opened further prospects for the cable system; those viewers who are unable to accommodate a receiving dish, or who wish to view programmes intended for other European countries (with or without English soundtrack) can be catered for, as the installation of receiving dishes for fringe satellite and DBS (direct broadcasting by satellite) reception is most economically done on a 'community' basis.

Provided the demand (and hence finance) were there, the cable scheme would enable television to become as locally-based as the current BBC and ILR district radio services, particularly relevant in the provision of text and data transmissions of local interest only; and in the potential for local advertising.

Cable types

The early cable transmission system consisting of twisted pairs carrying HF vestigial sideband TV signals is now obsolescent, though used in some districts for satellite programme relay.

The choice for multichannel TV carrying videotext (and possibly other data channels) lies between co-axial, standard telephone and glass-fibre optic cables, and each has various advantages and disadvantages. Co-axial cable is currently well established, and scores on the counts of easy interfacing with existing equipment and lower initial cost, at least on comparatively short runs. The optical-fibre technique requires more complex terminal equipment but has advantages in the areas of data-handling capabilities, immunity to electrical interference, security against 'tapping', and a small physical diameter, enabling a greater number of services to be laid in existing ducts, and a reduction in the cost of routing. Optical fibre has advantages over co-ax in terms of transmission efficiency, too; for a given data-rate the *repeaters* ('boosters' to overcome transmission losses) can be spaced at greater intervals than with the co-ax system. Regarding cost, glass is intrinsically much cheaper than copper and in volume production the glass-fibre technique may well show an overall cost advantage over copper cables.

It has been shown that for trunk lines optical transmission has much to recommend it, and British Telecom currently operates many glass-fibre optic links for transmission of television, audio, data and telephone traffic. Some can operate at very high data rates (140 Mbit/second) which confers the simultaneous ability to handle thousands of phone calls, or many broadcast-quality digital TV channels. It may be that the most economical way to implement a large cable system will be to adopt a 'hybrid' solution, with optical-fibre trunk routes to local distribution points, whence co-axial cables will 'spur off' to individual dwellings grouped around the fibre cable head. Alternatively, connection can be made to existing subscribers' telephone lines for conveyance of picture, data and sound.

Transmission modes

In the same way as air or space can be used to carry virtually any radio frequency using any of the several modulation systems outlined in Chapter 1, so it is (within reason!) with co-ax and fibre-optic systems. Obviously the 'launching' and 'interception' methods differ, and for fibre the basic carrier is light, rather than an electrical wave. Thus we can use baseband, AM, FM, PM or PCM in cable systems, at such carrier frequencies as are appropriate to the signal, the distance between terminals, and the transmission medium. Typically a co-axial cable will carry analogue signals modulated by FM or AM onto carriers in the VHF and (for short runs) UHF range. Fibre-optic cables will work from analogue-baseband to the 140 Mbits/second PCM mode described above. It should be remembered that the light-carrier in a fibre system (usually infra-red rather than visible light) is itself an electromagnetic wave with a frequency of the order of 3×10^8 MHz or 300 THz (see Chapter 2) so that the upper limit on the rate of data throughput in an optical fibre, is perhaps, limited by technology rather than physics! The main restriction on bandwidth in glass fibre links using modulation of a sub-carrier (sub-, that is, to the frequency of the light wave itself) is the effect of *fibre-dispersion,* which tends to slightly 'blur' in time the sharpness of received pulses, giving an *integration* effect to their shape. Fibre-optic cable is a very efficient carrier: a typical loss figure at 1550 nm light wavelength is 0.23 dB per km.

The network

There are two basic methods of cable distribution. The more traditional is the tree-and-branch system (see Figure 14.1*a*) in which all available programmes are continuously sent over the network in separate channels, with user selection by means of some form of switch at the receiving point. This was the *modus operandi* of the original radio and TV cable system, in which each household may be regarded as being on the end of a 'twig'.

The alternative and better system is termed a 'switched-star network', shown in Figure 14.1*b*. Here the available programmes are piped to a 'community-central' point analogous to a telephone exchange. The subscriber can communicate with the 'exchange' and request the desired programme(s) to be switched to his line for viewing, listening or recording. The advantage of the switched-star network is its *interactive* capability, whereby the subscriber can 'talk-back' to a local or central exchange; this opens the possibility of the 'single-fibre' household in which all communications services (radio, TV, telephone, Internet, banking, public-utility meter reading, text and data etc) come into the dwelling via a single link which can, by means of recording devices at either end, be utilised during off-peak and night hours. The switched-star configuration lends itself well to Pay-TV (either pay-by-channel or pay-by-programme) because security is more easily arranged. It is simpler to deny a programme to a non-subscriber by a central 'turn-key' process than by expensive signal *scrambling* and 'authorised unscrambling' systems of the sort described in the previous chapter.

Terminal equipment and propagation modes in glass-fibre cables

Since the basic transmission 'vehicle' in fibre-glass cables is light energy we must now see how the signal is launched into the cable and intercepted at the

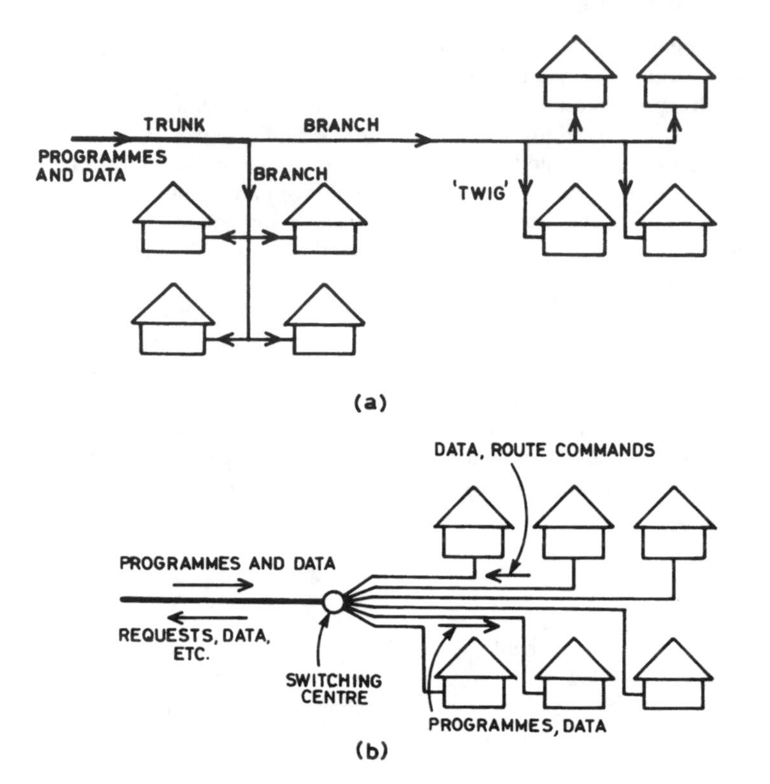

Figure 14.1 *Two methods of signal distribution: (a) the 'passive' trunk-and-branch system; (b) the 'interactive' switched-star network. These are discussed in the text*

receiving end. Depending on the distance to be covered, the sending device may be an LED or low-power semiconductor laser operating on a wavelength (infra-red) around 850 nm. The radiant energy in the sending device is surprisingly small, typically 200–300 μW for an LED and 1–3 mW for a laser. The fibre termination is an integral part of the light-source encapsulation for maximum coupling efficiency, permitting virtually all the light to be concentrated in the cable.

The receiving device is a light-sensitive diode, again intimately coupled to the fibre end. For low noise and highest possible sensitivity this pick-up device is 'tuned' to the light wavelength in much the same way as a radio set is tuned to an RF transmission. For short-haul reception a silicon P-I-N photodiode is generally used; with long-distance fibre cables greater sensitivity can be obtained by using an *avalanche* photodiode which combines the property of light detection with an internal amplification process.

The glass-fibre core, made of silicon dioxide (SiO_2) glass, is of very small diameter, typically 50–100 μm, surrounded by an intimately-bonded cladding layer of about 20 μm thickness. Further layers give strength and environmental protection, the outer jacket consisting of a tough waterproof polyurethane cover. The transmission of light along a glass fibre depends on the phenomenon of *total internal reflection* in which the light, when it encounters the inner surface of the fibre wall, is 'bounced' by the mirror-like wall surface back into the fibre. Light can enter the fibre end at any angle and the bounce-path in transmission can thus take several forms, as shown in Figure 14.2. Large bounce-angles give rise to a long path length (in terms of the passage along the entire cable) and are called high-order transmission modes; at lesser angles the total path length is shorter, described as low-order transmission mode. A beam which travels down the axis of the fibre takes the shortest possible path length in what is known as axial mode.

The nature of light propagation down a fibre depends on its diameter and on the difference in refractive index between the fibre core and its cladding material. Where a sudden change of refractive index is present at the fibre wall we have a *step-index* fibre in which several modes (high- and low-order) are taken by the light. An alternative form of construction is the *graded-index fibre,* where the interface between 'core' and 'cladding' represents a more gradual change in refractive index. This has the effect of making the light rays turn less sharply when they encounter the fibre wall and thus reduces reflection loss, as shown at the top right of Figure 14.2. Low-order modes predominate in such a graded-index fibre cable, and such high-order modes as are present travel faster along their longer path, reducing the fibre-dispersion effect described earlier. Graded-index fibres offer a low transmission loss and greater bandwidth than step-index types, at the expense of higher production costs and greater coupling losses at the junctions between the fibre and the sending and receiving devices.

Both step- and graded-index fibres operate in what may be called multimode with many possible light path angles within the fibre. If we can arrange a fibre to concentrate on the axial mode we shall significantly reduce the transmission loss. In this *monomode* system a high-grade glass core is used, with a small diameter in the region of 5 μm (about one-tenth of diameter of multimode cable cores). The light wavelength used here is longer, around 1.35 μm, and the much straighter light path gives very good transmission efficiency. Repeaters (regenerators) are therefore required at much longer intervals than in conventional fibre (and particularly co-axial) systems, and in a typical monomode transmission system repeaters can be as much as 30 km apart; this economy in equipment easily outweights the cost disadvantage of the monomode fibre cable itself.

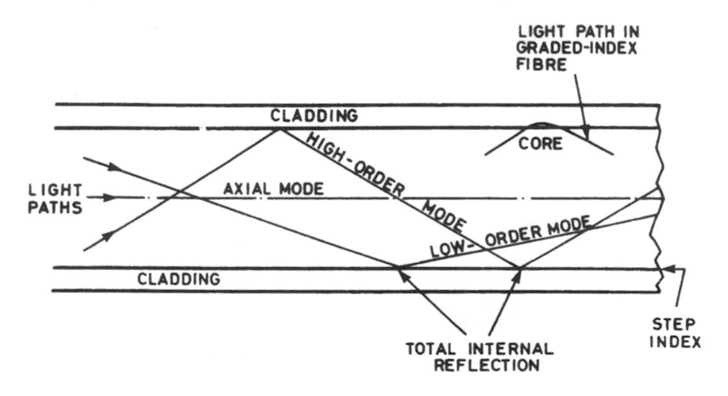

Figure 14.2 *Propagation modes in fibre-optic lightguides*

Repeater power

For repeaters generally, the operating power can be sent down the cable in co-axial systems, as described earlier for domestic masthead amplifiers. Fibre-optic cable plainly cannot carry DC, but conductive members can if required be incorporated in its protective sheath, or local power sources can be used, in view of the long intervals between repeaters made possible by fibre-optic technology. Many thousands of kilometres of fibre-optic cable are in use in the UK, primarily by British Telecom.

TV via the phone line: ADSL

In the UK alone there are over 30 million twisted-pair phone lines in operation between British Telecom exchanges and individual subscribers' premises. Most of them use copper wire, some aluminium, and all represent a ready-made medium for distribution of TV signals if they can be pressed into service in this application.

Originally the telephone lines were designed to carry simple command (dialling) pulses and frequency-restricted (300–3500 Hz) baseband voice signals, and their attenuation factor at higher frequencies is very great: that of a typical voice-grade telephone circuit at 200 kHz is 60 dB, and at 1 MHz is 110 dB. The *group delay* characteristic, however, is constant over a wide range; group delay is an expression of the ratio of phase change per unit of frequency. Under these circumstances pulse signals suffer relatively little distortion, opening the way to transmission of relatively high bit-rate digital datastreams over conventional phone cables. The system is called ADSL, Assymetric Digital Subscriber Line.

With the type of digital compression (MPEG-standard) described in Chapter 11, a bit-rate of 2 Mbit/second can be used to send an entertainment-standard (equal-to-VHS) down a telephone line, complete with high-quality digital stereo sound. At a bit-rate of 4 Mbit/second broadcast-standard pictures are possible, and in both cases provision is made for a relatively low bit-rate return channel via which the subscriber can specify and request the service, movie etc. required, and even to provide such features as freeze-frame, search forward or backward (cue/review) and slow motion. The spectrum allocations for this are given in Figure 14.3. There are several ways of configuring the control and

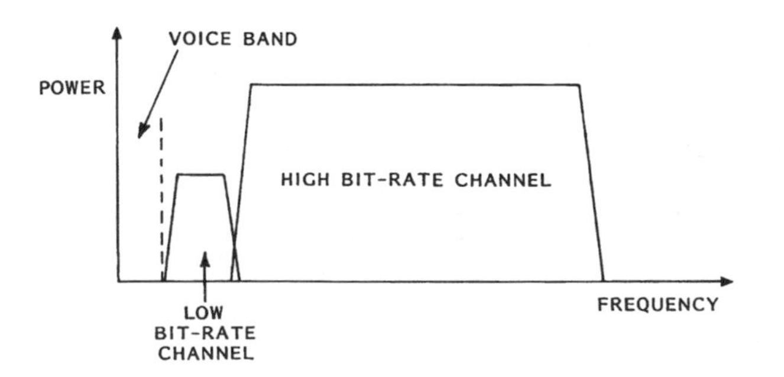

Figure 14.3 *Spectrum of digital TV signal in a narrowband telephone circuit*

picture/sound data, depending on the length and electrical characteristic of the line in use. Where required, it can be done *adaptively* by automatically selecting the best mode for optimum S/N ratio and minimum bit error rate. The data passes over the line in balanced form to minimise radiation and RF interference, coded in a format called 2B1Q wherein bit-pairs are individually converted to one of four quaternity symbols.

Modulation

Various modulation systems can be used to convey the data; an effective one is DMT (discreet multi-tone modulation) in which the data is distributed among many sub-channels in similar fashion to the OFDM (orthogonal frequency division multiplex) scheme described in Chapter 12. Figure 14.4 outlines the process: the serial input datastreams are first encoded in parallel form, then the resulting frequency samples are converted into time values in an FFT (fast Fourier transform) processor. Now the derived values are transcoded into serial form for application to a D-A converter before transmission. Noise and interference picked up by the signal in its passage is distributed over many channels by this technique, and the resulting minimal damage to individual pulse trains is amenable to repair by the sorts of techniques described earlier in this book in connection with digital sound and vision processing. Error control by cross-interleaving is incorporated into the transmission code to assist the process.

A typical ADSL set-up uses about 208 carriers, each QAM-modulated, and extending from about 200 kHz to the region of 1 MHz. At the subscriber's home is a band-splitter and a modem, and at the telephone exchange a modem and SAM: Subscriber Access Multiplexer, whence the signal passes into a cable network on its way to the service-provider. Unlike the ordinary telephone system with its line-grabbing and dial-up processes, ADSL is continuously alive and capable of two-way communication; data routing is directed by addressing information within the datastream, and each subscriber's terminal has its own individual address.

Video on demand

While satellite broadcast provides a form of video-on-demand (VOD) by running popular (usually subscription) movies at (e.g.) fifteen-minute intervals across many channels, a cable network – and particularly ADSL – has a unique ability to offer what might be called 'true' VOD. Here the subscriber can choose a film from a large catalogue, and request it to be sent down the line. The movies are stored on digital magnetic disks in central distribution centres, and the return path permits the sorts of replay tricks (cue, review, freeze-frame) as a domestic VCR affords.

Figure 14.4 *DMT modulation system for narrowband transmission channel. Adaptive separation techniques can be used to optimise signal transfer*

15 *Development of video tape recording*

The story of video tape recorders really began before the turn of the century with the experiments of Valdemar Poulson. By this time the relationship between electricity and magnetism was well understood, and the idea of impressing magnetic pulses on a moving magnetic medium was sufficiently advanced in 1900 to justify the US patent on Poulson's apparatus, the Telegraphone. The medium was magnetic wire rather than tape, and without any form of recording bias, and scant means of signal amplification, the reproduced sound signal was low, noisy, non-linear and lacking in frequency response. These problems of 'tape' and record/replay head performance are ones that have continually recurred throughout the history of sound and vision tape recording, as we shall see.

By the early 1930s, many advances had been made in the field. DC bias or pre-magnetism of the recording wire had been tried with better results, then overtaken by the superior system of AC bias, as used today. The magnetic wire gave way to steel tape 6 mm wide travelling at 1.5 metres per second, and performance became comparable with the contemporary disc recording system. Not 'hi-fi' by any means, but certainly adequate! The BBC adopted and improved the Blattnerphone system and in 1932 broadcast a programme of the Economic Conference in Ottawa, for which *seven miles* of steel tape was used, edited by means of a hacksaw and soldering iron! This era also saw the first crude forerunner of a servo system in the Marconi–Stille machine of 1934. The earthed tape was arranged to contact insulated metal plates when it became slack, the plates being wired to thyratron control valves. Relays in the thyratron anode circuits modified drive motor currents to regulate tape speed.

A great impetus was given to the industry when it became possible to coat a flexible insulated base with a finely-divided magnetic substance. This was achieved in Germany by Dr. Pfleumer and developed by Wilhelm Gaus under the auspices of AEG. This activity culminated in the successful demonstration at the 1935 Berlin Radio Exhibition of the first commercial sound tape recorder, the AEG Magnetophone, using a cellulose acetate tape coated with carbonyl iron powder. Performance of these sound recording machines steadily improved during the '30s and '40s to the point where at the end of the 1940s much radio broadcast material was off-tape, and indistinguishable from live programmes.

Once audio magnetic recording had become established in the radio industry, attention was turned towards the possibility of recording television images on tape. The problems were formidable, mainly because of the relatively large bandwidth of a television signal. Plainly, it would be necessary to increase the tape speed and one approach, by Crosby Enterprises in 1951, took the form of a 250 cm/second machine designed to record a monochrome picture whose frequency spectrum was split into ten separately-recorded bands with additional sync and control tracks. The same principle was embodied in a fearsome machine designed by RCA, in which the tape travelled at 600 cm/second to record on three separate tracks simultaneous R, G and B information for colour TV. Other designs involving longitudinal recording called for tape speeds approaching 1000 cm/second, amongst which was the BBC's VERA (video electronic recording apparatus) of 1956. Such machines were wasteful of tape and thoroughly frightening to anyone who happened to be in the room in which they were operating!

Already the seeds had been sewn of a new system, one which was to hold the key to the modern system of TV tape recording. This was simply the idea of moving the record or replay heads rapidly over the surface of a slowly-moving tape to achieve the necessary high 'writing' speed. Initially, the hardware consisted of a circular plate with three recording heads mounted near its edge at 120° intervals, their tips protruding from the flat surface of the faceplate. The 5 cm-wide tape was passed at 76 cm/second over the surface of the rotating plate, whose heads had an effective velocity of over 6000 cm/second, resulting in narrow arcuate tracks across the width of the tape. It was a step in the right direction, but results were poor for several reasons. The bandwidth of the TV signal being recorded was difficult to get on and off the tape due to noise and head-gap problems (we will meet these in detail in the next chapter), and the valve technology of the time did not really lend itself to such requirements as an ultra-wideband, high gain and stable amplifier.

Two more factors were required for success, and these were engineered by Dolby, Ginsburg and Anderson, of Ampex between 1952 and 1955. The problems associated with the tape-track configuration were solved by the use of a horizontal head-drum, containing four heads and rotating on a shaft mounted parallel to the direction of tape motion as shown in Figure 15.1*a*. The video heads lay down parallel tracks across the tape width, slightly slanted, (see Figure 15.1*b*) with each head writing about 16 television lines per pass. The tape-to-head speed (i.e. the *writing speed*) at an incredible – to us today – 40 metres per second was most adequate, with response beyond 15 MHz. The final hurdle was cleared with the introduction of an FM recording system; this involves frequency-modulating a constant-amplitude carrier with the picture signal before application to the recording head. Charles Ginsberg, one of the collaborators, once said that FM was proposed by the man who was assigned to design an AGC circuit for the AM system originally in use! The Ampex *Quadruplex* system was enthusiastically taken up by TV broadcasters, and rapidly became a world standard. For less exacting requirements and limited funds, a simpler recording system was really needed, and this gave birth to the helical-scan system. The idea of a rotating video-head drum containing one or two heads, each laying down one complete television field per pass, was first mooted in 1953. The Japanese Toshiba company was foremost in the field in the early days, though

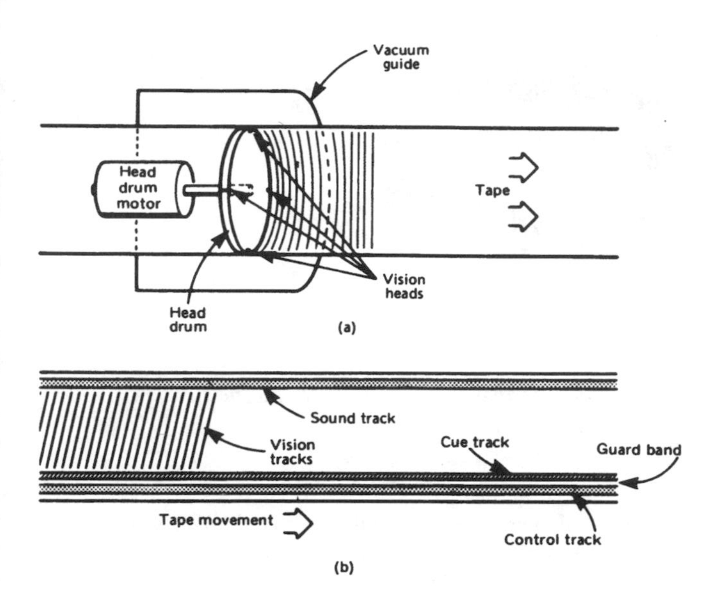

Figure 15.1 *(a) The transverse-scan system. The vacuum guide draws the tape into a curved profile to match the contour of the head drum. (b) Resulting tracks on the tape*

several other companies were working on the idea. By 1961 a handful of manufacturers were demonstrating helical machines, all with the open-reel system, each totally incompatible with all the others, and none having a performance which could approach that of the quadruplex system. The market for these machines was intended to be in the industrial and educational spheres, with very little regard as yet to the domestic market. If there was one thing these professional users wanted above all else, it was standardisation of the helical format to ensure software compatibility. Does this, 40 years later, have a familiar ring about it?

By the early 1970s, helical-scan machines had, by and large, settled into three camps. The Sony U-matic standard was well established, with a very good performance for its era, using 12.7 mm tape in a large cassette for easy handling. Most Far-Eastern manufacturers had thrown their weight behind the EIAJ (Electronics Industries Association of Japan) system, using a cartridge containing a single spool of tape. The third contender was Philips. Their 'VCR' machine, illustrated in Figure 15.2, was the first to be designed truly for the domestic as well as the educational and industrial markets. It took a relatively compact, fully-enclosed cassette and was styled for the home environment. Easy-to-operate controls, a simple timer facility and one-hour capability were offered, together with built-in TV tuner and RF modulator for use with a conventional TV set. This, the N1500, was a milestone in video recorder development.

The N1500 did not sell in very great numbers, in spite of the intensive effort made by Philips' engineering and publicity departments. Its release in 1972 coincided with an explosion in colour TV sales, and the average householder blew his savings on a colour receiver rather than a VCR. The maximum one-hour playing

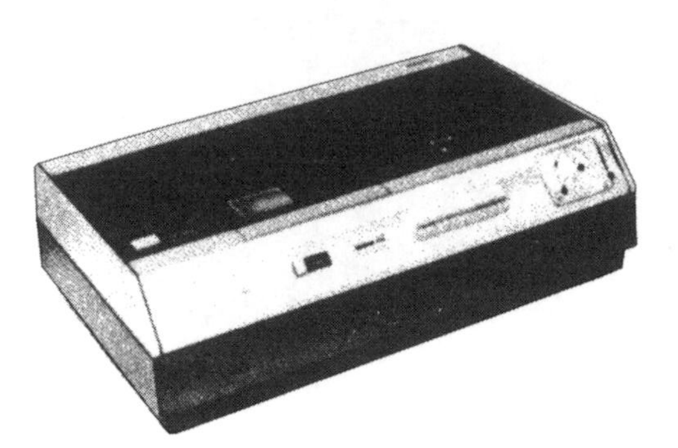

Figure 15.2 *The Philips VCR format was first in the domestic field. A late version of the first Philips machine, type N1500*

time, insufficient for a football match or a feature film, also held back sales and the introduction by Philips in 1976 of a similar machine, but with two-hour capability (N1700, VCR-LP) was a step in the right direction. The Philips monopoly was broken, however, in 1978 when the rival Japanese systems VHS (Matsushita) and Betamax (Sony) appeared on the shelves of high street shops.

All three systems used cassettes, a thick double-stacked type for the Philips VCR and VCR-LP format, and smaller, lighter, co-planar cassettes for the VHS and Betamax systems. All the early machines of each format had piano-key controls similar to those of audio tape recorders. The rather awkward and expensive Philips VCR cassette, together with its relatively short playing time, spelt the early demise of this format, and by 1980 VHS ruled supreme in the UK, with Betamax in second place and rapidly gaining ground. Sales of domestic VCRs rapidly increased, and a wide range of software appeared alongside. The pace of machine development quickened, and in 1980 piano-keys had given way to light-touch sensors and remote control, sophisticated timers and programmers had appeared, and a form of freeze frame had been introduced. Another significant event of 1980 was the appearance of a new helical VCR format, the Philips/Grundig Video 2000 system, for use with the VCC (video compact cassette). The Compact Cassette designation indicated the intention that it should become as popular as the universally-used audio compact cassette. In the event V2000 format was not successful, and production had ceased by early 1986. It used an advanced DTF (dynamic track following) tracking system, very similar to that of the later and more successful Video 8 format, whose *ATF* feature will be described later.

Specialised ICs had been making steady inroads into domestic VCR design, LSI (large scale integration) and microprocessor devices becoming commonplace in contemporary machines. These made possible advanced remote control systems, comprehensive timers and programmers, trick-speed, still-frame and visual search features. They also led a trend away from mechanical complexity in the tape deck and towards electronic control of direct-drive systems; this considerably simplified the mechanics of the tape transport, threading and head drive systems, while retaining a relatively low electronic component count.

Advances had been made, too, in the field of portable VCR equipment. Purpose-designed battery operated machines became available in VHS, VHS-C and Video-8 format for outdoor location work. Those currently on offer incorporate the camera section and videorecorder in one unit – *camcorders*. In this realm there was strong competition between the VHS camp whose champions are JVC and Panasonic, and the Video 8 protagonists, led by Sony of Japan and having in its ranks many companies with backgrounds in the world of conventional and cine photography. Details of VHS, VHS-C and Video-8 formats will unfold throughout the book, and the relative merits of the formats (including the high-definition variants, S-VHS and Hi-8) will be discussed in Chapter 25.

The mid-nineties saw the advent of the digital format DVC, primarily intended for camcorders, and taking advantage of MPEG technology of the sort described in Chapter 12. For the first time here was a format backed by virtually all the manufacturers in the 'domestic' field, and DVC-format products, with compatibility across the board, became available in a wide range of types and prices, with competition working well for the consumer. The late 1990s saw the introduction by Sony of *Digital-8* format, a hybrid capable of dual-standard operation with Video-8/Hi-8 and digital recording and playback. JVC began marketing D (Data)-VHS in 1999, a 'homedeck' format for use with DTV television systems, and disc-based digital camcorders made their first appearance, from Hitachi, at about the same time. We shall look at the operation of digital tape systems in Chapter 24. At the turn of the new century computer-type hard disc drives were incorporated into DTV set-top boxes for recording (*auto*-recording) of TV programmes. Chapter 24 will describe how home video footage is recorded in digital form on computer disks, as well as the editing systems associated with this mode of data storage.

16 *Magnetic tape basics and video signals*

All magnetic materials, videotape coatings amongst them, may be regarded for practical purposes as consisting of an infinite number of tiny bar magnets, each with its own north and south poles. This is a simplification, but suits our purposes well. In the natural state these bar magnets are randomly aligned within the material so that their fields cancel one another out, and no external magnetic force is present. Thus the contents of a box of steel nails, for instance, will have no particular attraction for each other. To magnetise the material, be it a blank tape or a solenoid core, we have to apply an external magnetic force to align the internal magnets so that they sit parallel to one another, with all their N poles pointing in the same direction. When the external field is removed, most of the magnets remain in alignment and the material now exhibits magnetic properties of its own.

Remanent magnetism

If the relationship between externally applied force, or *flux* and retained flux in the material, were linear, the business of tape recording would be much simpler. Unfortunately this relationship, called the *transfer curve*, is very far from linear, as Figure 16.1 shows. Here we have a graph with the magnetising force (H) plotted along the horizontal axis, and the stored flux density (B) on the vertical axis. Our starting point (with the new box of nails!) is in the centre, point 0. Here, no external magnetising force is present, and the magnetic material's internal magnets are lying in random fashion – hence zero stored flux. Let's suppose we now apply a linearly-increasing external flux. The *flux-density* (strength of magnetism) in the material would increase in non-linear fashion as shown by the curve 0-T. At point T the material has reached magnetic saturation and all its internal magnets are rigidly aligned with each other – no increase in applied magnetic force will have any effect. If we now remove the magnetising force, bringing the H coordinate back to zero, we see that a lot of magnetism is retained by the specimen, represented by point U. This is the *remanence* of the material, the 'stored charge' as it were. For recording tape, it needs to be as high as possible.

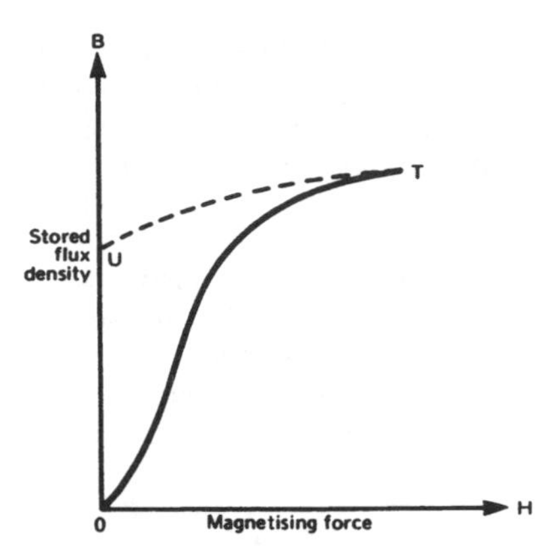

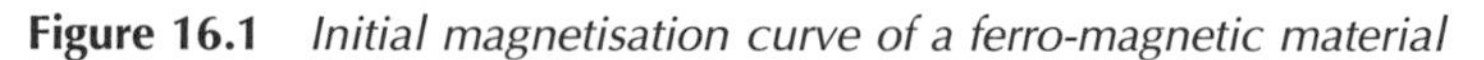

Figure 16.1 *Initial magnetisation curve of a ferro-magnetic material*

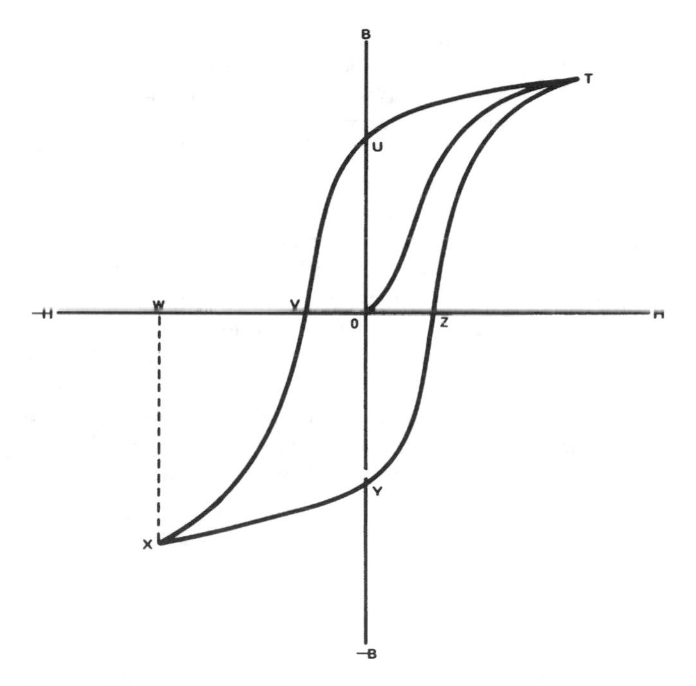

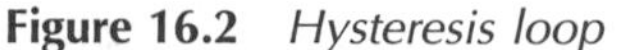

Figure 16.2 *Hysteresis loop*

Hysteresis

Figure 16.2 is an expansion of Figure 16.1 to take in all four quadrants. We left the material at point U with a stored flux. To remove the flux and demagnetise the material (we'll call it tape from now on) it's necessary to apply a negative magnetising force, represented by 0–V, whereupon stored flux B returns to zero. Further negative applied force (V–W) pushes the stored flux to saturation point in the opposite direction, so that all the magnets in the tape are once again aligned, but all pointing the other way, represented by point X. Removal of the applied force takes us back to the remanence point, this time in the negative direction – point Y. To demagnetise the tape a positive applied force 0–Z is required, an increase of which will again reverse the stored flux to reach saturation once more at point T. The diagram is called a magnetising hysteresis curve, and one cycle of an applied magnetising waveform takes us right round it. It can now be seen how the 'degaussing' or *erasing* process works. Here we apply a large alternating magnetic field, sufficient to drive the material to saturation in both directions. The field is then allowed to decay linearly to zero, creating smaller and smaller hysteresis loops until they disappear into a dot at point 0, and the material is fully demagnetised.

Transfer characteristic and bias

Figure 16.3*a* is based on the previous diagram, but shows the initial magnetising curve in solid line (T0X) with the hysteresis curve in dotted outline. If we apply a magnetising force, 1H, and then remove it, the remanent flux falls to a value 1′. A larger magnetising force 3H will, if applied and removed, leave a remanent force 3′, and so on, up to saturation point 8H and remanence point 8′. The same applies in the negative direction, and plotting remanence points against applied force we get the curve shown in Figure 16.3*b*. This is the transfer characteristic. It is very non-linear at the middle and ends, but for a typical recording type, will contain reasonably linear sections on each flank, PP and QQ. If we can bias the recording head to operate on these linear parts of the transfer characteristic, the reproduced signal will be a good facsimile of that originally recorded. DC bias (Figure 16.4) puts the head into *one* of the linear sections, but the recorded signal will be noisy and inadequate; AC bias, shown at the top of Figure 16.4, allows the head to operate in two quadrants, with superior results. We shall see that the chrominance signal in a VCR is recorded along with a bias signal, which is in fact the FM carrier for the luminance signal.

Head-tape flux transfer

The recording head consists of a ferrite 'ring', with its continuity broken by a tiny gap. A coil is wound around the ring, and when energised it creates a magnetic field in the ring; this is developed across the head gap. As the tape passes the gap, the magnetic field embraces the oxide layer on the tape and aligns the 'internal magnets' in the tape according to the electrical signal passing through the head. Provided that some sort of bias is present (Figure 16.4) the relationship between *writing current* and flux imparted to the tape is linear, so that a magnetic facsimile of the electrical signal in the head is stored in the tape, as shown in Figure 16.5*a*. The tape passes the head at a fixed speed, so that low frequencies will give rise to long 'magnets' in the tape, and high frequencies short ones.

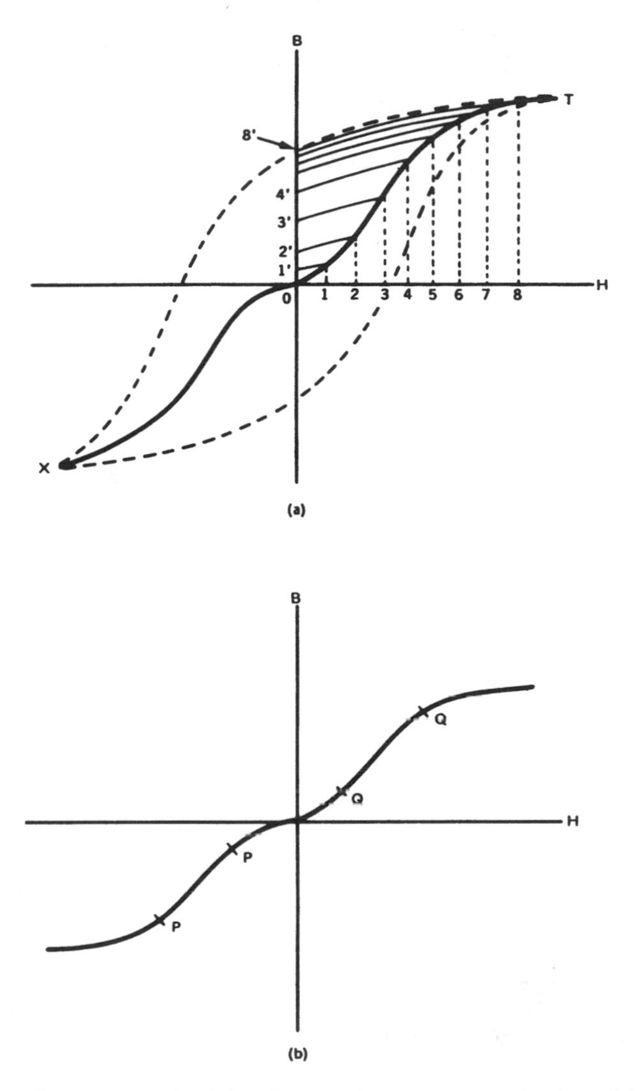

Figure 16.3 *Deriving the transfer curve: at (a) is plotted the remanent flux for eight linear steps of applied magnetising force; (b) shows the resulting transfer curve*

Head gap and writing speed

The linear relationship between head field strength and stored flux in the tape, described above, holds true when the wavelength to be recorded on tape is long compared to the width of the head gap. However, when the wavelength of the signal on the tape becomes comparable with the head-gap width, the flux imparted to the tape diminishes, reaching zero when the recorded wavelength

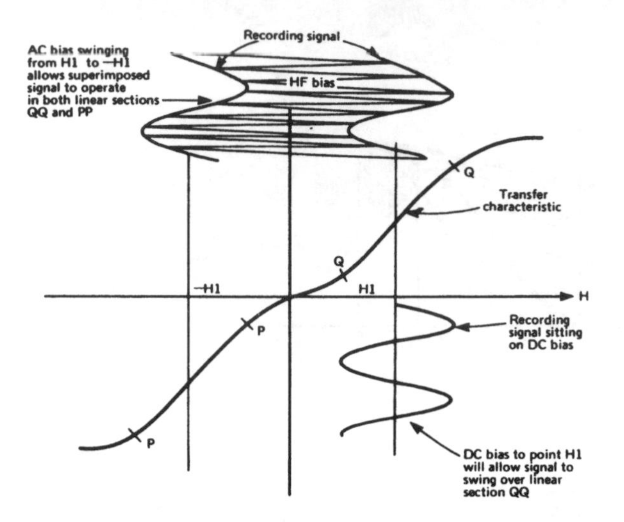

Figure 16.4 *The effect of AC and DC bias*

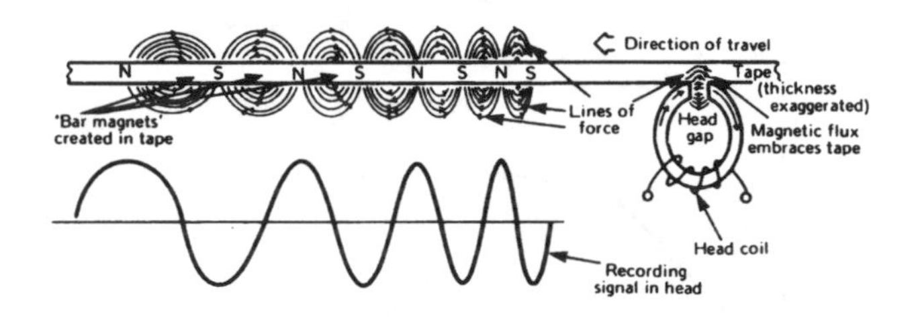

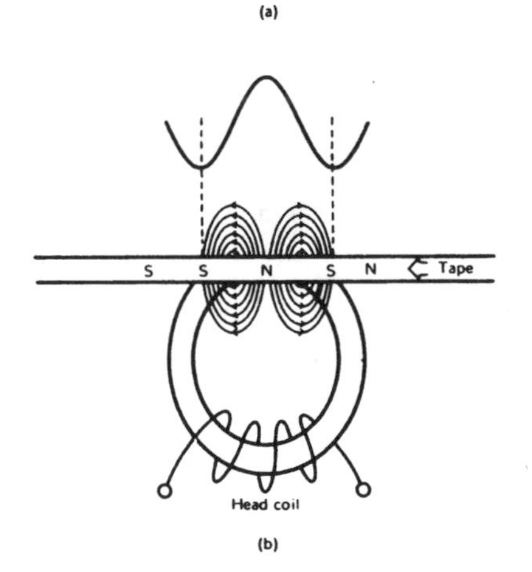

Figure 16.5 *Storing flux on the tape: at (a) the flux appearing across the head gap is penetrating the oxide surface to leave magnetic patterns stored on the tape; (b) shows the effect when one complete cycle of the recorded waveform occupies the head gap – no signal transfer will take place. This point represents fex, extinction frequency*

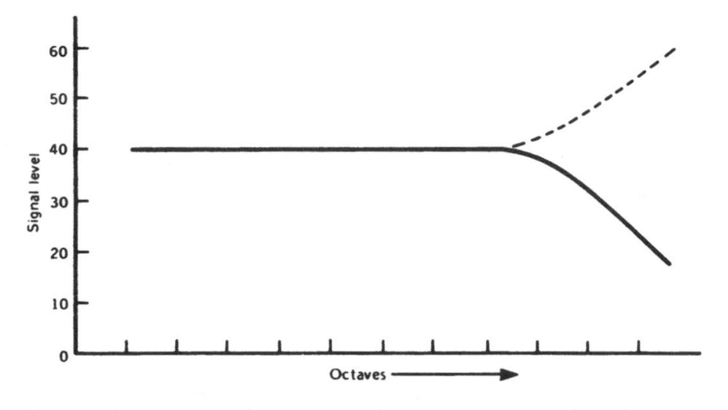

Figure 16.6 *Losses in the recording process. The dotted line shows a compensating 'recording equalisation' curve*

(or rather the non-recorded wavelength!) is equal to the width of the head gap. This is illustrated in Figure 16.5*b*, where it can be seen that during the passage of a single point on the tape across the head gap, the applied flux has passed through one complete cycle, resulting in cancellation of the stored flux in the tape.

This point is known as the extinction frequency (*f*ex) and sets an upper limit to the usable frequency spectrum.

For video recording we need a large bandwidth, and a high *f*ex. This can be achieved by reducing the head-gap width or alternatively increasing head-to-tape, or *writing* speed. There is a practical limit to how small a gap can be engineered into a tape head, and currently this is about 0.3 μm, less than half a micron (1 micron = 10^{-6} metre). For domestic applications, where full broadcast-bandwidth signals are not required, a writing speed of about 5 m/second is required with such a head gap. How this is achieved will be explained in due course. Other HF losses also occur during recording. The head is by definition inductive, so losses will increase with frequency. Eddy currents in the head will add to these losses, as will any shortcomings in tape-to-head contact. High frequencies give rise to very short 'magnets' in the tape itself, and it is the nature of these to tend to demagnetise themselves. For all these reasons, the flux imparted to the tape tends to fall off at higher frequencies, as in the solid line of the graph shown as Figure 16.6. To counteract this, recording equalisation is applied by boosting the HF part of the signal spectrum in the recording amplifier, as per the dotted line.

This is called recording equalisation, and its aim is to make the frequency/amplitude characteristic of the signal stored on the tape as flat as possible.

Replay considerations

As with any magnetic transfer system, the output from the replay head is proportional to the *'rate of change'* of magnetic flux. Thus, assuming a constant flux density on the recorded tape, the replay head output will double for each doubling of frequency. A doubling of frequency is called an *octave* and a doubling of

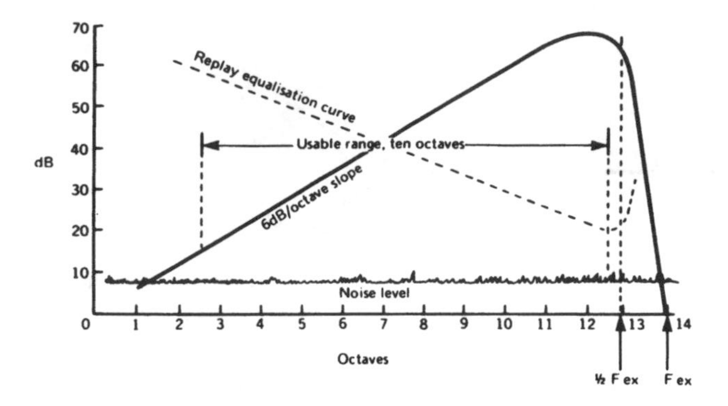

Figure 16.7 *Playback curve for a tape with equal stored flux at all frequencies. For a 'level' output the gain of the replay amplifier must follow the dotted replay equalisation curve*

voltage represents an increase of 6 dB. This holds good until the extinction frequency, *f*ex, is approached, when the head output rapidly falls towards zero. This is shown in Figure 16.7. The upper limit of the curve is governed by the level of the signal recorded on the tape, limited in turn by the tape's magnetic saturation point. At the low-frequency end, the replay head output is low due to the low rate of change of the off-tape flux. At some point, it will be lost in the 'noise' off-tape, and this will occur at about 60 dB down from peak level. Thus, even with playback equalisation (represented by the dotted curve in Figure 16.7), the *dynamic range* of the system is confined to 60 dB or so. With the unalterable 6 dB/octave characteristic, we are limited, then, to a total recording range of ten octaves. This is inherent in the tape system and applies equally to audio and video signals. Ten octaves will afford an audio response from 20 Hz to 20 kHz, which is quite adequate. TV pictures, however, even substandard ones for domestic entertainment, demand an octave range approaching 18 and this is plainly not possible. No wonder they had so much trouble in the pioneering days!

Modulation system

To be able to record a video signal embracing 18 octaves or more it is necessary to modulate the signal on to a carrier and ensure that the octave range of the carrier is within the capabilities of the tape recording system. While the carrier could be AM (amplitude modulation), the FM (frequency modulation) system has been adopted because it confers other advantages, particularly in the realm of noise performance. An FM signal can be recorded at constant level regardless of the modulating signal amplitude, so that head losses and effects of imperfect head-to-tape contact are less troublesome. To achieve a picture replay with no perceptible background noise (snow), the S/N ratio needs to be over 40 dB, and this can be achieved by an FM recording system in a domestic VCR. Professional and broadcast machines can do much better than this!

FM basics

An FM system, familiar to us in VHF sound broadcasts, starts with a CW (continuous wave) oscillator to generate the basic carrier frequency. The frequency of the oscillator is made to vary in sympathy with the modulating signal, audio for VHF sound transmitters, video for VCR recording systems and satellite broadcasts. For any FM system the *deviation* (the distance that the carrier frequency can be 'pulled' by the modulating signal) is specified. In VHF sound broadcasting it is ±75 kHz, giving a total frequency swing of the carrier of 150 kHz. In a VCR, carrier frequencies are specified for zero video signal amplitude – represented by the bottom of the sync pulse – and full video signal amplitude, i.e. peak white.

All modulation systems generate sidebands, and those for FM are more complex than occur in an AM system. In fact an FM system theoretically generates an infinite number of sidebands, each becoming less significant with increasing distance from the carrier frequency. Figure 16.8 shows the sidebands of a VHF-FM sound broadcast transmission. The modulating frequency is 15 kHz, and the sideband distribution is such that the first eight sidebands on either side of the carrier are significant in conveying the modulation information. Thus to adequately receive this double-sideband transmission we need a receiving bandwidth, a window, of 240 kHz or so – in practice 200 kHz is sufficient, and this is the allocated channel width.

Modulation index

The example given, 200 kHz bandwidth for transmission of a 15 kHz note, seems very wasteful of spectrum space, and certainly will not do for our tape system in

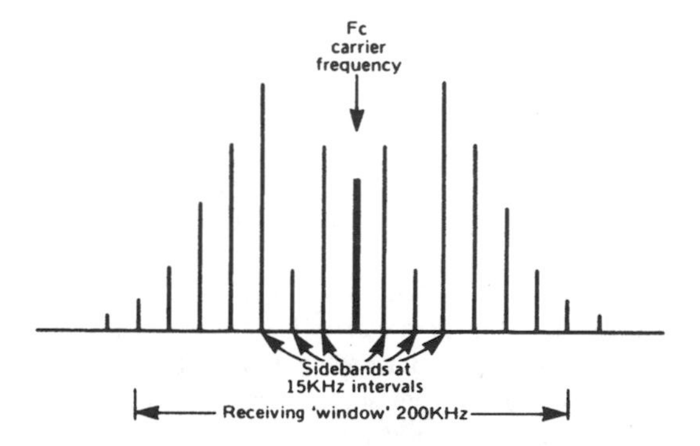

Figure 16.8 *Sidebands of a VHF-FM sound transmitter*

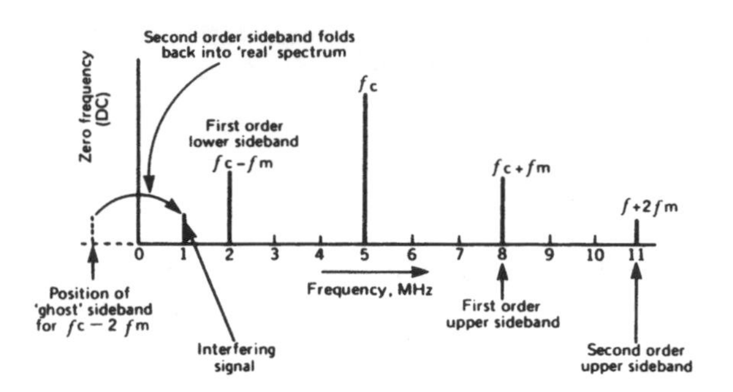

Figure 16.9 *The effect of folded sidebands. Carrier frequency is 5 MHz, modulating frequency 3 MHz*

which elbow-room is very limited! In the above example, the *modulation index*, given by the formula

$$\frac{\text{carrier deviation}}{\text{modulating frequency}} \quad \text{is} \quad \frac{75}{15} \quad \text{or} \quad 5.$$

If we can reduce the modulation index, the significant sidebands draw closer to the carrier frequency, and at modulation indexes below 0.5, the energy in the first sideband above and below the carrier becomes great enough for them to convey all the necessary information in the same way as those of an AM signal. In video tape recording we go a step further and use only one sideband along with a part of the other, rather similar to the vestigial sideband scheme used with AM television broadcast systems.

By using a low modulation index, then, the sidebands of the FM signal can be accommodated on the tape. FM deviation has to be closely controlled and carrier frequencies carefully chosen to avoid trouble with the sidebands, which if they extend downwards from the carrier to a point beyond zero frequency will not disappear, but 'fold back' into the usable spectrum to interfere with their legitimate fellows, leading to beat effects and resultant picture interference. The effect of a folded sideband is shown in Figure 16.9.

FM video

Taking the VHS system as an example (all the analogue formats use similar frequencies and parameters), the video signal is modulated on to the FM carrier according to Figure 16.10.

Sync-tip level gives rise to a frequency of 3.8 MHz, black-level 4.1 MHz, mid-grey 4.45 MHz and peak white 4.8 MHz. Thus deviation is limited to a total of 1 MHz, and with a restricted video bandwidth of about 3 MHz, modulation index is about 0.3. The resultant spectrum of the on-tape signal is shown in Figure 16.11, with a full lower-sideband, vestigal upper sideband (limited by the system frequency response and the approach of *f*ex) and a carefully arranged gap between

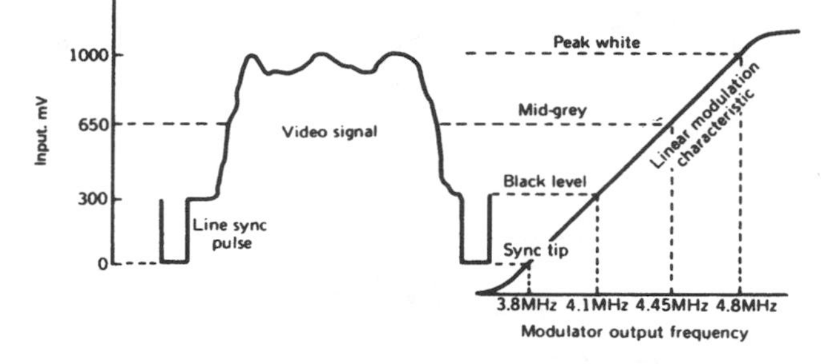

Figure 16.10 *FM modulation characteristic for a VCR*

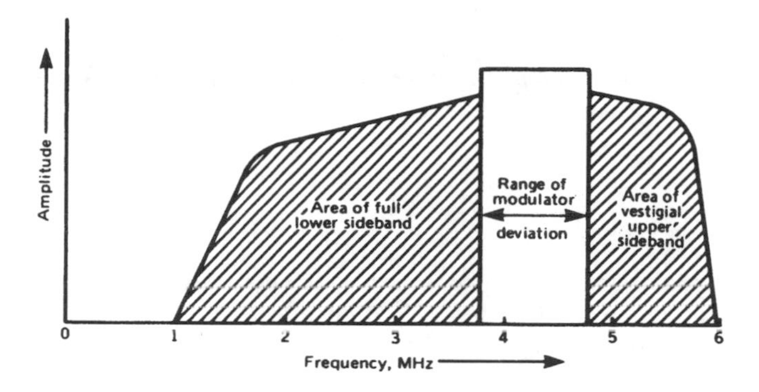

Figure 16.11 *Spectrum of luminance video signal on tape*

0 and 1 MHz into which (as later chapters will show) is shoe-horned the chrominance, and where applicable ATF (automatic track-finding) control signals. If we regard the FM luminance spectrum as extending from 1 MHz to 6 MHz, it is now occupying less than three octaves, well within the capabilities of the system. The S-VHS parameters will be discussed later.

Pre-emphasis

Noise is the enemy of all recording and communications systems, and in FM practice it is common to boost the HF components of the modulating signal prior to the modulation process. FM radio uses this technique, called pre-emphasis, and so do we in VCR FM modulation circuits. The effect of this after demodulation (in the post-detector circuit of a radio or the playback amplifier of a VCR) is to give a degree of HF lift to the baseband signal. In removing this with a filter, gain is effectively reduced at the HF end of the spectrum, with an accompanying useful reduction in noise level. The idea is shown in Figure 16.12, which illustrates the filter charactertistics in record and replay, along with the effect on the noise level.

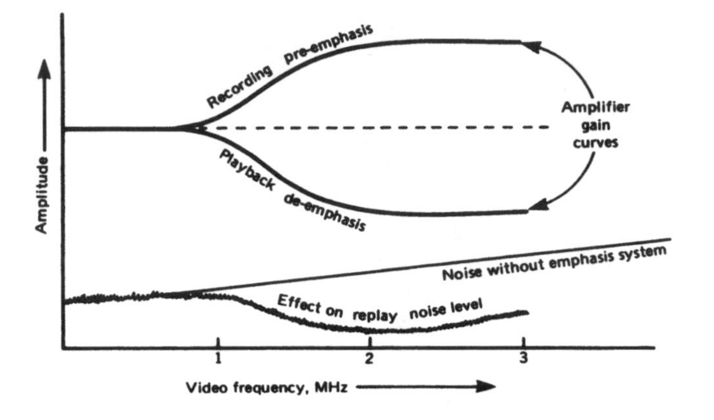

Figure 16.12 *The effect of pre-emphasis and de-emphasis on playback signal-to-noise ratio*

Summary

The FM system is fundamental to the video recording process. Here is a summary of the main points, as applied to domestic VCRs. The video signal is bandwidth-limited in a low-pass filter to restrict it to the band 0–3 MHz. This luminance signal (chroma components are removed by the filter) is pre-emphasised and applied to a voltage-controlled oscillator to provide an FM signal with a total swing of about 1 MHz, which is then limited and carefully controlled in amplitude to fall in the middle of the linear portions of the tape transfer characteristic. The FM oscillator frequency is chosen to minimise sideband interference and beat effects, and the FM signal is applied to the video heads after equalisation to compensate for HF losses. The signal is recorded on the tape at a writing speed of around 5 m/second, and the recorded tape carries a vestigial sideband signal ranging up to 6 MHz or so.

17 *Video tape: tracks and transport*

In Chapter 15 we saw that the necessary high head-scanning speed was first achieved by fast longitudinal recording techniques, then in the successful Quadruplex system by transverse scanning of the tape by a four-head drum with the video heads mounted at 90° intervals around its periphery. The transverse-scan method has a lot going for it! The very high writing speed confers great bandwidth, enabling full broadcast-specification pictures to be recorded and replayed. The relatively wide tracks are almost at right-angles to the tape direction, so that any jitter or flutter in the tape transport has little effect on the timing of the video signal, merely causing slight momentary tracking inaccuracies, easily catered for by the wide track and the FM modulation system in use.

The cost of transverse-scanning machines is high because of the need for a complex head-drum system and precision vacuum tape guides. To maintain the necessary intimate contact between video head and tape, all VCRs have their head tips protruding from the surface of the drum so that they penetrate the tape and create a local spot of 'stretch.' Hence the need for a precise vacuum guide at the writing/reading point in a transverse machine to maintain correct tape tension. These techniques are not amenable to domestic conditions or budgets, and the alternative and simpler helical system has undergone much development. It is now used exclusively.

Principle of helical scan

In a helical scan system, several problems are solved in one go, but other shortcomings are introduced. The idea is to wrap the tape around a spinning head drum, with entry and exit guides arranged so that the tape path around the head takes a form of a whole or part helix. The principle is shown in Figure 17.1 where a video head in the course of one revolution enters the tape on its lower edge, lays down a video track at a slant angle and leaves the upper edge of the tape ready to start again at the tape's lower edge on the next revolution. During the writing of this one track, the transport mechanism will have pulled the tape through the machine by one 'slant' video-track width, so that track no. 2 is laid alongside the

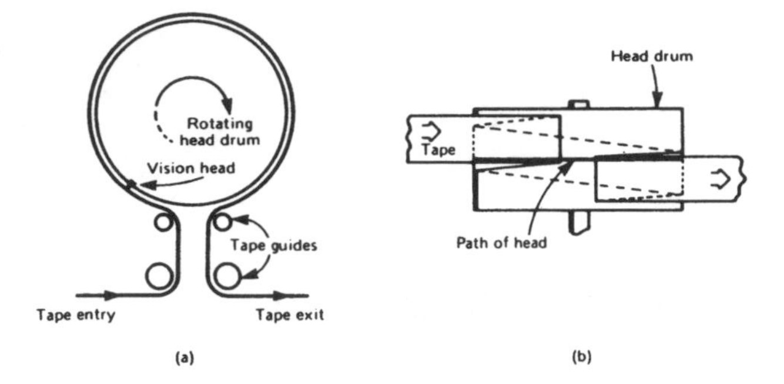

Figure 17.1 *(a) The Omega wrap. This diagram shows the arrangement for a single-head machine; the path of the head is the heavy line in (b)*

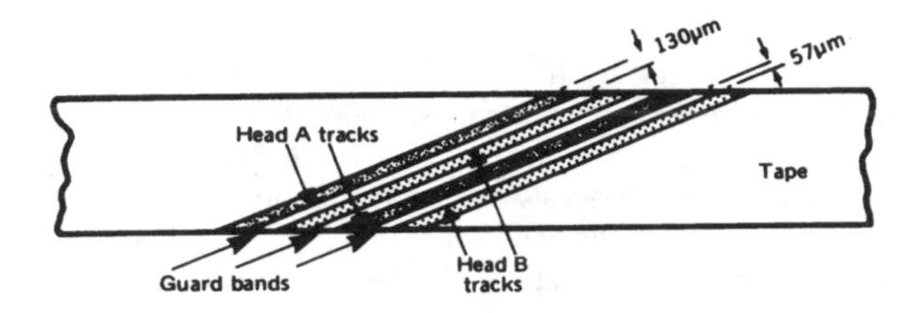

Figure 17.2 *The diagram of Figure 17.1 drawn from the point of view of the tape*

first; each track is the width of the video head. Figure 17.2 shows how this works and gives an elementary impression of the track formation.

Let's look at the strengths and weaknesses of the helical format. Tape tension and hence head-tip penetration is now governed by the tape transport system rather than a precision guide arrangement. One or two complete television fields can be laid down per revolution of the head drum, so that the problems of matching and equalising heads (to prevent picture segmentation) disappear. In a single-head machine such as we have described, no head switching during the active picture period is necessary. Two-head helical machines have a simple head-switch or none at all; what price do we pay for these advantages? The main penalty is timing jitter in the recorded and replayed video signals. Because the tracks are laid at a small angle to the tape direction (about 5° off horizontal for VHS) they may be regarded as virtually longitudinal, so that the effects of the inevitable transport flutter, variations in tape tension, bearing rumble etc. will be to introduce minute timing fluctuations into the replayed signal. This effect cannot be eliminated in any machine, and causes problems with colour recording and certain types of TV receiver, as will become clear in Chapters 18 and 27.

In practice, two video heads are used in the drum of a helical VCR, pictured amongst others in the photo of Figure 17.3. The two-head system means that the tape needs only to be wrapped around half the video head drum perimeter, with one head joining the tape and beginning its scan as the other leaves the tape after

Figure 17.3 *Video head drum. This photograph of the underside shows five heads with a 'blank' (for mechanical balance) near the top. The heads at one o'clock and seven o'clock are double types for video, the others for Hi-Fi audio (two o'clock and eight o'clock) and flying erase near the bottom*

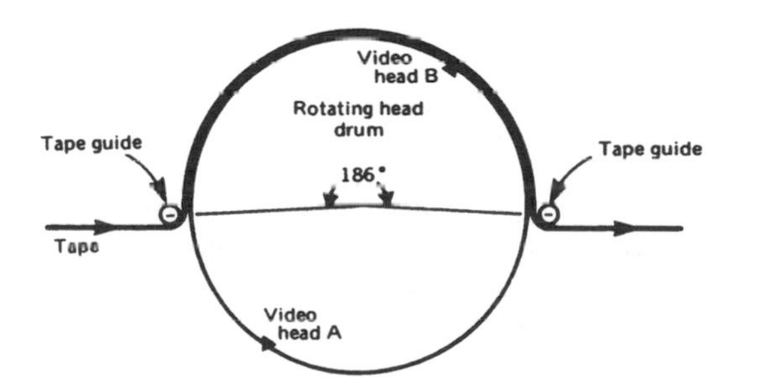

Figure 17.4 *Omega wrap for a two-head drum. The tape occupies rather more than half a turn of the drum*

completing its stint. To give a degree of overlap between the duty-cycles of the two heads, the tape wrap is in fact about 186°, slightly more than half a turn. This is known as an omega (Ω) wrap, outlined in Figure 17.4.

Track configuration

A typical track layout for a two-head helical VCR appears in Figure 17.5. Here we can see the video tracks slanting across the tape, shaded for head A, white for head B. At the edges of the tape, further tracks are present: the lower carrying a control track (serving a similar purpose to the sprockets in a cine film, and described later) and the upper carrying the sound track. Sound is recorded longitudinally in the same way as in an audio recorder, but with limited frequency

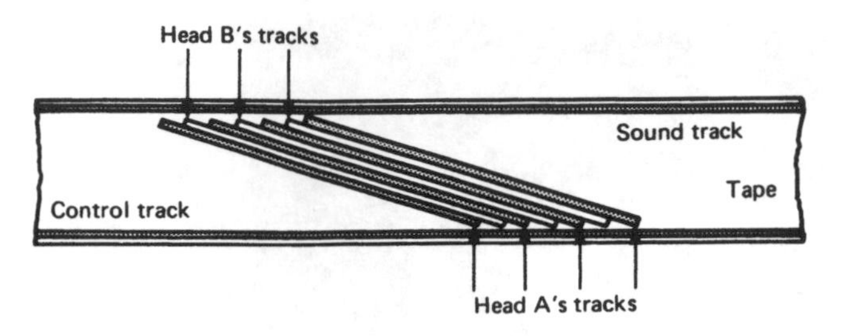

Figure 17.5 *Video tracks written by a two-head system, with each head writing alternate tracks. The relative positions of the sound and control tracks are also shown (VHS format)*

response due to the low linear tape speed in the sorts of VCR we are dealing with. Budget machines only have this mono longitudinal sound track; other VCRs are additionally fitted for hi-fi sound. This superior system is described in Chapter 19. Tape is 12.65 mm (VHS) or 8 mm (Video 8) wide. It progresses at speeds between 23.4 and 10.06 mm/second, depending on format and mode.

Now let us examine the practicalities of the mechanical arrangements of helical scan machines.

Tape threading

One characteristic of all domestic VCRs is the storage and transport of the tape in an enclosed cassette. This not only prevents contamination of the tape and physical damage to it, but also makes easy the loading and operation of the machine. Unlike an audio cassette, where the capstan, pinch-wheel (pressure roller) and heads 'come to visit', as it were, the video cassette system requires that a loop of tape be drawn from the cassette and wrapped around the video head drum, stationary heads, tape guides and capstan assembly. This is called threading, and of the several ways of going about it, we shall describe two.

Domestic VCRs use, as we have seen, a co-planar cassette similar in form to the compact audio type, but larger. To load the machine, the cassette is fed horizontally into a carrier which is itself then moved downwards into the machine. This action also opens the hinged front flap of the cassette. A pair of posts penetrate into the cassette shell (see Figure 17.6*a*) and when the thread mode is initiated they move away from the cassette, drawing out a loop of tape in 'M' formation. At the limit of their travel, the posts locate in 'V' notches mounted vertically at each side of the head drum to give the required 186° head wrap. This completes the threading operation, and the posts now form guide rollers on the tape path, as can be seen in Figure 17.6*b*.

An alternative arrangement is used in some types of VHS decks, notably those designed by Philips. Here (Figure 17.7) the cassette closes over four tape guides and a small pinch roller, all initially positioned *inside* the tape loop across the front of the cassette shell. When threading commences they move away from the cassette in opposite directions. The entry guide, fixed to a loading ring, moves to the left and then in a loop around the head drum so that the tape embraces the

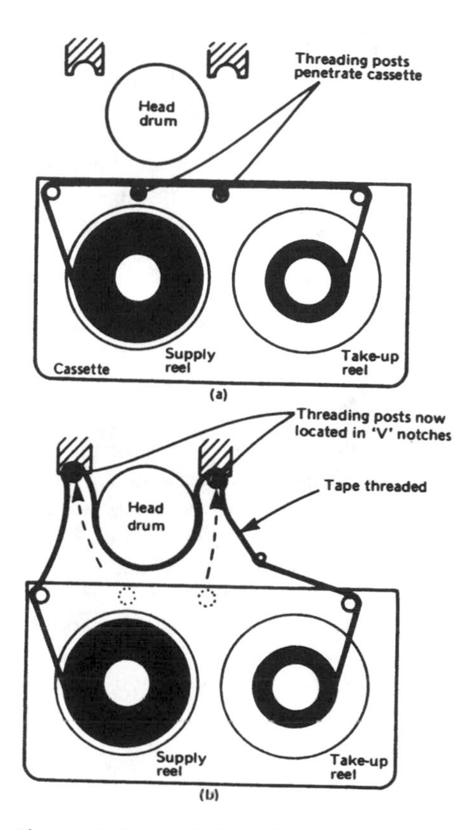

Figure 17.6 *VHS threading system: (a) shows start of the threading and (b) threading completed*

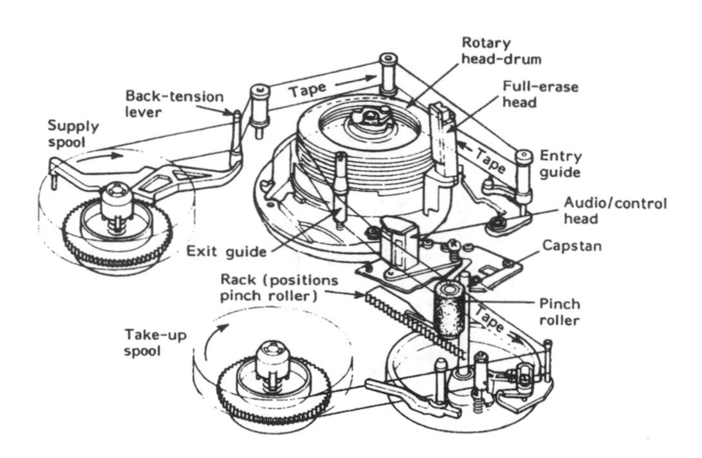

Figure 17.7 *U-wrap tape threading and path in domestic VHS VCR*

drum in a 'U'-wrap. The outward path of the tape is held clear by two additional posts which move into position as threading progresses. The guide locates in a 'V'-notch at the far side of the head drum to define the tape path at the entry-point of the tape wrap.

Meanwhile the pinch roller, mounted on a toothed rack, moves away from the cassette to the right, taking a loop of tape with it. It comes to rest against the capstan shaft, upon which it is pressed by a spring-loaded lever, again brought into position by the loading mechanism. In this configuration the audio head stack (see later) is positioned directly in front of the cassette.

Tape path

For 'homebase' VHS VCRs, the mechanical and electrical components on the tape deck do not vary fundamentally between any of the formats or manufacturers, with the exception of the Philips deck just described. Let's follow the passage of the tape on its involved journey from the supply spool to the take-up spool. Figures 17.8 and 17.9 show a typical deck, with the tape threaded and ready to go. The supply spool is on the left-hand side. On emergence from the cassette, the tape tension is checked by the tension arm, which brakes the supply spool to maintain tape tension constant. On most machines this is a purely mechanical operation, though some designs use electrical circuits in a *tension servo.* The tape next encounters the full-width erase head, which in record mode is energised with

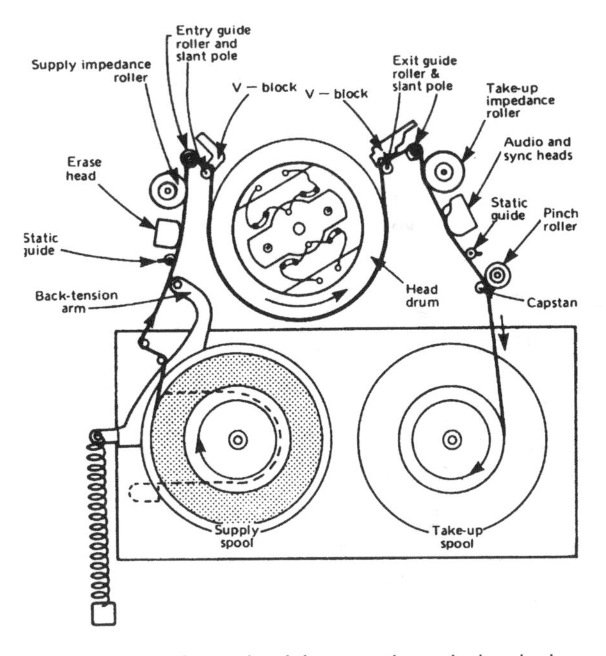

Figure 17.8 *The path of the tape through the deck*

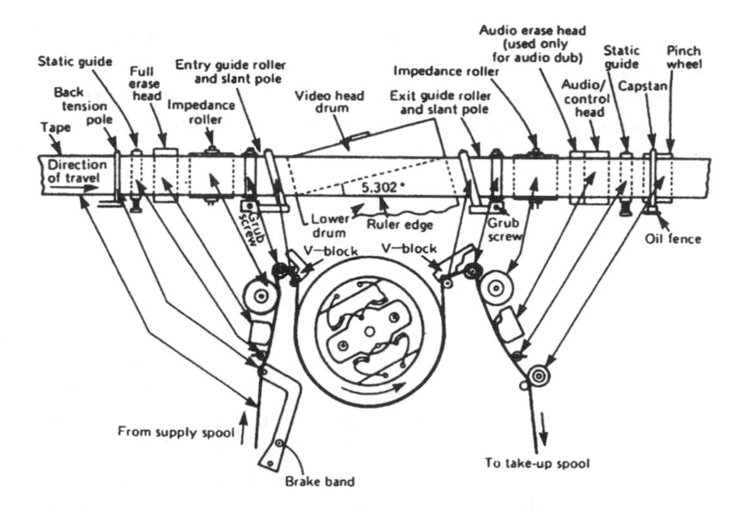

Figure 17.9 *Tape path and 'linear' sequence of deck components*

a high-frequency, high-amplitude CW signal to wipe clean all recorded tracks, video, sound and control. Its mode of operation is the same as the erase head of an audio machine. From here, in some machines, the tape passes around a *supply impedance roller* to iron out any speed or angle fluctuations imparted by the tension arm or the tape's passage over the erase head. An entry guide roller, part of the threading assembly, passes the tape to the slant pole. The latter is inclined from the vertical, and the effect of this is to create more tension on the top edge of the tape than at the bottom. The result is that the tape ribbon is 'nudged' downwards so that it sits firmly on the critically-positioned ruler edge around the video head drum.

The head drum itself is tilted at an angle of about 5° so that the head-wrap angle of the tape is correct with the tape moving parallel to the horizontal deck surface. On leaving the head drum the tape is again nudged downwards by an inclined exit guide pole to ensure that it is correctly seated on the ruler edge all the way round the 186° wrap of the head drum assembly. The tape now passes over the exit guide roller/threading post and may then be steadied by a take-up impedance roller. Now the tape encounters the audio/sync head, a single assembly with separate heads lined up with the upper and lower edge of the tape. A further guide or roller keeps the tape aligned on its way to the capstan, the prime mover of the tape transport system.

The capstan is a precision-machined shaft, motor-driven under the influence of a servo system and a stable timing reference. Holding the tape tightly in contact with the capstan is the pinch roller, which is disengaged during stop, pause, and fast transport modes. Finally the tape passes back into the cassette and on to the take-up spool, which is gently driven by a slipping clutch or direct-drive motor. The sequence of deck components is illustrated in the 'linear' diagram at the top of Figure 17.9. In a given format, all models by all manufacturers will conform to this 'linear' diagram so that tapes recorded on any machine will play back on any other. This compatibility is an essential feature, demanding that all machines agree with the parameters laid down in the format specification. The most critical and significant parameters are: video head positioning with respect to height and angular mounting on the head drum; head-to-ruler edge angle on the video head drum; drum exit to sound/sync head spacing; linear tape speed; and of course the track configuration, which we will shortly examine in more detail.

Half-loading

Many VCRs have facilities for real-time counting and index-search systems, which are required to operate in fast-forward and rewind modes. In those machines which do not perform these fast transport functions with the tape fully laced around the head drum, it is necessary to draw a small loop of tape out of the cassette and keep it stretched tight across the audio/control head face so that the control track can be read continuously. It's achieved by one or two *half-loading* arms which, like the guides of Figure 17.6, penetrate the cassette, behind the tape ribbon, as it is lowered onto the deck. During fast-transport mode the half-loading arm(s) moves out from the cassette to route the tape across the ACE head as shown in Figure 17.10. The working principle of real-time counting and tape indexing will be described in Chapter 21.

Other deck components

Before we leave the deck, however, there are several other devices to mention, mainly concerned with the *systems control* or safety aspect of the VCR. The most

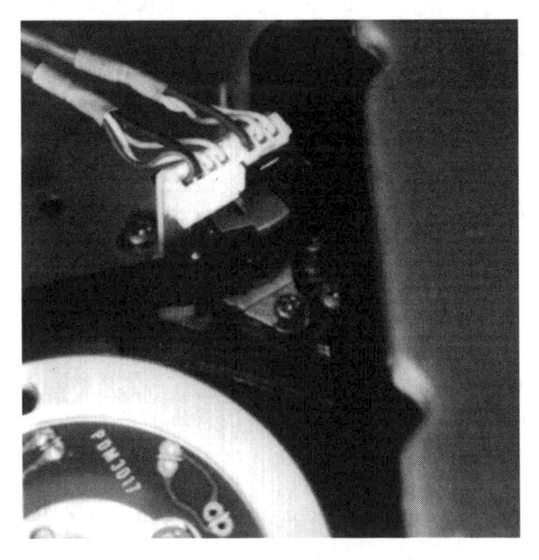

Figure 17.10 *The half-loading arm (centre of picture) maintains control-track readout during fast tape transport. In FF and REW modes the tape is not laced around the head drum as shown here*

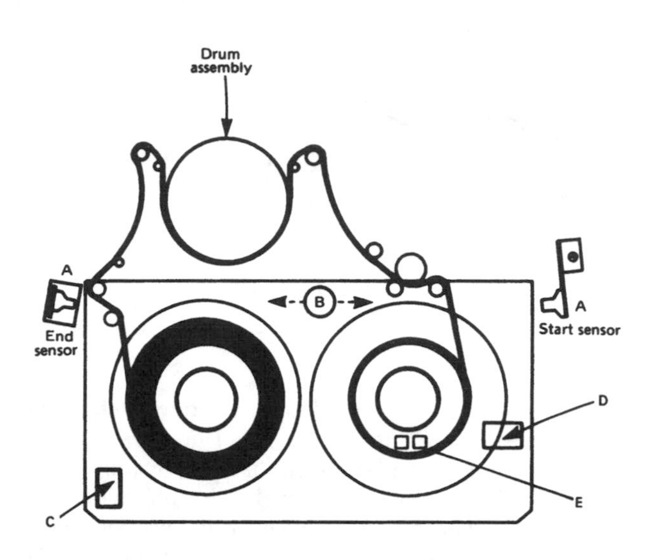

Figure 17.11 *System control sensors on typical VHS tape deck*

important of these are shown in Figure 17.11 starting with the end-of-tape sensors (AA). These prevent physical damage to the tape by inhibiting transport when the end of the tape is reached in either direction, and permitting further motion only in the 'safe' direction. The sensors are optical, depending on a centrally-mounted infra-red LED emitter, B, to illuminate photo-electric sensors through clear leader tape at each end of the tape. In some machines the passage of the cassette itself through these light paths is used in place of 'D' as a 'cassette-in' indicator.

Under the cassette carriage there are two microswitches, C and D. One is the 'cassette in' indicator which inhibits mechanical action in the absence of a cassette. The other is a 'tab detector' which prevents accidental erasure if the safety tab has been removed from the cassette in question. More details on these will be found in Chapter 21. Sensor E detects spool rotation on the take-up side. If it ceases, slack tape can build up, so the control system shuts down the deck. Some VCRs have a similar sensor on the supply reel.

Scanning systems

In the original VCR plan, a guard band was left between video tracks on the tape. This was true of the first machine to appear on the domestic scene, the Philips N1500. Linear tape speed here was over 14 cm/second, and the track configuration is shown in Figure 17.12. It can be seen that each video track is spaced from its neighbours by an empty guard-band, so that if slight mistracking should occur, crosstalk between tracks could not take place. Each video track was 130 µm wide and the intervening guard bands 57 µm wide. This represents relatively low packing density of information on the tape, and it was soon realised that provided the

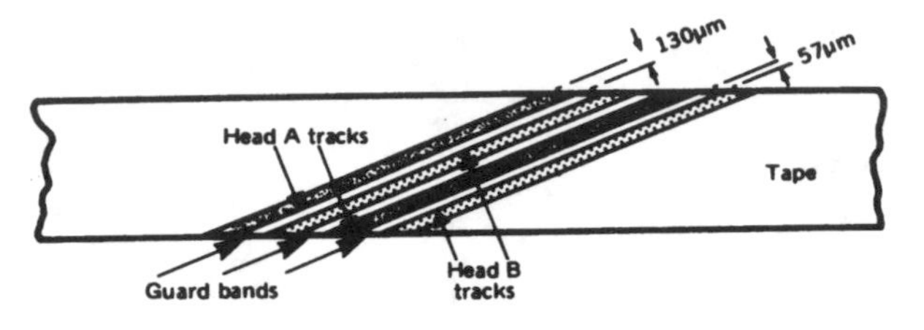

Figure 17.12 *Guard-band recording: the track formation for VCR format as recorded by the Philips N1500 machine*

tape itself was up to it, a thinner head could be used to write narrow tracks; if the linear tape speed was also slowed down the tracks could be packed closer together, eliminating the guard band. Using both ideas, tape playing time for a given spool size could be doubled or trebled. First, though, the problem of cross-talk had to be solved. Even if the mechanical problems in the way of perfect tracking could be overcome so that each head always scanned down the middle of its intended track, crosstalk would occur due to the influence of adjacent tracks, and the effect on the reproduced picture would be intolerable. A solution to this problem was found in the form of azimuth recording.

The azimuth technique

For good reproduction from a tape system it is essential that the angle of the head gap on replay is exactly the same as was present on record, with respect to the plane in which the tape is moving. In an audio system the head gap is exactly vertical and at 90° to the direction of tape travel. If either the record or replay head gap is tilted away from the vertical, even by a very small amount, tremendous signal losses occur at high and medium frequencies, the cut-off point travelling further down the frequency spectrum as the head tilt or *azimuth error* is increased. If the same head is used for record and replay (as is usually the case in audio tape recorders) the azimuth error will not be noticed, because there is no azimuth difference between record and replay systems. A pre-recorded tape from Granny in Scotland will not be up to much, however!

This azimuth loss effect, bad as it is for Granny, is the key to successful recording and replay of video signals without a guard band. Let's designate our video heads A and B, and skew A's head gap 15° clockwise and B's head gap 15° anticlockwise as in Figure 17.13*a*. This imparts a total 30° difference in azimuth angle between the two heads, and the result is video tracks on the tape like those in Figure 17.13*b*.

With the built-in error of 30°, head A will read virtually nothing from head B's tracks, and therefore the guard band can be eliminated. This was the *modus operandi* of the Philips VCR-LP format, using the same cassette and virtually the same deck layout as the original VCR format, but with linear tape speed reduced by 50 per cent, and video track width down to 85 μm. It worked, and the two-hour machine was a reality.

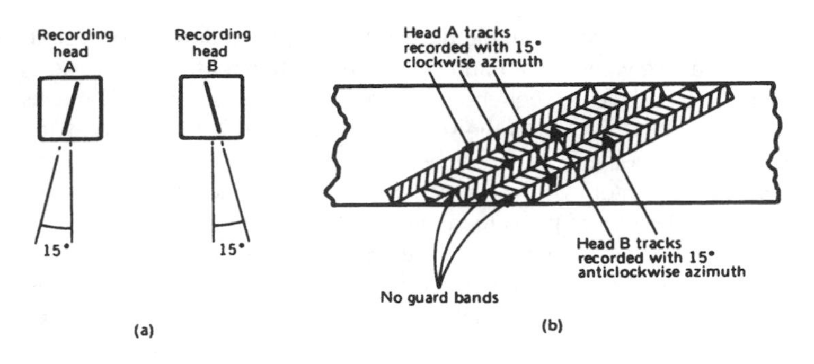

Figure 17.13 *Azimuth offset. The head gaps are cut with complementary azimuth angles so that the guard bands of Figure 17.12 can be eliminated*

Subsequent formats use a smaller azimuth tilt: 6° for VHS and 10° for Video 8. The offset *between* heads is double this figure in each case.

Compatibility

The format specification sets out the parameters of the recorded signal on the tape and it is important that each machine in record and playback conforms exactly to these parameters. If a VCR is in a worn or maladjusted state, electrically or mechanically, it may well record and play back its own tapes satisfactorily, but mistracking and other problems will arise when it is called upon to replay tapes from another machine, or when its tapes are replayed elsewhere. Possibly the most common cause of incompatibility in a VCR is mechanical misalignment of deck components, particularly head-drum entry/exit guides, dirt build-up the head drum ruler-edge guide and physical displacement of stationary heads and pinch roller.

To check mechanical and electrical alignment a precision test tape is used, recorded on a 'design centre' machine under carefully controlled conditions. Also required is a series of mechanical jigs produced by the VCR manufacturer, as described in Chapter 27. A properly aligned machine will be compatible with others of the same format, needing at most an adjustment of the tracking control during replay. The function and need for this control will be described in Chapter 20.

Television recap

Before we go on to describe the laying-down of the video tracks in detail, let us briefly recap on the characteristics of the UK television system, known as CCIR System I. For the purposes of this chapter we need only examine the luminance, or monochrome component, of the signal; the colouring signals will come under scrutiny later!

As is well known, the picture is made up of 625 lines, each drawn from left to right of the screen as we view it. For bandwidth conservation reasons (see page 2) the lines are not transmitted in sequence, but in interlaced fashion. This means that line 1 is traced out at the top of the screen, then a gap is left before drawing line 2. Below line 2 another gap is left between it and line 3, and so on, all the way down to the bottom of the TV screen. By the middle of line 313, we are the centre of the bottom of the screen, and at this point the scanning process is suddenly terminated, recommencing at the top of the screen. So far we have traced out one *field* of $312\frac{1}{2}$ lines in a period of one fiftieth of a second or 20 ms. During this time the instantaneous brightness of the scanning spot along each line has been changing in sympathy with the video signal to build up the picture. The scanning lines for the first field are shown in solid line in Figure 17.14.

The lines of the second field (nos 313 to 625) are slotted into the gaps between the lines of the first field as the vertical timebase commences its second journey down the screen, shown by the dotted lines of Figure 17.14. The second field lasts another 20 ms, and the two fields combined make up a television *frame,* one of which is completed every one twenty fifth of a second, or at 40 ms intervals. Each field contains $312\frac{1}{2}$ lines, and the duration of each line is 64 µs.

The television waveform contains two sorts of synchronising pulses, one at 64 µs intervals to define the starting point of a new line and a more complex one at 20 ms intervals to signify the beginning of a new field. These are illustrated in Figure 17.15*a* and *b* respectively, and are separated from the vision signal in the TV receiver to initiate the flyback or retrace strokes of the line and field timebases. Thus the video signal is 'chopped' as it were, at line and field rate, and a sharply-defined pulse inserted as a timing reference. In a VCR the timing reference at field rate is a useful marker, and is used to set the video head position and define the length of a video track on the tape. We can now relate the TV picture to the magnetic 'signature' it writes on the video tape.

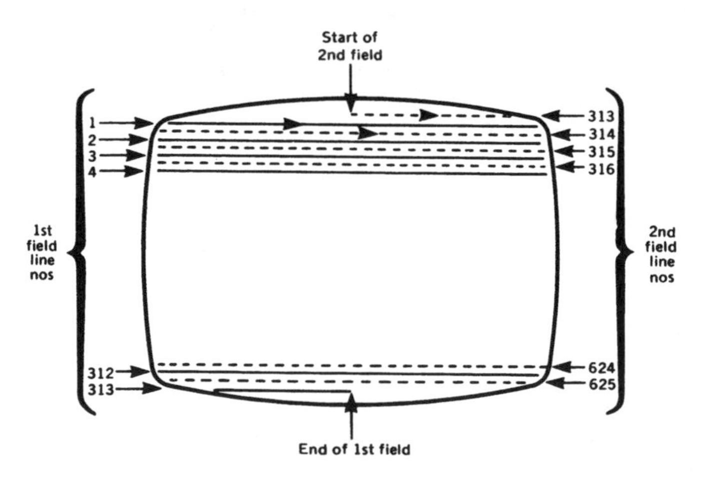

Figure 17.14 *Interlaced TV scanning, in which the lines of field 2 are traced in the gaps between those of field 1*

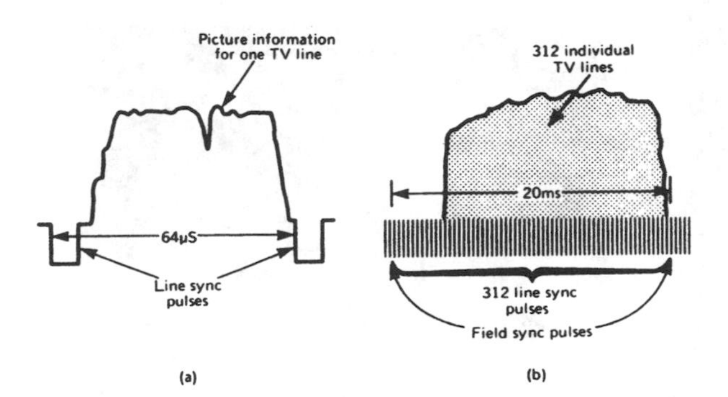

Figure 17.15 *Line and field synchronising pulses in the television waveform*

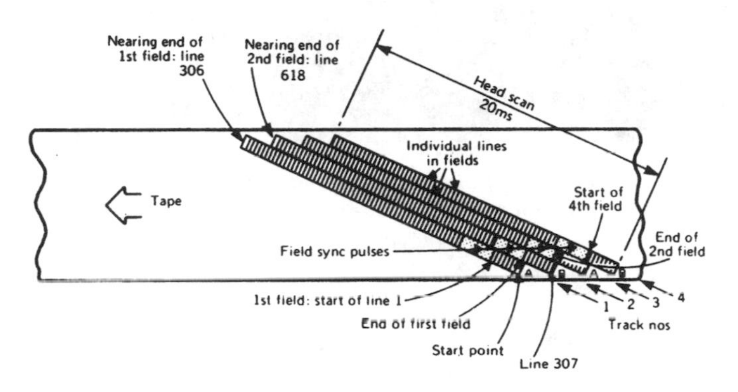

Figure 17.16 *The positions of TV lines and fields in the recorded tape track. Each head writes one field of video information*

The video track

Figure 17.16 shows four adjacent tracks on the tape. Track 1 is laid down by head A, and the 'phasing' of the spinning video head drum is arranged so that the A head enters on to the tape and starts to record just before a field sync pulse arrives. It will write about seven lines of picture before recording the field sync pulse, and then go on to write the rest of the lines in the field. By the time line 306 has been recorded, head A is leaving the top of the tape, having recorded one field of $312\frac{1}{2}$ lines; head B has entered onto the tape and is about to record track 2, consisting of the next sync pulse and field; and the head drum has turned through half a revolution, or 180°. During this time the capstan has pulled the tape through the machine just far enough to ensure that track 2 lays alongside, and just touching, track 1. Track 3 is laid down by head A again, track 4 by head B, and so on.

We can see, then, that signals recorded towards the lower edge of the tape correspond to those in the top half of the picture and *vice-versa*. Thus a tape damaged by creasing along the top edge may be expected to give a horizontal band of disturbance in the bottom half of the reproduced picture. Around the

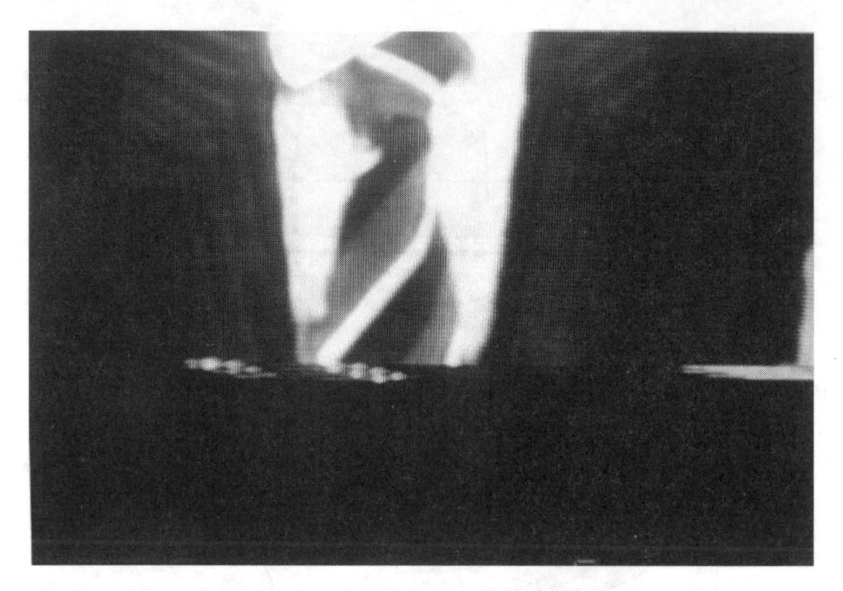

Figure 17.17 *The video-head switchover period generates a picture disturbance over the bottom few lines of the tape-replay picture. Normally hidden below the visible image, here it is revealed by adjustment of the monitor's picture height control*

period of the head changeover point, both heads are at work for a brief instant, one just about to run off the top of the tape, and the other having just entered at the bottom. Thus there is an overlap of information. All VCRs incorporate a head switch which electronically switches between the heads at the appropriate time, just before the field sync pulse. As should now be clear, this takes place at the very bottom of the picture, and any picture disturbance due to head changeover during record and replay is hidden by the slight vertical over-scanning which takes place in a correctly-adjusted TV. Figure 17.17 shows the head changeover on a TV whose picture height has been reduced to demonstrate the effect.

Still-frame considerations

In domestic VCRs the video head drum and the tape move in the same direction, so that with the head drum spinning anticlockwise (as they do in all formats) the tape is pulled through the half-wrap round the head drum in an anticlockwise sense. This, in conjunction with a special grooved surface on the periphery of the head drum, ensures a minimum of friction between tape and drum.

From our studies so far, and the video track diagrams already described, it might reasonably be supposed that if the tape transport were stopped while video head rotation continued, the heads would repeatedly scan the same track over and over again to produce a good still-frame picture. In fact it does not happen, for reasons which will become clear.

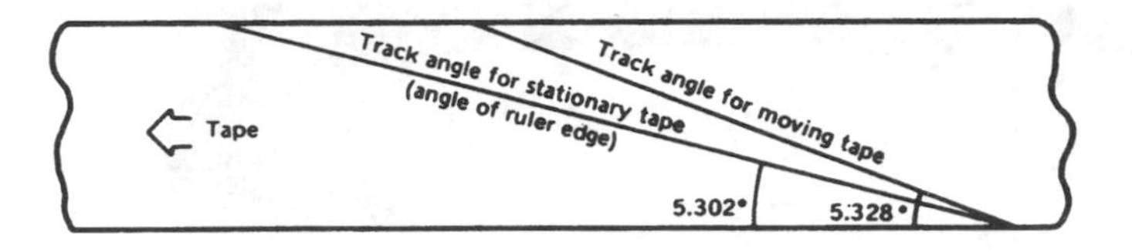

Figure 17.18 *The effective video track angle changes when tape transport is stopped for reasons explained in the text*

Referring again to Figure 17.9, the angle between the plane of the head path and the ruler edge around the head-drum mounting is 5.032° (VHS system) and when the tape is stationary the video heads move across the tape at this precise angle. Once the tape is moving, however, the tape and head velocities subtract, because the head and tape are moving in the *same* direction. Think of the tape as a slow lorry being overtaken by a fast car representing the video head; the car's speed *relative to the lorry* is less than if the lorry were stationary. This effective slowing of the writing speed means that in the fixed time available (20 ms, one half-turn of the head drum) the track angle will be steeper than the 5.302° set by the guide, and the recorded track length shorter than when the tape is stationary. The relative track angles and lengths for stationary and moving tape are shown in Figure 17.18, where it can be seen that the track angle has changed to 5.328° when the tape is moving in record and replay. Although the angles and parameters quoted are for VHS, the same principle applies to all formats. Let's see what effect this has on still frame reproduction.

The video track, recorded at an angle of 5.328°, is now stationary and being scanned by a head moving across the tape at 5.302°. The difference is not very great, but with video track width of the order of 49 μm it is sufficient to cause the head to diverge from its intended track on to an adjacent one, which, as we have already seen, is recorded with the 'wrong' azimuth angle. This results in mistracking and a band of noise on the reproduced picture, as shown in Figure 17.19. Where the noise band occurs depends entirely on the precise stopping point of the tape. Figure 17.20 shows three possibilities, with their effects on the still-frame picture. In simple VCRs where a stop-motion facility is provided, this unavoidable noise band precludes any serious use of the feature, and it is referred to as 'picture pause' rather than 'still frame.'

Picture freeze

Second-generation VCRs had special facilities for the display of noise-free still frames. Before the advent of tilting-drum technology (see below) other solutions to the problem of mistracking were found.

One approach is to make the video heads wider, so that in the stop-motion mode the replay head is 'reading' a sufficiently broad path to embrace the change in track angle without undue loss of signal. The idea is shown in Figure 17.21 where a head 59 μm wide is able to keep in sight the 49 μm video track throughout the field period. It must be appreciated that in stop motion both heads trace

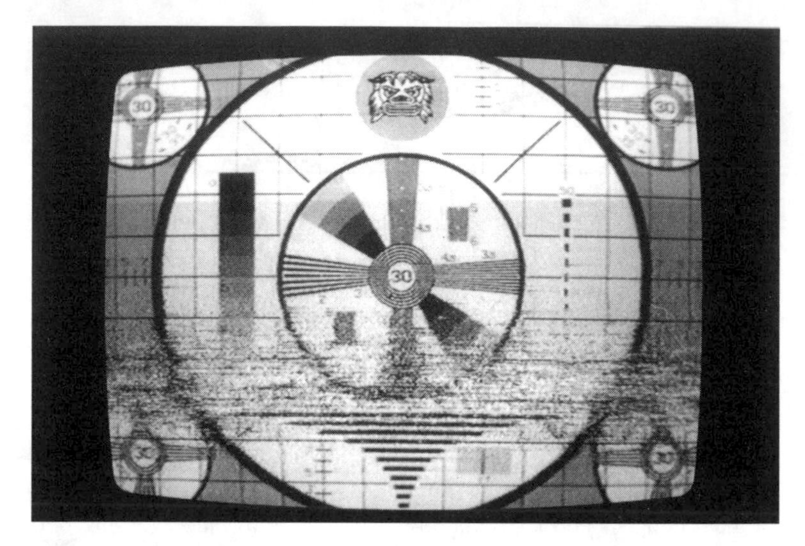

Figure 17.19 *Head-mistracking effect on playback picture*

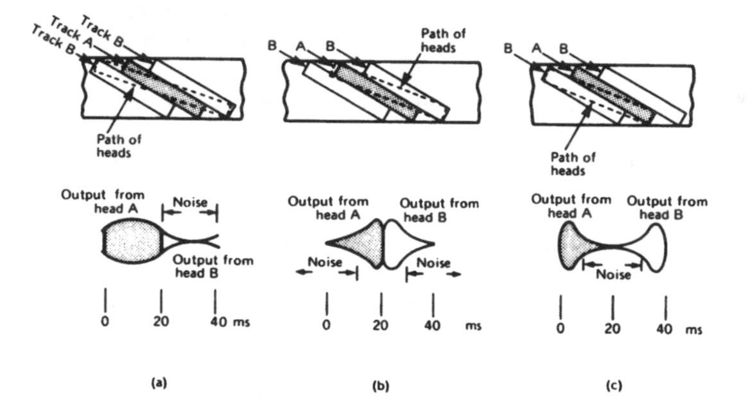

Figure 17.20 *The position of the mistracking bar depends on the exact point, relative to the heads' path, at which the tape stops*

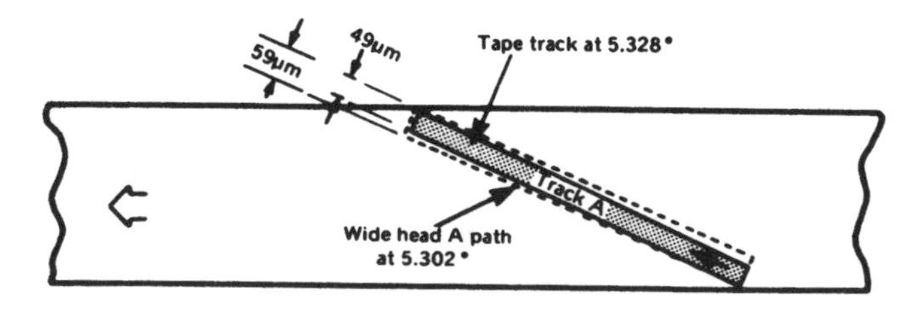

Figure 17.21 *A special wide video head scanning a video tape track during 'stop motion'*

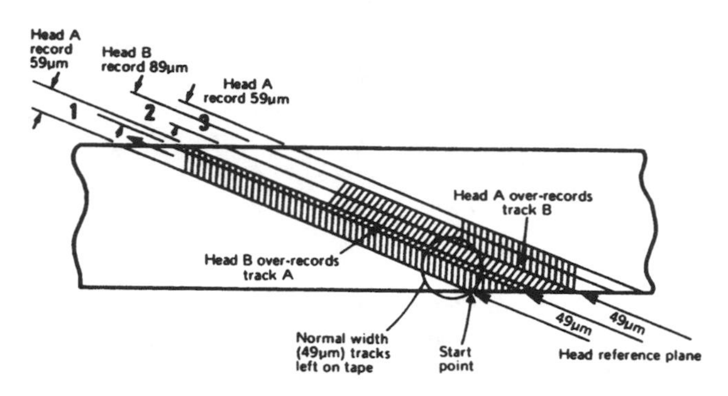

Figure 17.22 *Laying down standard-width tracks with wide recording heads. The excess track width is 'wiped off' by the next head sweep*

the *same path* across the tape, so that if track 1 (head A's province) is being scanned, head B will produce no output from it due to the azimuth error (refer again to Figure 17.20). To obtain a useful output from head B, then, it needs to see some of its own track 2, and to this end it is made even wider than head A, in fact 79 µm. Even when travelling along the same path as head A, then, it reads enough of the adjacent track 2 to provide a usable output, though the picture quality in still frame is, not surprisingly, inferior to that on normal replay. The extra-wide heads are also necessarily used during record to lay down normal compatible 49 µm tracks according to the format specification. This is achieved because the lower edges of both heads are on the same reference plane, and each new track recorded by either head will wipe off, or over-record the excess width of its predecessor, as shown in Figure 17.22.

Three-head and four-head drums were introduced to improve still-frame reproduction. The extra heads are wide, and optimised for 'freeze' reproduction; a further advantage was that both 'trick' heads could now be cut with an azimuth angle to scan *the same track*. Figure 17.23 shows one such approach. Electronic field-store memories can be used for consummate freeze-frame pictures in home VCRs.

Tilting head drum

For many years it was thought that the achievement of good still-frame (and 'trick-play') pictures by mechanically tilting the head-drum was not reliably possible because of the close tolerances involved. It was achieved, however, by JVC with the *Dynamic Drum* mechanism shown in Figure 17.24. The whole drum assembly tilts, under servo control, to line up the head-sweeps with the tape tracks in freeze-frame and picture-search modes. In record and normal replay modes the lower drum block sits on four fulcrum points. For still- and search modes the two tilt screws are rotated by a motor so that the drum see-saws on two fulcrums, coming to rest at the point where the off-tape signal indicates that tracking is optimum.

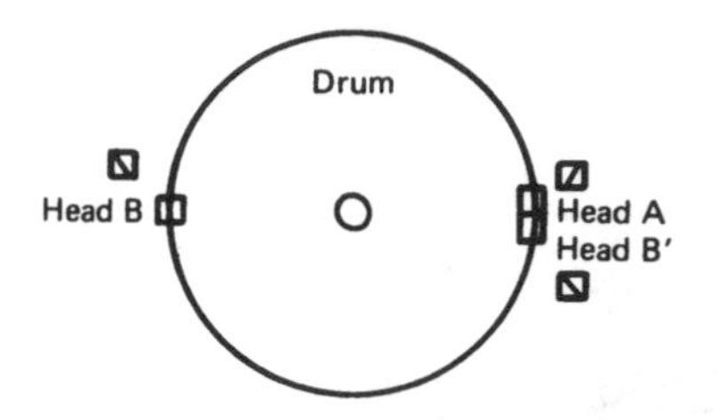

Figure 17.23 *One form of three-head drum. In record and normal playback heads A and B are used. For still-frame, heads B and B′ scan the* same video track *for noise-free and jitter-free reproduction*

The total movement is just a few tens of microns, and the mechanical arrangement is very stable and accurate.

Miniature VHS head drum

The 62.5 mm drum diameter of the standard VHS specification is a great handicap in portable video equipment. To achieve a deck size small enough to be accommodated in a light camcorder a small head drum is used: it is 41.3 mm in diameter. To permit record and playback of *standard* VHS tracks some complexity in the mechanics and electronics of the machine is unavoidable, and Figure 17.25 shows the essence of the arrangement. The travel of the tape guides follows a longer and more sinuous path than before, to wrap the tape around 270° of the periphery of the small drum, which rotates at 2250 rpm. This high speed is calculated to sweep a single drum-mounted head along the entire length of a *standard* VHS video track during its contact with the wrapped tape. The inclined and continuous ruler-edge around the lower drum assembly maintains the tape at the normal 5.302° angle to the head-sweep path.

Plainly, one pair of heads will not suffice to work this system. At the point when one head is leaving the tape wrap, there needs to be another just 90° ahead, the point where it is just entering the tape wrap. This ensures continuity of signal feed onto the tape, whose linear speed around the drum conforms to standard VHS specifications – 2.34 cm/second for SP mode, 1.17 cm/second for LP mode. If the head which has just left the tape is writing or reading 'A' tracks the one ahead of it and the one behind it must be 'B' heads, with azimuth angles cut accordingly. Hence the A-B-A-B configuration of the four heads around the drum in Figure 17.25. Each head scans every *fourth* track on the format-standard tape.

At any given moment only one of the four heads will be active in record or playback, and since two others will be in contact with the tape at this time a four-phase head-switching system is required during both record and replay. Figure 17.26 shows the switching sequence and the time relationship of the video signal to the active period of each head. The switching system is the same in record and playback modes, though of course the routing of the video FM carrier is opposite.

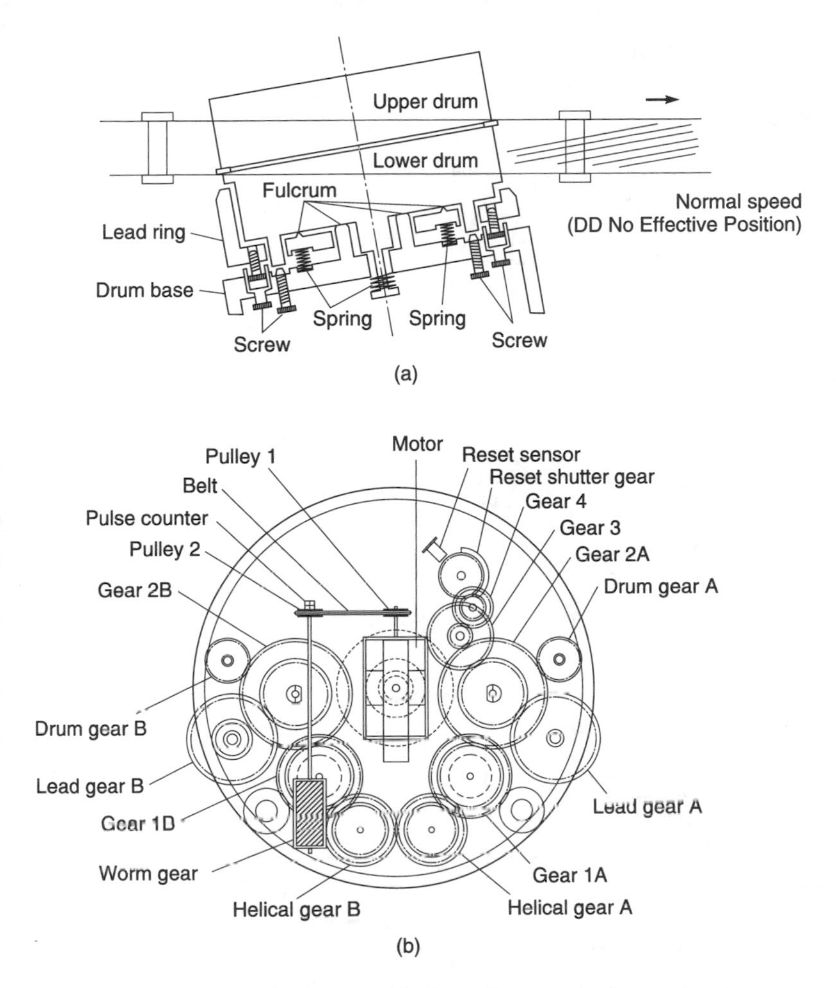

Figure 17.24 *Tilting drum assembly by JVC. At (a) is shown the drum mounted on fulcrums; at (b) the gear train (driven by the central motor and belt) which drives the tilting screws*

It can be seen, then, that four heads are necessary in a 'small' VHS head drum to do the work of the two in a conventional drum. If separate heads are provided for SP and LP eight heads are required on the drum, though they can be mounted in four chips, each carrying two windings and two head gaps. Some VHS camcorders are additionally fitted with a flying erase head, giving an effective total of *nine* heads around the drum periphery, with a multi-winding rotary transformer to couple the recording signals to the heads.

Video-8 format was designed from the outset for a two-head 40 mm-diameter drum in a small camcorder. Even so, further miniaturisation has led to the use in Video-8 and Hi-8 camcorders of the same small-head technique, with a drum of 26.6 mm diameter.

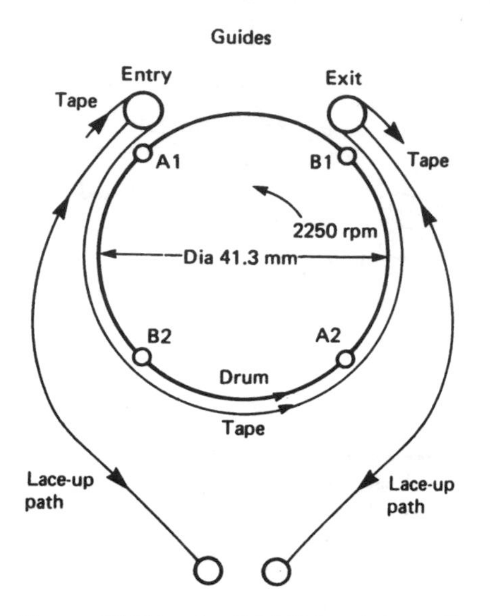

Figure 17.25 *Small VHS head-drum. Four heads are required to read and write standard tracks*

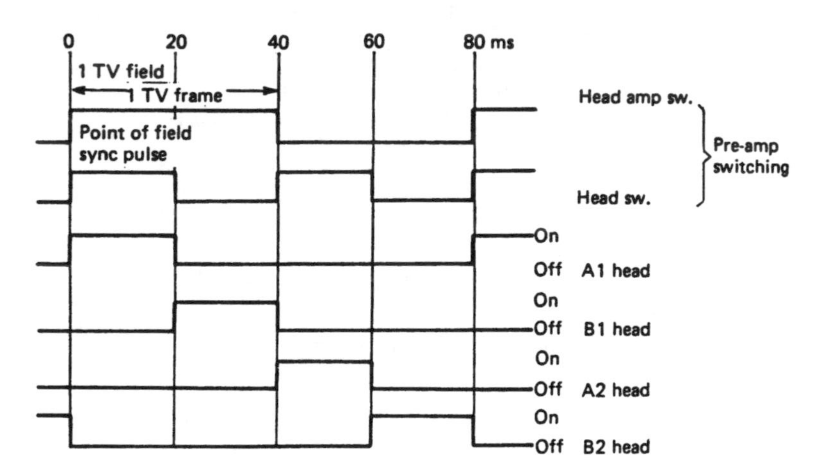

Figure 17.26 *Sequential switching for the heads in Figure 17.25. Switching is applied on record and playback*

Tracking

We have seen that during record the phasing of the head drum position relative to the incoming video ensures that the field synchronising pulse is laid down at the beginning of each head scan of the tape. During replay it is necessary to set and maintain the relative positions of the tape and head drum so that each head scans down the centre of its own track. The record phasing and replay tracking is carried out by the VCR's servo systems, which have a chapter to themselves later in this book. Suffice it here to say that on record a *control track* is recorded along the edge of the tape as a timing reference. This takes the form of a 25 Hz pulse train and is used on replay to set and maintain the relative position of the tape tracks and head drum.

Automatic tracking systems

The system of marking the physical position of each video track on tape by control-track signals is a long-established one, and lends itself to various methods of quick programme finding by modifying the control track pulses temporarily at the beginning of each new recording. The VISS and VASS search systems used in VHS machines detect the change of control-track pulse formation in search, FF and REW modes to stop transport. More of this in Chapter 21.

Servo control by a separate tape track, however, requires provision and careful alignment of a separate pulse record/replay head; and a tracking control which in many cases is a manual (user-operated) type. Correct tracking with this arrangement is also dependent on correct tape tension and precision alignment of tape guides. For these reasons and others, control-track systems are vulnerable to tracking errors, especially in LP modes where the video tracks on tape are narrow and less tolerant of head-path errors.

In VHS VCRs of recent and current design, automatic tracking is achieved by sampling the amplitude of the off-tape FM carrier signal and applying it to the servo circuit to 'trim' the tracking for optimum results. Even so, a manual override is provided, most often used in replay of Hi-Fi sound cassettes which are 'difficult', perhaps due to slight head misalignment in the duplicating deck on which they were recorded.

Automatic track finding

Automatic track finding (ATF) is a more advanced form of tracking, in which head-guidance signals are continuously recorded *in the video tracks.* The concept was first introduced in the now-defunct Philips/Grundig V2000 format, where it was known as DTF (dynamic track following) and was used with a piezo-bar mounting system for the heads, whose position could be set by applying DC deflection voltages to the piezo bars.

Video-8 format uses an ATF system, the essence of which is a pilot tone which is recorded with the picture throughout every video track. Four pilot tone frequencies are used: f1, 101.02 kHz; f2, 117.19 kHz; f3, 162.76 kHz; and f4, 146.48 kHz. They are added to the luminance FM record signal and recorded in the sequence f1, f2, f3, f4 in successive head sweeps, see Figure 17.27. Relatively low frequencies like these are almost unaffected by any azimuth offset of the replay head, so pilot-tone crosstalk from adjacent tracks is easily picked up by the

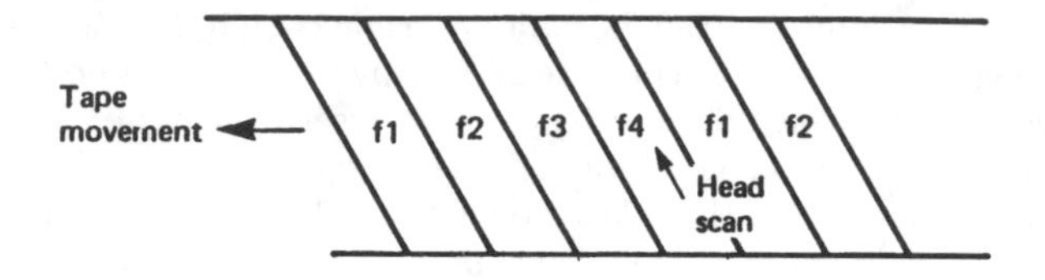

Figure 17.27 *Tone sequence laid on tape for ATF*

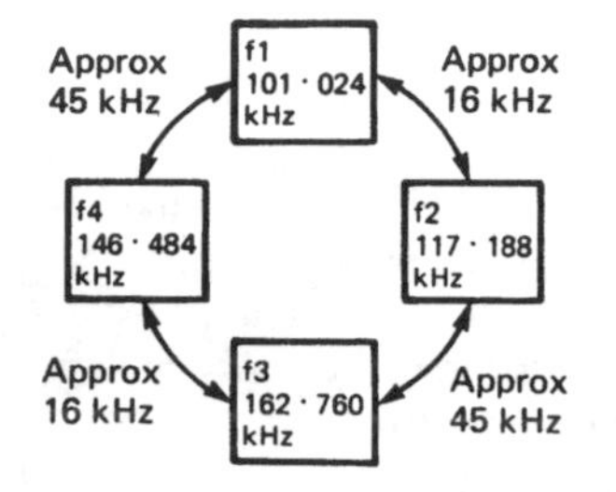

Figure 17.28 *ATF tones have a carefully-selected frequency relationship*

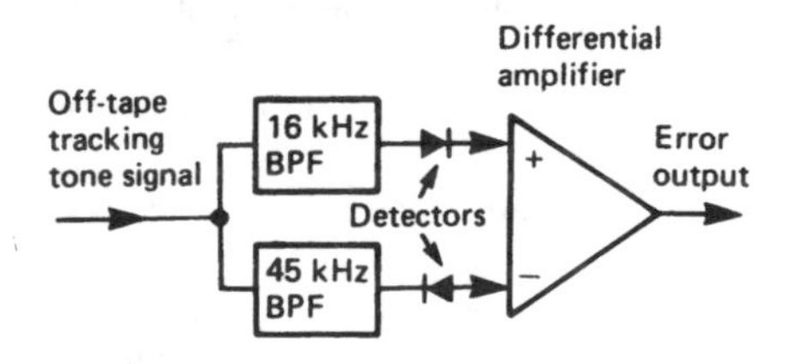

Figure 17.29 *Basic arrangement for derivation of an error voltage from inter-track pilot beats*

video heads during playback. The pilot-tone frequencies are chosen to have specific relationships as Figure 17.28 shows. The beat frequencies which arise when pilot tones from adjacent tracks are mixed are always 16 kHz or 45 kHz. These beat products are used to steer the head path/tape track alignment for optimum tracking: when the levels of 16 kHz and 45 kHz beat signal are equal the replay head must be scanning along the dead centre of its video track, indicating optimum tracking.

There are several ways in which the ATF pilot tones can be processed during playback. A simple one, illustrated in Figure 17.29, utilises two bandpass filters to pick off and separate the 16 kHz and 45 kHz beat products so that they can be separately detected and measured. The DC outputs are applied to the differential inputs of an operational amplifier whose output forms the error signal. This error output can be used to phase-lock either the capstan or head-drum servo to give accurate and continuous tracking correction with built-in compensation for 'mechanical' errors, tape-stretching etc.

ATF playback

While Figure 17.29 shows a simplistic approach to replay ATF processing, the actual system used is more sophisticated. Pilot tones are generated during playback, again changing on a track-sequential basis triggered by head tacho pulses. This time the tones are sequenced in reverse order, however: see Figure 17.30. The top waveform represents the head-drum flip-flop signal, high for head A, low for head B. The off-tape pilot tone (REC pilot) is shown below, and at the bottom the newly-generated local pilot tone (REF pilot) in the reverse order f4, f3, f2, f1. The diagram shows ideal tracking conditions, in which the off-tape tones switch in synchronism with the REF pilot tones. At every fourth field two f1 tones appear simultaneously at a mixer, whose output consequently drops to zero. During the other (properly-tracked) fields the beat product is either zero (f3/f3) or 29 kHz (f2/f4), the latter being outside the band of interest and thus rejected.

If the tape speeds up – see Figures 17.27 and 17.30 – the REC pilot moves to the left, permitting REC f2 to appear during the period of REF f1 and giving rise to a 16 kHz beat product. As other tracks are scanned all the beat products (f3/f4, f4/f3, f1/f2) are 16 kHz.

Conversely, when the tape slows down the REC pilot pattern moves to the right. Some REC f4 now appears during the REF f1 period to produce a 45 kHz beat product. Similarly during subsequent scans REC f1 beats with REF f4, f2 with f3 and f3 with f2, producing a 45 kHz beat output in each case.

Thus a fast-running tape always results in a 16 kHz output from the mixer; and a slow-running tape always produces a 45 kHz output. By using suitably tuned bandpass filters to select these products, and feeding their outputs to separate peak-detectors an ATF error signal is produced for subsequent smoothing and passage to the capstan servo control input. In practice the replay ATF processing circuit is also provided with artifices to detect 'false lock' conditions and to permit locking of mistracking-noise bars during search modes.

Video-8 tape-signal spectrum

The DTF pilot tones are recorded at the lowest part of the frequency spectrum as shown on the left of Figure 17.31. Here they do not interfere with the 'signal' components of the tape recording. The other parts of the tape-signal spectrum will be dealt with later: luminance and chroma in Chapter 18 and audio in Chapter 19.

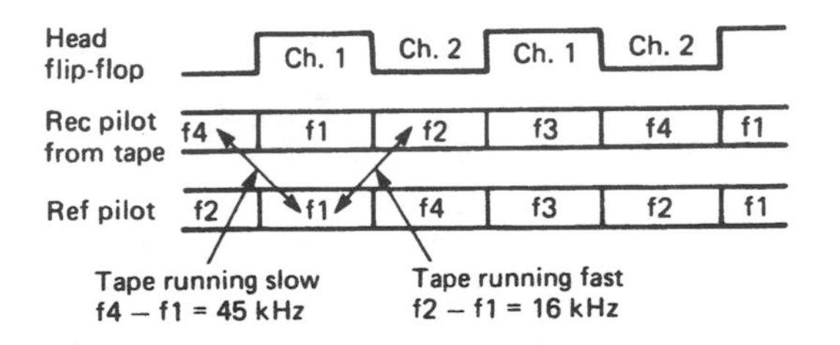

Figure 17.30 *Replay beat-tone generation: off-tape pilot frequencies are compared with a locally-generated tone sequence*

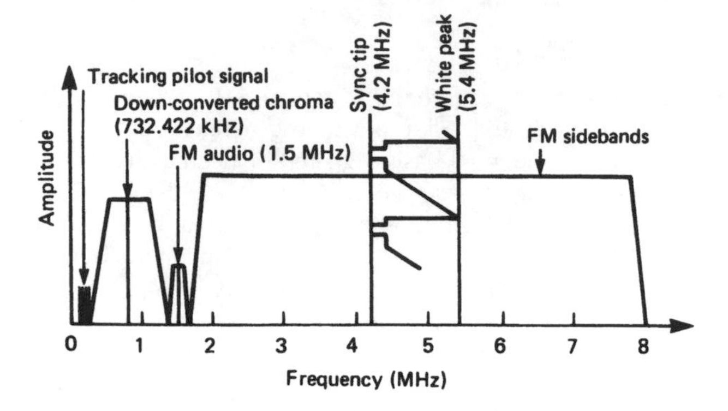

Figure 17.31 *The spectrum of signals on tape – Video-8 format. That for Hi-8 format is given in Chapter 25*

18 Signal processing: video

In previous chapters, we have seen some of the problems of tape recording video signals, and have examined how these are overcome by the use of high writing speeds and FM modulation. We have seen, too, how home VCRs achieve an acceptable performance at low cost by signal bandwidth restriction. This is by no means the only shortcoming of the domestic machine! In spite of these, the playback picture from a domestic VCR in good condition is very acceptable; in this chapter and the next we will examine not only the basic circuitry used to process TV signals within the VCR, but also the special circuits and processses, some of which owe their origins to broadcast technology, which compensate for the deficiencies inherent in a cassette VCR system, and enhance performance.

The luminance (or black-and-white) signal is dealt with separately from the chrominance (or colouring) signals in home VCRs. For luminance the basic idea is to modulate the signal on to an FM carrier for application to the recording head, and demodulate it to baseband during the replay process. As the simplified block diagram of Figure 18.1 shows, however, there are several other processes undergone by the luminance signal, and these will be described in turn.

AGC (automatic gain control) and bandwidth limiting

The basic luminance signal that we wish to record may come from a TV camera or other local video source, or more likely, a broadcast receiver built into the VCR. In either case it will be positive-going for white and will probably contain a chrominance signal modulated on to a 4.43 MHz carrier. It is important that the signal recorded on the tape is within the limits of the recording system, so the luminance signal is first passed through an AGC amplifier with a sufficiently wide range to compensate for signal inputs of varying amplitudes. This works in a similar

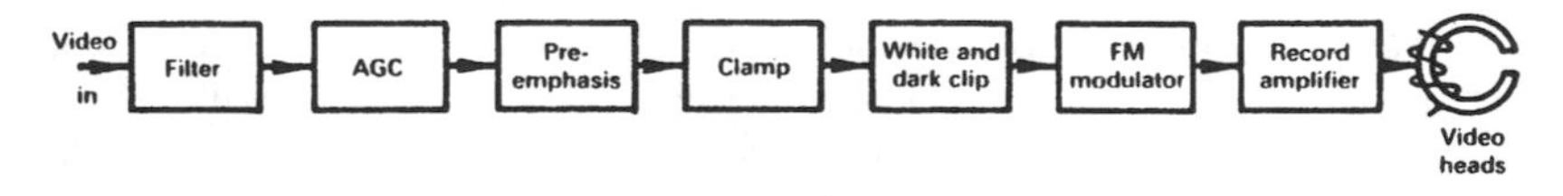

Figure 18.1 *The stages in the luminance-recording process*

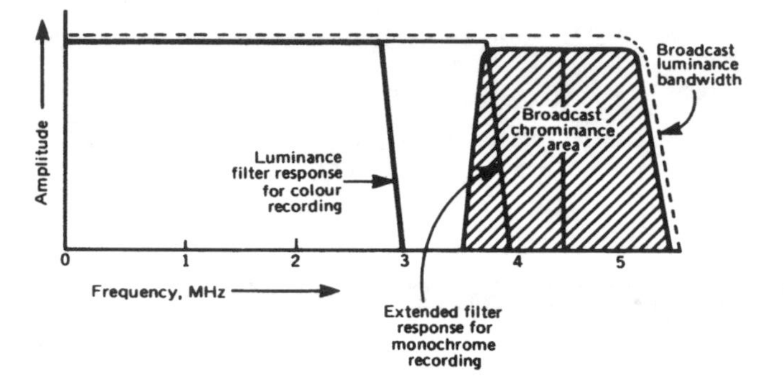

Figure 18.2 *VCR luminance-recording filter characteristics relative to the spectrum of the composite video signal as broadcast*

manner to the AGC system of a radio or TV, by sampling its output level to produce a DC control potential, and applying this to an attenuator at the amplifier input. Thus the output from the AGC stage will be at constant (say) 1 V amplitude.

In low-band (e.g. VHS) VCRs, luminance bandwidth is restricted to about 3 MHz, give or take a hundred kHz or so between the formats. If higher frequencies than this are allowed to reach the modulator they will make mischief with sidebands, as explained in Chapter 16. A low-pass filter with a quite sharp cut-off around 3 MHz is incorporated in the record signal path, then, and this also eliminates all the chrominance components of the signal, which are based on a subcarrier of 4.43 MHz. When recording in monochrome more bandwidth can be allowed, occupying the space normally reserved for the chrominance signal, and many machines have an automatically-switched filter characteristic for monochrome and colour recordings, as shown in Figure 18.2.

Pre-emphasis

Chapter 16 showed how pre-emphasis is used in an FM system to reduce noise. What is required is a boosting of high-frequency (HF) signals before the FM modulator, and this is achieved by a circuit like that in Figure 18.3*a*. Here we have a common-emitter transistor with a small capacitor (C1) in its emitter circuit. At high frequencies, the capacitor's reactance will become comparable with the emitter resistor R1, and the negative feedback due to the latter component will diminish, resulting in greater output from the stage (i.e. at the collector) at those frequencies. A typical pre-emphasis curve is shown in Figure 18.4*a*, in which 10 dB of pre-emphasis is given to frequencies above 1 MHz.

In some machines, particularly Video-8 types (which make use of very narrow video tracks) a further, *non-linear*, pre-emphasis circuit is used. This applies a degree of HF boost which is dependent on signal level (greater for small signals, less for large signals). A family of curves for this system is given in Figure 18.4*b*. Although this calls for more complex circuits (the de-emphasis characteristic has to be non-linear to compensate) it does offer a useful reduction in noise level on replay.

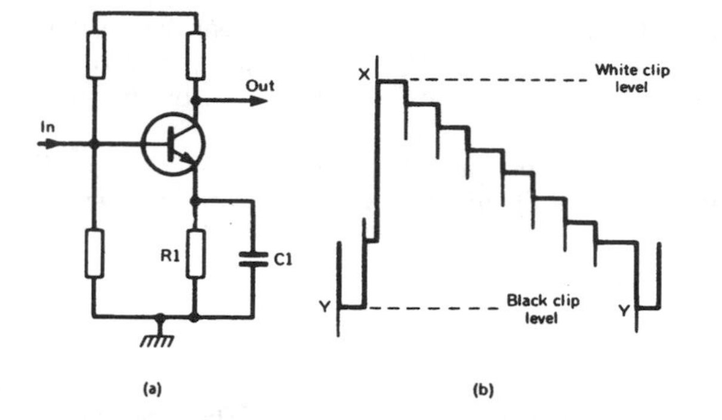

Figure 18.3 *(a) Rudiments of pre-emphasis circuit, and (b) its effect on the luminance staircase waveform*

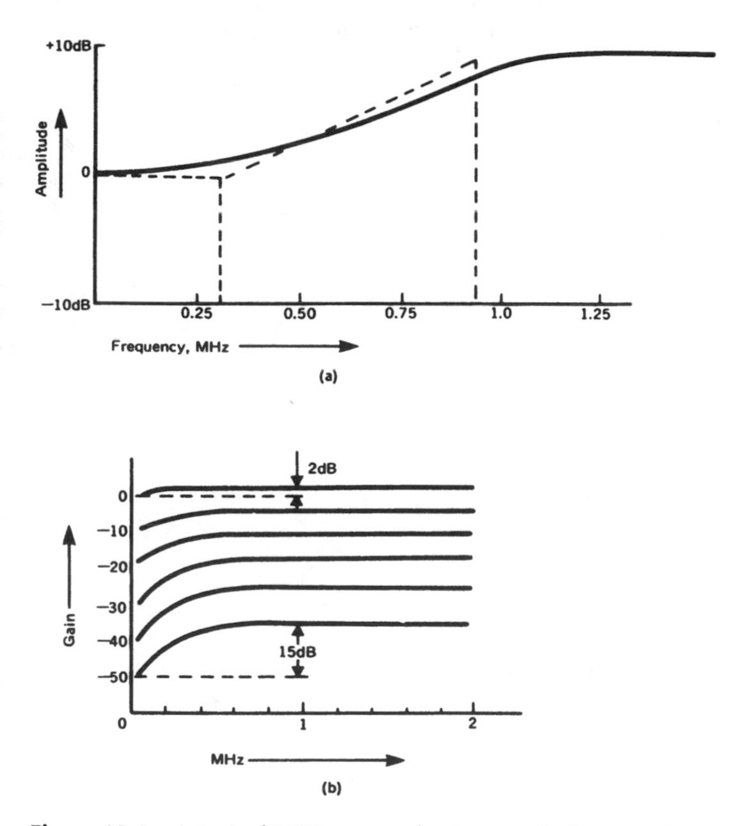

Figure 18.4 *A typical VCR pre-emphasis curve is shown at (a). In (b) appears a family of curves for a non-linear pre-emphasis system, in which the degree of HF 'lift' varies from 2 to 15 dB, depending on signal level*

Because the steep rising and falling edges of a luminance transient, or step, look like a high frequency they are effectively differentiated in the pre-emphasis process, so that the 'risers' of a luminance staircase waveform take on the spiky characteristic shown in Figure 18.3*b*.

Where the spikes exceed peak white level as defined by the output of the AGC stage, we are once again in danger of driving the FM modulator into excessive deviation, and this must be prevented.

White and dark clip

The next process, then, is a clipping action on the video signal. The danger areas for overmodulation are those shown at points X and Y in Figure 18.3*b*, and the dotted lines indicate the levels at which the signal must be clipped. This action takes place in an IC, where a suitably biased diode, transistor or pair limits the maximum amplitude of the signal in both directions.

FM modulator

The modulator's output needs to have equal mark-space radio regardless of deviation, and deviation must be linear with respect to modulating (video) voltage. The output must not contain harmonics which would generate harmful sidebands, and frequency stability has to be of a high order. A constant output level is also required over the whole of the deviation range, but this can be achieved by limiting stages after modulation, as we shall see. The curve of Figure 18.5 shows the relationship between modulating voltage and output frequency for a typical system.

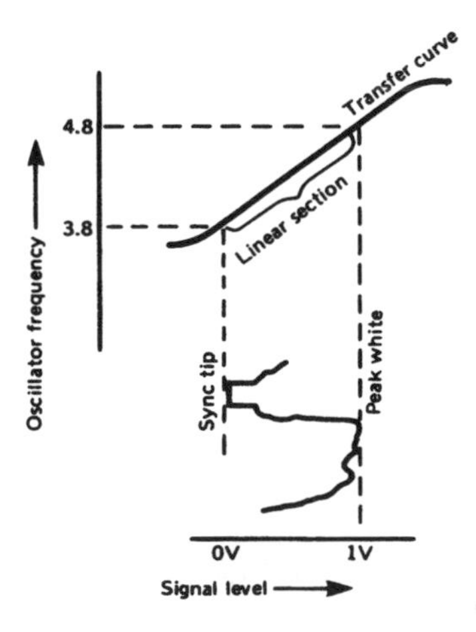

Figure 18.5 *The transfer curve for an FM modulator. It has to be very linear between the frequencies corresponding to sync-tip and peak white. For the VHS system illustrated, these are 3.8 to 4.8 MHz*

Limiting and head drive

A degree of amplitude modulation can occur in the FM conversion and coupling processes, and it is important that this is ironed out before the signal is applied to the recording heads. To illustrate the effect, Figure 18.6 shows a clipper made of two diodes, and its effect on the waveform, whose richness in harmonics will call for low-pass filtering downstream. In practice, of course, the clipper is incorporated with other functions inside an IC.

The recording amplifier has two basic functions: to provide power amplification to drive the recording heads, and to apply recording equalisation to compensate for the falling response of the video heads and head/tape interface with increasing frequency. The circuit diagram of Figure 18.7 shows the arrangement. The first stage after the limiter, Tr1, provides equalisation by giving a degree of lift to the higher frequencies. Frequency-selective negative feedback is again used, the

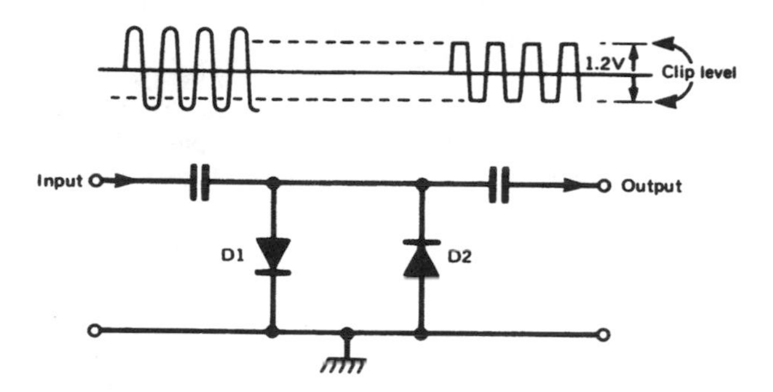

Figure 18.6 *Simple diode amplitude clipper*

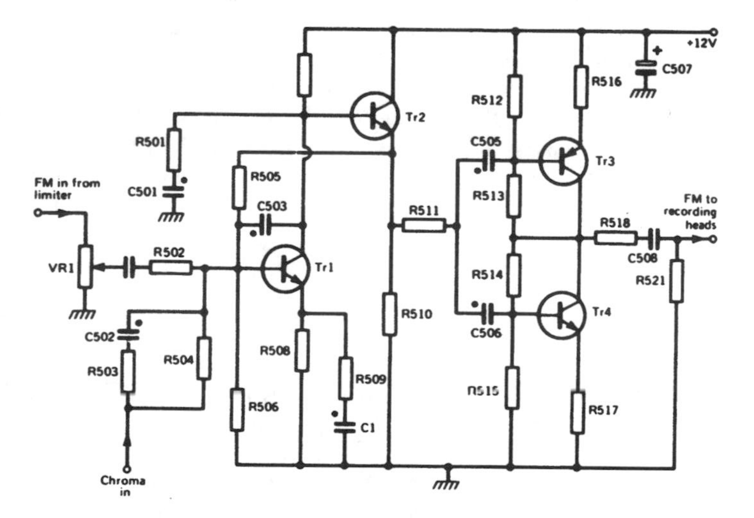

Figure 18.7 *An FM recording amplifier by Panasonic*

operative component being C1. An emitter follower stage Tr2 acts as a buffer to drive the complementary output pair Tr3/Tr4. These work in similar fashion to audio and field output stages in radio and TV sets, but at a much higher frequency.

The level of the FM luminance signal fed to the recording heads is critical, and is set at an optimum level for the type of videotape in use. The constant-level carrier acts as a recording bias signal for the colour signal which accompanies it on to the tape, and the critical FM luminance *writing current* is set by VR1 in Figure 18.7 to bias the recording level to the centre of the linear flanks of the transfer characteristic shown in Figure 16.4. Again in current VCRs the circuit function shown in Figure 18.7 is integrated into a chip, and has no manual adjustment: the diagram, however, provides an illustration of the principle of operation.

Transferring the video signal to a pair of heads rotating at 1500 rpm is something of a problem, and the later recovery of the tiny playback signal an even greater one! Any form of brushgear is impractical from the noise point of view, so rotating transformers are used, with the primary winding stationary and the secondary winding rotating with the heads. These roles are reversed during playback, of course. Two separate and independent transformers are used, having their own windings and ferrite cores; signal transfer takes place across a very narrow air-gap between the two ferrite discs. Multi-head VCR designs use separate and concentric transformers. Except for small-drum VHS systems (Chapter 17) head switching is not carried out on record, both heads being driven together via the rotary transformers from the recording output stage.

IHQ

'Intelligent high quality' is the name given to a concept developed by Akai for home VCR use. In similar fashion to high-quality audio-cassette recorders, it involves a short test-recording just before the 'real' recording takes place; its purpose is to establish the magnetic characteristics of the tape in use. In a fully-automatic sequence taking a few seconds, a test signal is recorded on the tape, which is then rapidly rewound and replayed for analysis in a measuring circuit. The data thus obtained is memorised and used to set the response of a filter in the luminance record amplifier stage which optimises the shape of the lower sideband recorded on tape. The effect is one of equalisation of the energy in the spectrum of signals on tape, leading to enhanced detail and lower noise levels in the reproduced picture.

Replay circuits

During replay, the off-tape signal from the video heads is very small, and to maintain the necessary >40 dB S/N ratio in the reproduced picture, low-noise amplification is necessary. The main replay processes after the preamplifiers are head switching, equalisation, limiting, drop-out compensation, demodulation to baseband, de-emphasis, crispening and amplification, after which the luminance signal is restored to its original form, usually 1 V peak-to-peak, negative-going

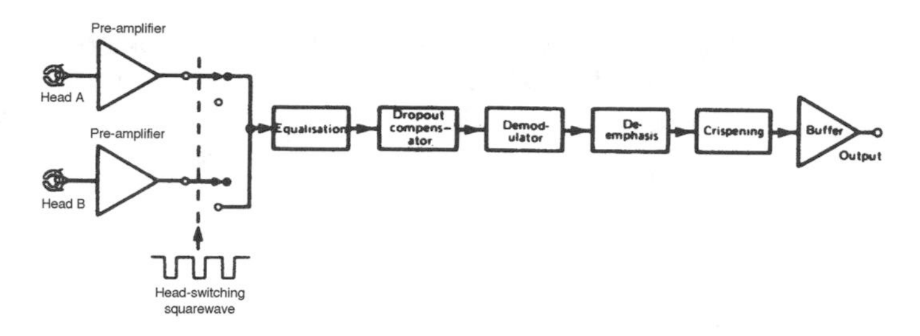

Figure 18.8 *The replay chain for the luminance signal*

syncs. A block diagram showing the order of the replay circuits is given in Figure 18.8. We will examine these in turn.

Head preamplifier and switching

The RF signal from the rotary transformers contains signal components ranging from 100 kHz to over 8 MHz. These are at high impedance and very vulnerable to noise pick-up. The head pre-amplifier is built into the IC which performs head drive on record; it has internal play/record switches to separate the functions. Although both heads are driven together during the record process, on all formats head switching is carried out on replay. This serves the dual purpose of eliminating noise from the inoperative head (remember only one of the pair is scanning the tape at any given moment) and sharply defining the head changeover point to minimise picture disturbance during the switch. As shown in Figure 18.8 a squarewave pulse train (derived from a *tachogenerator* on the head drum and often called SW25) is used to switch between heads.

Equalisation

Chapter 16 explained why during replay the signal from the tape head falls at the rate of 6 dB per octave. HF losses due to the approach of *f*ex are largely compensated for by a resonance circuit associated with the replay head. To even-out the response on replay, further equalisation is provided, and in practice it takes the form of a 'boost' in the 2–3 MHz region, the major area for the lower FM sideband of the luminance signal. This is often catered for by the provision of a reactive component in the collector circuit of the preamplifier or a following stage within the IC.

Limiting

For correct operation of any FM demodulator, it is important that its input signal contains no amplitude variations. For this reason, several stages of limiting are usually provided in the path of the FM replay signal to clip it to a constant level.

The form of diode clipper shown in Figure 18.6 (D1 and D2) is suitable for this. In current VCR designs, limiting takes place (along with the other replay processing) within an IC.

Replay picture optimisation

It is on the playback RF carrier that the Nokia ASO (active sideband optimisation) system works. One cause of poor high-frequency response in a tape replay system is the random phase distortion which occurs at sudden frequency transitions in the off-tape signal, representing vertical edges and outlines of picture objects. The ASO concept depends on a small LC delay network, working in conjunction with a pair of clamp diodes. This L/C/D network has the effect of speeding up the transition between the old and new levels in the demodulated video signal to give 'cleaner' and better defined edges.

The action of the Akai IHQ system (whose function in record mode we met a few paragraphs ago) during replay is based on a noise-reduction system, whereby the degree of noise cancellation is made dependent on the S/N level of the replayed luminance signal. This adaptive noise-cancellation scheme optimises playback quality from a wide range of tape types and signal qualities, ranging from a low-noise recording on a high-grade tape to a well-worn library cassette.

Dropout compensation (DOC)

A videotape is not the perfect medium that we would like it to be! The magnetic coating is not completely homogenous, and with video track widths smaller than the diameter of a human hair even a microscopic blemish in the magnetic coating will delete some picture information. As the tape ages, slight contamination by dust and metallic particles, and oxide-shedding effects, will aggravate the situation. The effect of these tiny blemishes is a momentary loss of replay signal known as a *dropout*. Unless dropouts can be 'masked' in some way a disturbing effect will take place in the form of little ragged black or noisy 'holes' in the reproduced picture.

In practice, the video information on one TV line is usually very much like that on the preceding line; so that if we can arrange to fill in any dropout 'holes' with the video signal from the corresponding section of the previous line, the patching job will pass unnoticed. What's required, then, is a delay line capable of storing just one TV line of 64 μs duration, so that whenever a dropout occurs we can switch to the video signal from the previous line until it has passed. For delay-line bandwidth reasons, this is difficult to achieve at video baseband frequencies, so it is carried out on the FM signal before demodulation. To avoid disturbance on the picture the switching has to be very fast, virtually at picture-element rate. One form of DOC circuit is shown in block-diagram form in Figure 18.9. The FM replay signal takes three paths, the upper of which is via demodulator 1 and on through the following replay circuits; this is the path taken by the signal under normal circumstances, that is, when no dropouts are present. The bottom path

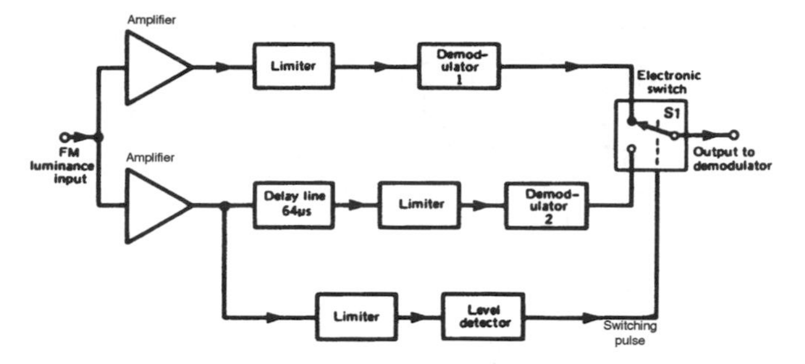

Figure 18.9 *Operating principle of a dropout compensator. Demodulator 2 is operating on an FM signal one TV line earlier than that in demodulator 1*

consists of a level detector which monitors the FM signal from the playback heads. When a dropout comes along, the FM signal falls below the detector's threshold level and the detector output falls to zero. This operates a fast diode changeover switch S1, which now selects the middle path via demodulator 2, whose FM input is exactly 64 µs (one TV line period) 'old' due to the delay line, and thus corresponds to the FM replay signal at the same point on the previous line. This signal is held by the switch until the dropout has passed and full FM input is restored at the DOC input, when the detector output reappears and the changeover switch drops back to the output of demodulator 1.

Recycling DOCs

The system outlined above breaks down when any dropout exceeds one TV line duration, and this happens often – the tape area occupied by one line is microscopic. If the dropout period exceeds one line, then, both demodulators in Figure 18.9 will be looking at noise, and a disturbance will be visible on the TV screen. To prevent this, further steps must be taken. In early designs, a 'grey oscillator' was used, consisting of a circuit, active after the first dropout line, which generated a CW signal corresponding to mid-grey in terms of off-tape frequency. The resulting neutral tone inserted in the picture was subjectively less noticeable than the dropout it replaced.

Later designs perform better by recirculating the last 'good' TV line around the delay line and reading it out continuously for the duration of the dropout. This gives better reproduction than the grey oscillator, and a recirculating DOC, based on an IC, is shown in Figure 18.10. The replay FM signal enters the chip at pin 2 and emerges, after amplification, at pin 3. After passing through a switched filter the signal re-enters the chip at pin 5 and passes through pole A of the dropout switch to emerge at pin 9, *en route* for the FM demodulator. A second output of the amplified RF signal appears at pin 4, routed through the limiter between pins 11 and 12, then into the dropout detector on pin 14. The dropout switching signal appearing at pin 13 is amplified in the source-follower X8 and reapplied to pin 10 of the chip, whence it operates the changeover switch to select pole B. Here the delayed signal is available, and it is routed not only to the demodulator via IC

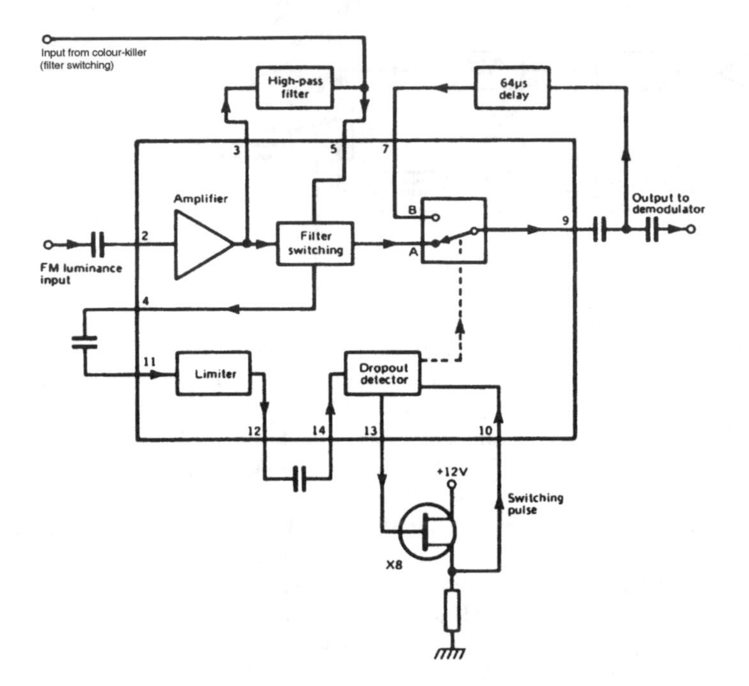

Figure 18.10 *A recirculating dropout compensator*

pin 9, but back to the delay line input for recycling. This circuit requires only one demodulator.

Demodulation

Several methods of FM demodulation are available, and such configurations as the ratio detector and Foster-Seeley discriminator have been used for many years for FM sound recovery in radio and TV receivers. More up-to-date designs use a PLL (phase locked loop) or quadrature demodulator for better performance, but the low modulation index used in the home-VCR system (operating frequency 3–5 MHz, modulating bandwidth 0–3 MHz) requires a different technique.

The principle of a suitable type of demodulator is illustrated by the waveforms in Figure 18.11. At (a) is shown the off-tape FM carrier after amplitude limiting. A form of differentiator derives a 'spike' from each transition in the waveform as illustrated in (b). Now, in effect, the spikes are full-wave rectified, clipped in an amplitude-limiter and then further amplified to render the pulse train shown in waveform (c). The frequency-doubling effect greatly assists with filtering after the demodulator. All that is now required is a low-pass filter, across whose integrating capacitor appears the dotted waveform of Figure 18.11c, a facsimile of the original video waveform, now at baseband.

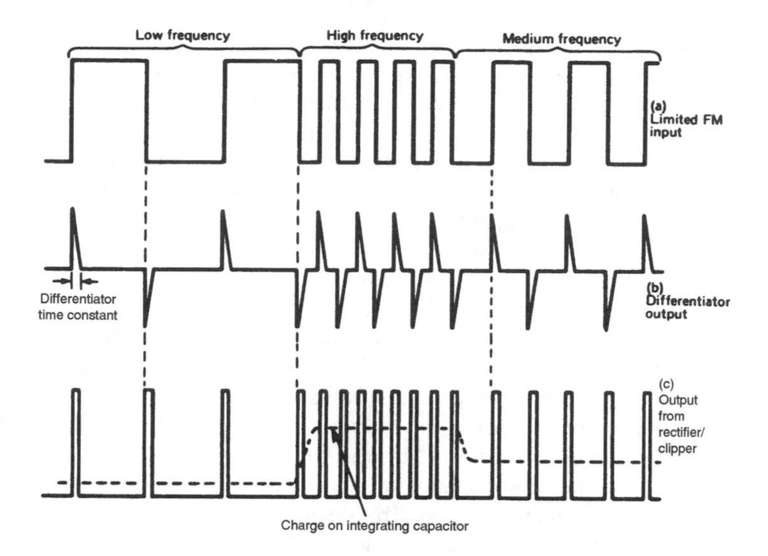

Figure 18.11 *Waveforms for a video FM demodulator. The frequency of the incoming squarewave (a) is effectively doubled at (c)*

The demodulator circuit is carefully designed to offer linear operation throughout the deviation range, and the filter/integrator effectively rejects the residual (doubled) carrier ripple. The luminance signal we have now recreated consists of composite video (i.e. with sync pulses present) with a bandwidth of about 3 MHz, but as yet without a chrominance component.

Crispening

The 3 MHz capability of a standard VCR means that replayed pictures will lack the sharpness and definition of an off-air picture; and fine detail, for example the frequency gratings of a test pattern, will not be reproduced. Several other factors are present, though, and these are worth examining. Most video recording involves colour programmes, and colour receivers incorporate notch filters in their luminance amplifiers which impair definition above 4 MHz. At about the same frequency physical limitations of the colour display tube come into play. The pitch of the phosphor-dot matrix on many tube face plates is such that a picture element *(pixel)* smaller than the area of one RGB dot-group cannot be reproduced, and this is a limiting factor for luminance definition, especially in smaller colour screen sizes. Very little of the content of an average picture consists of fine repetitive detail, and the subjective effect of limited HF response is a blurring of sharp vertical edges.

The crispening circuit goes some way to compensate for HF roll-off by artificially 'sharpening up' vertical edges in the picture. The technique has been used in broadcast and CCTV for many years to compensate for the finite scanning spot size in cameras and FSS systems. In such applications it is known as aperture correction because it has the effect of limiting the scanning spot size to that of a smaller 'aperture' inserted into the tube, and enhancing definition thereby.

The effect of poor HF response is to slow down the rise and fall times of the luminance waveform so that a sudden transition from white to black (or vice-versa) is reproduced as a waveform (a) in Figure 18.12. There are many types of crispening circuit, and a block diagram of an easy-to-understand one appears in Figure 18.13. The luminance step waveform (a) is fed into a short delay line to emerge slightly later at the centre tap, waveform (b). Waveform (a) is also differentiated, producing a negative-going spike, waveform (c). After inversion this spike appears in positive-going form, waveform (d). After a further slight delay waveform (e) appears at the end of the delay line and is differentiated to a spike, waveform (f). Thus we have three carefully timed waveforms (b), (d) and (f) passing into the add matrix. Its output is shown as waveform (g), showing the *pre-shoot* effect of waveform (d) and *overshoot* effect of waveform (f).

The slightly delayed signal now appears to have sharper transitions, and the process works on positive-going steps too. The delay time, represented by the two vertical lines in Figure 18.12 is carefully arranged along with the differentiator time-constants to give optimum correction at a chosen frequency – study of Figure 18.12 shows that above this optimum frequency the crispening circuit actually reduces definition! The degree of pre-shoot and overshoot is sometimes adjustable by a pre-set control labelled *picture soft/sharp*.

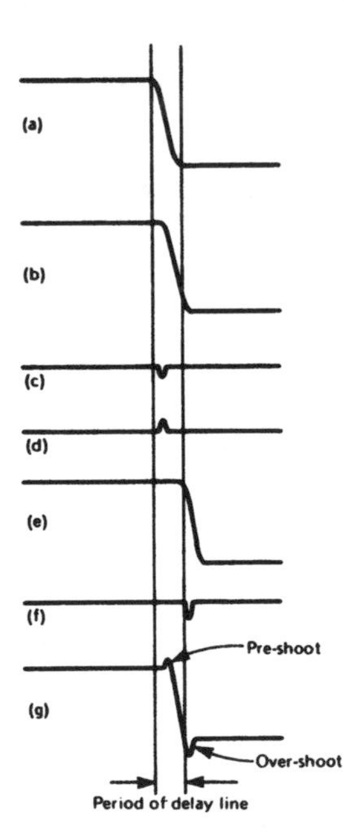

Figure 18.12 *Waveforms for the crispening circuit of Figure 18.13. Their derivation is explained in the text*

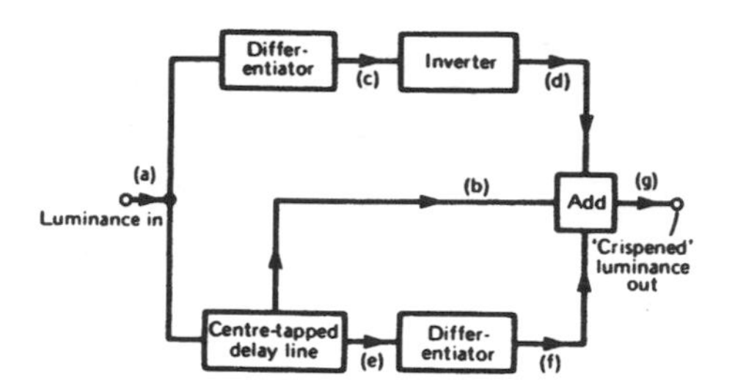

Figure 18.13 *One form of crispening, or aperature-correction circuit*

Luminance delay

The luminance signal processing is now almost complete. Before the playback luminance signal is mixed with the off-tape chrominance signal, however, it is passed through a short delay line. This component serves the same purpose as the luminance delay line in a colour TV, and is necessary because of the differing bandwidths of the luminance and chrominance signal paths. The chrominance signal, being of relatively narrow bandwidth, has a slower rise time than the accompanying luminance signal, and if uncorrected this would have the effect of printing the chrominance component slightly to the right of the luminance image on the TV screen. The delay line, then, slightly retards the timing of the luminance signal to bring the two into step, ensuring correct *registration.*

After amplification, chroma addition, and conversion to low impedance, the vision signal is ready to pass out of the VCR. It is applied to the 'video out' socket in 1 V peak-to-peak form, and also to the RF modulator (along with the sound signal) where it is impressed on to a UHF or VHF carrier for injection into the aerial socket of a TV receiver.

S-VHS

The 'super' variant of VHS format overcomes the definition limitations of conventional formats. Here the FM carrier frequency deviates between 5.4 and 7.0 MHz, a 60 per cent expansion. The luminance sidebands are correspondingly wide; the entire luminance FM 'package' occupies a bandwidth of over 8 MHz, compared with 5 MHz in the original plan. S-VHS cassettes have similar running time and appearance to standard types, but contain cobalt-oxide coated tape and can reproduce pictures with 400-line resolution. A diagram of the S-VHS recording spectrum is shown in Figure 25.3.

Chrominance processing

To understand the colour circuits of a VCR, it is essential to have a working knowledge of the way in which chrominance signals are encoded in ordinary broadcast transmissions. Full details are given in Chapters 3 and 4, and revision of those sections of the book is recommended to readers as a basis for study of the colour section of a VCR. One of the advantages of home VCRs is that at no point is the colour signal demodulated – it remains in subcarrier form throughout the system, giving little opportunity for phase (thus hue) errors.

Phases and vectors

The subject of phase relationships and vector diagrams is not a widely-understood one. We have already touched on the subject of phase in describing the colour-encoding process, and it will crop up many times in our study of VCR chrominance and servo systems. The word 'phase' describes the timing relationship between two waveforms, usually – but not necessarily – at the same frequency. In Figure 18.14*a* is shown a sinewave which we will call the *reference*; *b* shows an identical sinewave with the same timing as *a* and this is said to be 'in phase' with the reference, as shown by the vector diagram alongside. One complete trip round the vector diagram (360°) represents one cycle of the reference signal, so that waveform *c*, being half a cycle behind the reference signal, is phase-retarded by 180°, or half a circle; it is said to be *anti-phase*. It will be noticed that waveform *c* is an 'upside-down' version of the reference *a*, so that inverting a waveform *if it is symmetrical about the zero line over one cycle* (i.e. a sine-wave or square-wave) will give the same effect as a phase change of 180°. The waveforms involved do not have to be sinusoidal of course; *d* shows a square-wave in phase with the reference and *e* a triangular wave in antiphase, or at 180°.

When two waveforms have a phase relationship of 90° or 270° they are said to be in *quadrature* and their vectors are at right-angles to each other; this quadrature relationship is a useful one, because it means that when the reference waveform is at its peak, its quadrature-companion is passing through zero, so that by 'sampling' at the correct instant we can separate the information carried by each of two mixed signals in quadrature – without crosstalk. This, of course, is the principle used in the synchronous demodulators of colour decoders in TV sets. Waveform *f* in Figure 18.14 is in quadrature with the reference *a*.

Rotating vectors

If we introduce a second signal at a frequency slightly lower than that of our reference, the situation will be as in Figure 18.14*g*. Here the waveforms start in phase, but the lower frequency (*g*) will fall behind the reference as time goes on, so its phase lag will increase with time, giving rise to the clockwise-rotating vector shown. After a period, the second signal will be exactly one cycle behind the reference, and the two will be momentarily in phase again, with the vector having completed one full clockwise circle. For a frequency higher than the reference signal, it can be seen that the vector will rotate anticlockwise; in both cases, the speed of rotation of the vector will be proportional to the difference in frequency between the reference and the sample.

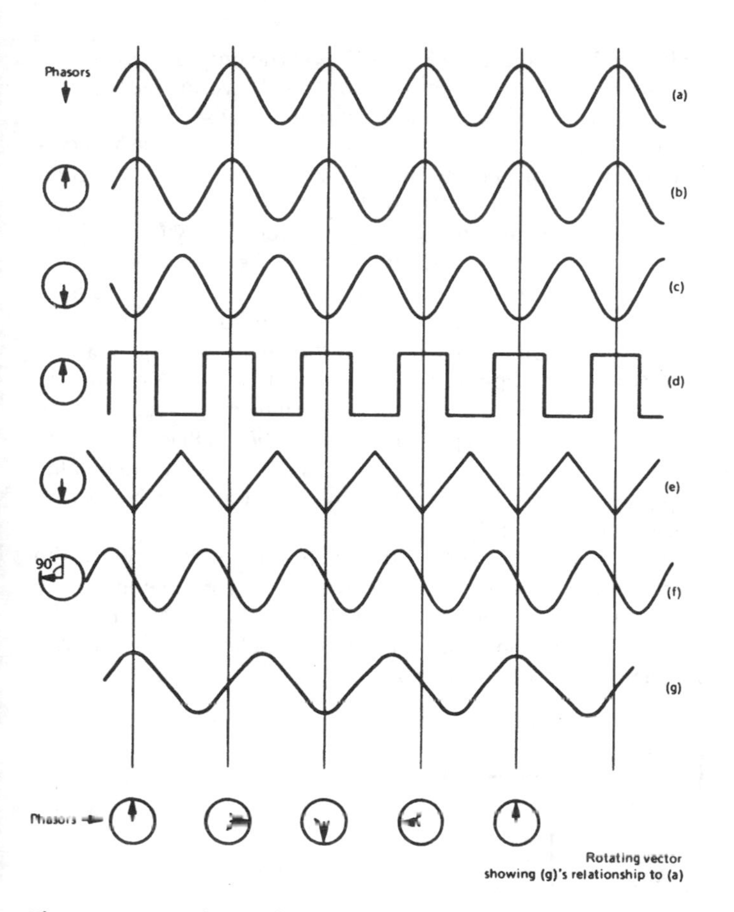

Figure 18.14 *Phase relationships between various waveforms: (a) is the reference and the 'clocks' on the left show vector relationships. Waveform (g) is at a frequency lower than that of the reference (a), giving rise to the clockwise vector rotation shown at the bottom*

If the frequency of the sample is constantly changing with respect to the reference, but by less than one cycle, the result is a 'phase jitter', represented by a constantly-changing vector angle, but confined to less than 360° (one full turn). This is what happens, for instance, in a colour TV system, where the reference signal is a stable subcarrier source, and the 'sample' is the transmitted chroma signal.

Phase-locked loop (PLL)

A common 'building block' in VCR systems is the phase-locked loop. It takes the form of an oscillator and phase detector, and is designed to produce an output signal with a fixed phase relationship to a reference signal; a familiar example is the flywheel line timebase circuit of a TV receiver.

A simple example of a PLL is shown in Figure 18.15. The process starts with a phase detector, or discriminator, consisting of a device with two inputs, X and Y. When the timing of input pulse X is earlier than that of input pulse Y, the detector will produce a positive DC voltage at its output Z. If, alternatively, pulse Y occurs before pulse X, output Z will move negatively. When both input pulses are coincident in time (i.e. *in phase*) output Z is zero. Thus the output Z of the phase detector indicates by its polarity the direction of phase error, and by its amplitude the amount of phase error. Not surprisingly, this is called the error voltage!

The second half of the PLL is a voltage-controlled oscillator (VCO). This is an oscillator whose output frequency – and phase – can be controlled by a DC voltage. It may consist of a crystal oscillator (as in the reference generator of a colour TV decoder), or an LC or flip-flop oscillator. Its output forms one input to the phase detector, and its frequency-controlling input is the error voltage from the phase detector. Thus is the 'loop' set up, and whenever the oscillator and input signals drift apart in timing, the discriminator generates an error signal to pull the oscillator back into phase lock with the control pulse, and again achieve zero phase difference between its X and Y inputs.

The oscillator is then said to be 'slaved' to the reference, or 'master', signal. The oscillator of a phase-locked loop does not necessarily run at the same frequency as the control pulse; with a suitable divider in the feedback path from the oscillator to the phase detector, an output locked to a multiple of the 'master' frequency can be obtained, as shown in Figure 18.16.

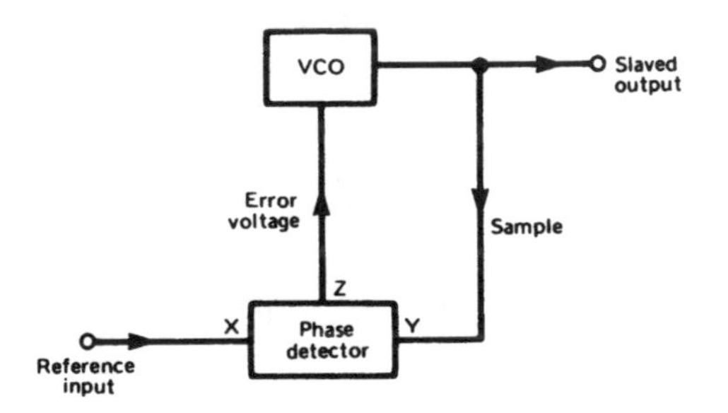

Figure 18.15 *Basic operation of a phase-lock-loop*

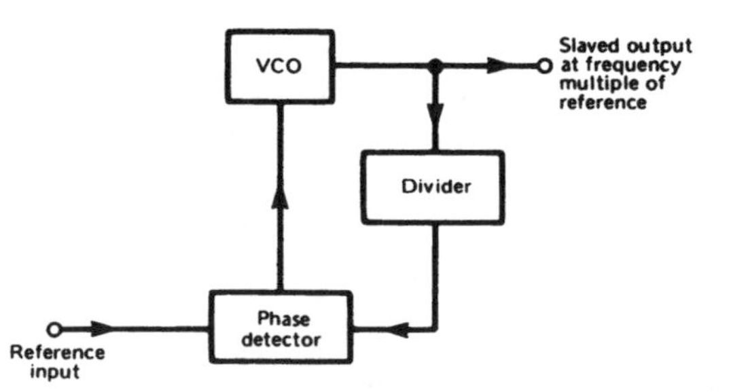

Figure 18.16 *Phase-lock loop with divider. The output becomes locked to a multiple of the reference frequency*

Colour signals in the VCR

If we had a perfect video tape recorder there would be no reason why we could not take the composite video signal, chrominance and all, and record it on the tape as an FM signal, demodulate it during replay and recreate the original video signal as described for luminance in the first part of this chapter. As we've already seen, a home VCR is far from perfect in its performance, and the direct recording idea will not work for two good reasons.

The bandwidth limitations of our VCRs have already been described, and with an upper cut-off frequency around 3.5 MHz, the colour signals at 4.43 MHz are 'left out in the cold' somewhat! Even in S-VHS format, whose signal passband approaches 5 MHz, it is still not possible to directly record an encoded colour signal. Because the VCR relies on mechanical transport, perfectly smooth motion cannot be achieved. Inevitable imperfections in bearing surfaces and friction drives, the elasticity of the tape and minute changes in tape/head contact all combine to impart a degree of timing jitter to the replayed signal. In good audio tape systems this jitter is imperceptible, and for luminance replay its effects can be overcome by the use of timebase correctors in professional machines, or fast-acting TV line-synchronising systems with domestic VCRs. The latter class of machine, in good condition, may be expected to have a timing jitter of about 20 µs over one 20 ms field period. Let's examine the effect of this on the chrominance signal.

We have seen that the hue of a reproduced colour picture depends entirely on the phase of the subcarrier signal. 'Phase' really means relative timing, so if the timing of the subcarrier signal relative to the burst is upset, hue errors will appear. One cycle of subcarrier occupies 226 ns, and a timing error of this order will take the vector right round the 'clock' of Figure 18.17, passing through every other hue on the way! A timing error of 100 ns (one ten-millionth of a second) will turn a blue sky into a lime-green one, and the jitter present in a mechanical reproduction system will make nonsense of the colours in the picture. For acceptable results,

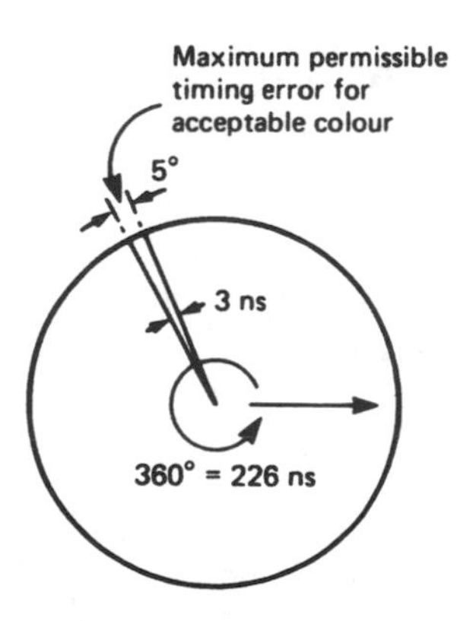

Figure 18.17 *Hue is dependent on phase angle, and only a very small phase change is permissible in a VCR system*

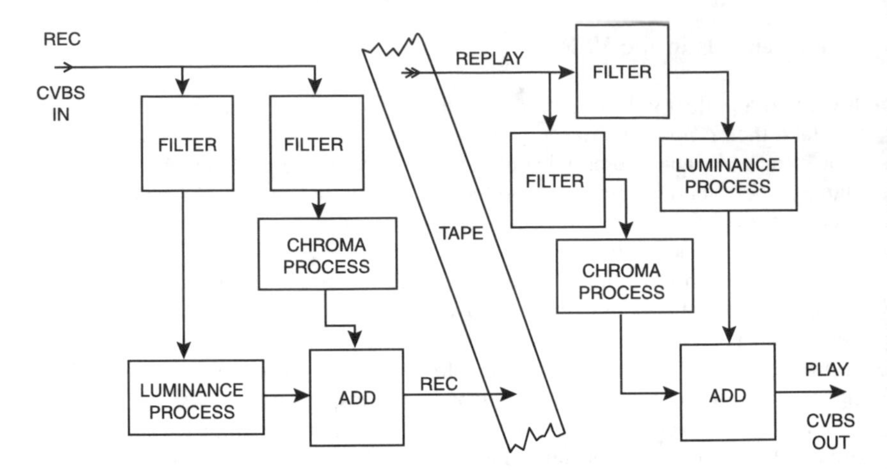

Figure 18.18 *'Streaming' of Y and C components of the video signal for tape recording*

subcarrier phase errors need to be held within 5° or so, representing a permissible maximum timing jitter of ±3 ns as shown in Figure 18.17. The physical causes of jitter cannot be eliminated so an electronic method of subcarrier phase correction is necessary. To achieve this the luminance and chrominance signals are separately treated; Figure 18.18.

Colour under

Let's first examine the problem posed by the high colour subcarrier frequency. Several solutions are possible, including decoding the chrominance signal to baseband (i.e. primary colour or colour-difference signals) before committing them to tape, then recoding them to PAL standard during the replay process. This would require a lot of circuitry, and be vulnerable to hue and saturation errors. A better solution is to keep the PAL-encoded signal intact, bursts, phase modulation and all, and merely convert it to a suitable low frequency for recording. During playback the composite chrominance signal can be re-converted to the normal 4.43 MHz carrier for recombination with the luminance component into the standard form of colour video signal, compatible with an ordinary TV set.

Space has to be found for the colour signal in the restricted tape frequency spectrum, however, and it is made available below the luminance sideband, where a gap is purposely left in the 0–1 MHz region. This area is kept clear by tailoring the luminance record passband and the FM modulator deviation to avoid luminance sidebands falling into it during colour recording. The colour signal, then, is down-converted and bandwidth-limited so that it occupies a frequency band of roughly 0–1 MHz. It is a double-sideband signal, so that the carrier frequency needs to be centred on about 500 kHz and the sidebands limited to 500 kHz or so. The spectrum of the colour-under signal is shown in Figure 18.19; compare this with Figure 18.2.

Down-conversion

The process for down-converting the chrominance signal is shown in Figure 18.20. The double-sideband signal is first separated from luminance information in a high-pass filter, then its bandwidth is limited to around 1 MHz by a bandpass filter. This curtailment of the sidebands of the modulated signal has the effect of restricting the chroma signal detail to around 500 kHz, less than half that of the broadcast signal. The chroma signal now passes into a mixer where it beats with a stable locally-generated CW signal at around 5 MHz. This is the familiar heterodyne effect, and the mixer output contains components at two frequencies, those of the sum and difference of the two input signals. A low-pass filter selects the required colour-under signal, and rejects the spurious HF product.

The colour-under signal contains all the information that was present in the 4.43 MHz chroma signal except the finer colour detail; thus chroma amplitude and phase modulation are preserved, as well as the phase and ident characteristics of the burst, albeit based on a much lower subcarrier frequency. Figure 18.21*a* shows an oscillogram of the chrominance component of a

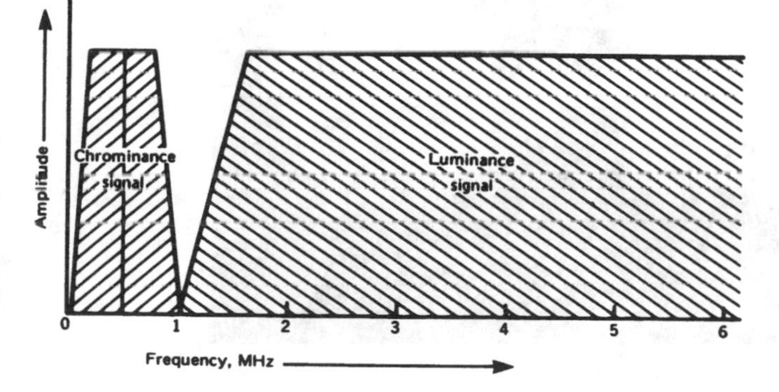

Figure 18.19 *Spectrum of luminance and chrominance signals on the tape. The luminance signal is now in FM form, and the chroma signal still at baseband, but with limited bandwidth and a lower carrier frequency*

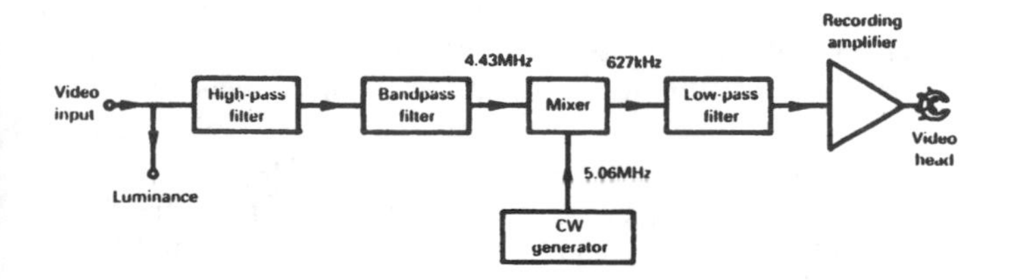

Figure 18.20 *The basic colour-under system. A heterodyne technique is used to convert the chroma signal to a new, low frequency*

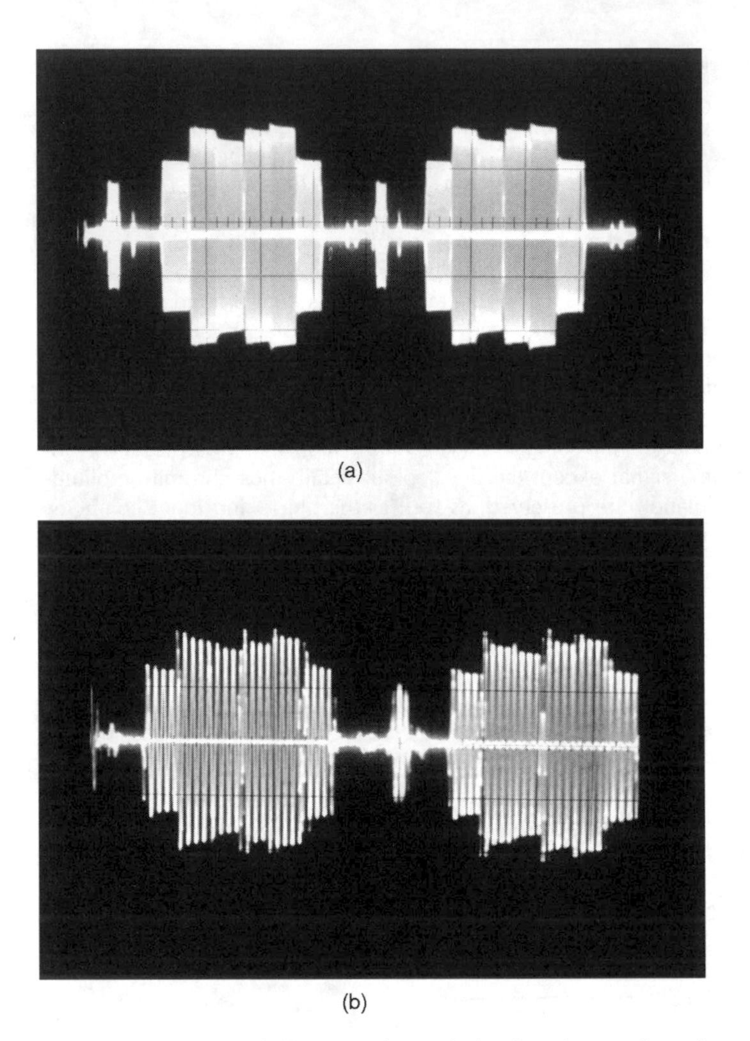

Figure 18.21 *(a) Oscillogram of encoded colour-bars as broadcast, based on a subcarrier frequency of 4.43MHz. (b) down-converted colour-bars, now at a carrier frequency about one-eighth of that in (a)*

colour-bar signal in 4.43 MHz form as broadcast, while in Figure 18.21*b* appears the same signal after down-conversion to a base frequency of around 500 kHz. Looking at the coarse structure of the waveform at *b* it is hard to believe that all information is preserved intact, especially the crucial burst signal, which appears as little more than two cycles of oscillation – it is so, however!

Up to now we have been quoting subcarrier frequencies in round numbers. Format specifications quote the colour-under frequency very precisely, and taking VHS as an example, the local CW signal is at a frequency of 5.060571 MHz, which when beat against the 4.433619 MHz broadcast subcarrier provides a colour under centre frequency of 5.060571 – 4.433619 = 0.626952 MHz, or

626.952 kHz. Due to broadcast subcarrier modulation, sidebands extend for roughly 500 kHz on each side of this carrier.

For reasons which will become clear later, the local CW signal at 5.06 MHz needs to be locked (i.e. have a fixed phase relationship) to the line syncs of the recorded signal and this is achieved by a PLL (phase-locked loop).

Chroma recording

The down-converted colour-signal based (for standard VHS format) on 627 kHz, and containing amplitude and phase modulation, is recorded directly on to the tape. Its amplitude is carefully controlled to achieve maximum modulation of the tape without non-linearity due to magnetic saturation. In this respect it is similar to audio tape recording, with the necessary HF bias being provided by the FM luminance record signal – the two are added in the recording amplifier, giving rise to the waveform of Figure 18.22, where can be seen the recording-head signal for a colour-bar signal. The chrominance signal is superimposed on the constant-level FM luminance carrier, and this composite waveform is fed to the heads for recording on the tape.

Chrominance replay

During playback the chrominance signal appears at the replay amplifier in the colour-under form described above, and with a waveform like that in Figure 18.22. Impressed on to it are the timing variations, or jitter, introduced by the mechanical record and replay transport system and these have to be eliminated. First, though, the chroma signal is separated from the luminance

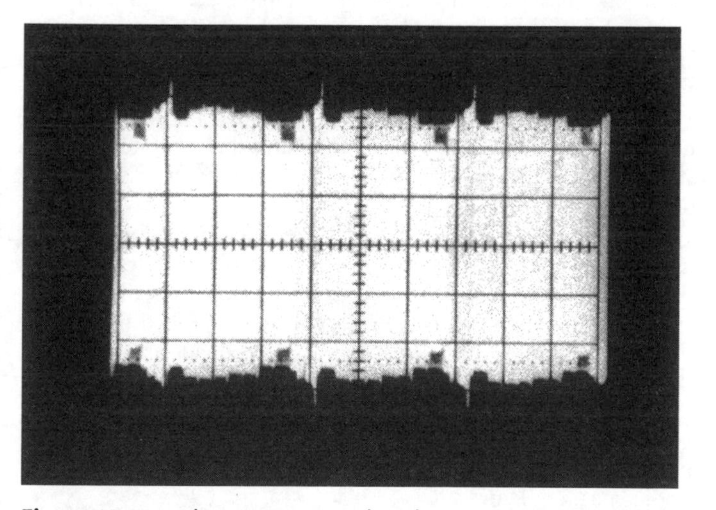

Figure 18.22 *The composite colour bar signal (FM luminance plus down-converted chroma) as applied to the recording heads*

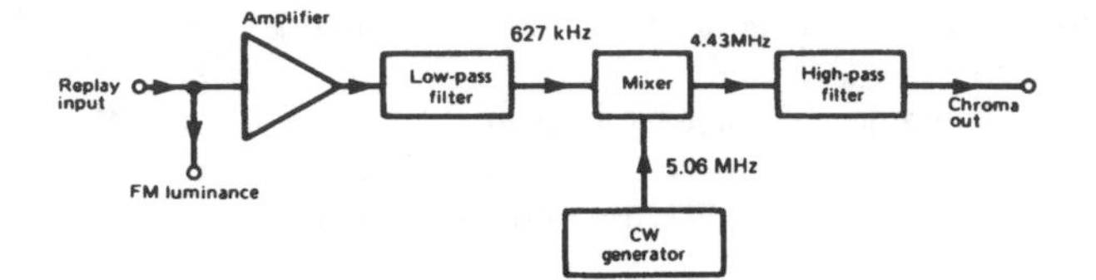

Figure 18.23 *Basic chroma replay process of up-conversion*

FM signal in a low-pass filter. Up-conversion to correct subcarrier frequency (4.43 MHz) must also be carried out, and we'll take this first.

The 627 kHz is heterodyned with a local 5.06 MHz CW signal to produce a beat frequency of 4.43 MHz, selected by a suitable bandpass filter. The idea is shown in Figure 18.23. Dropout compensation and crispening are not relevant to the relatively coarse and 'woolly' colour replay signal, and the recreated 4.43 MHz chroma signal is added to the luminance component and passed out of the machine.

De-jittering

As we have seen, the up-conversion process during replay involves a heterodyne system. If we can ensure that the 'local oscillator' is modulated by just the same timing errors as the off-tape chroma signal, the jitter on the replayed waveform will be cancelled out. The heterodyne process works by producing a *difference frequency* between two inputs (in this case the local 5.06 MHz CW signal and the off-tape chroma signal at 627 kHz). If both inputs contain the same jitter component they will move together in frequency and phase, so that the difference between them will be constant. This 'constant difference' is in fact the 4.43 MHz output signal, which will thus be unaffected by timing errors. On replay, then, the 5.06 MHz CW signal is known as the *jittering reference*. Its derivation will now be described.

The luminance signal recorded on the tape contains line sync pulses, and when these are recovered from the luminance FM demodulator they will contain the same timing errors, or jitter, as the chrominance signal, having been recorded and played back under the same circumstances and at the same time. If we separate them from the luminance part, they can be used to generate a suitable jittering reference. Figure 18.24 shows how. Here we have a stable crystal oscillator running at 4.435 MHz as one input to a mixer, and a 625 kHz signal as the other. The 625 kHz signal is generated by a VCO (voltage-controlled oscillator). Now 625 kHz is exactly 40 times line frequency, so if we divide the oscillator output by 40 in a *counter* we have a line-rate signal suitable for comparison with off-tape line sync in a phase detector. This is another example of a PLL and its effect is to impress the jitter signal on to the VCO output. This jittering 625 kHz signal when beat against the stable 4.43 MHz crystal output will produce a jittering 5.06 MHz reference as required.

Let's take an example to illustrate the action of the circuit. Assume that a timing error has momentarily increased the off-tape chroma carrier frequency by 40 Hz,

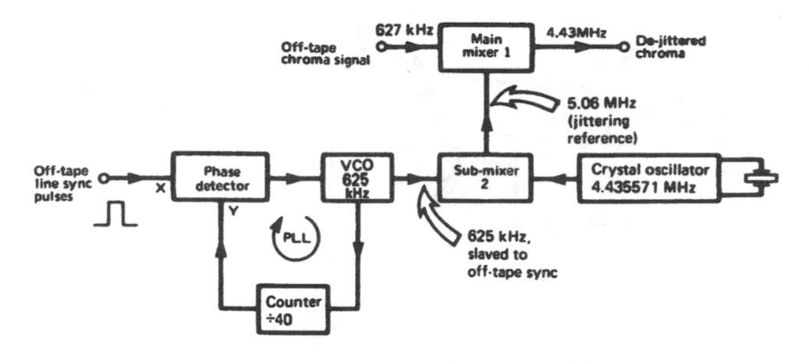

Figure 18.24 *A jittering reference generator using off-tape line sync and a phase-lock loop*

bringing it up to 626.992 kHz. Because colour-under frequency is 40 times line rate, the off-tape line sync pulse frequency will simultaneously rise by 1 Hz to 15.626 kHz, and this will appear at point X on Figure 18.24. The PLL will push up the VCO frequency to rebalance its phase detector and achieve 15.626 kHz at its Y input, and VCO output will rise to 625.040 kHz. The sub-mixer 2 will beat this against the steady 4.435571 MHz crystal frequency to produce an output of 5.060611 MHz. This passes into the main mixer 1 where it beats with the off-tape chroma carrier at 626.992 kHz to produce a final output frequency of 5.060611 − 0.626992 = 4.433619 MHz as was present with no timing error. Although this example describes a *frequency* error, phase errors (timing errors of less than one down-converted subcarrier cycle) are dealt with in the same way.

It is now plain why the colour-under signal is phase-locked to off-air line sync during record – the sync and colour signals are being recorded simultaneously, so that jitter cancellation works for both record and playback processes, the key in each case being the closely accompanying line sync pulse. The same PLL and 40-times divider is used in record and playback. The block diagram of Figure 18.25 shows the basic processes in chrominance recording, playback and jitter-compensation in a domestic VCR.

There are several other methods of deriving a jittering reference, all involving an off-tape signal of one kind or another. In the *burst-lock* system the colour-burst signal itself is used, being gated out of the replay chrominance signal and used to slave a local oscillator. Another system uses a pilot tone, laid down on the tape during record, to achieve the same end. Home formats, however, use the line-sync reference system described above, in conjunction with a 'tighter' jittering reference derived from the off-tape chroma sub-carrier signal itself.

ACC and colour-killers

Colour TV sets incorporate an ACC (automatic colour control) system to prevent tuning and propagation errors upsetting the luminance/chrominance ratio. This is necessary in a VCR for the same reasons, so that during record the ACC circuit monitors the amplitude of the colour-burst signal (transmitted at a constant level)

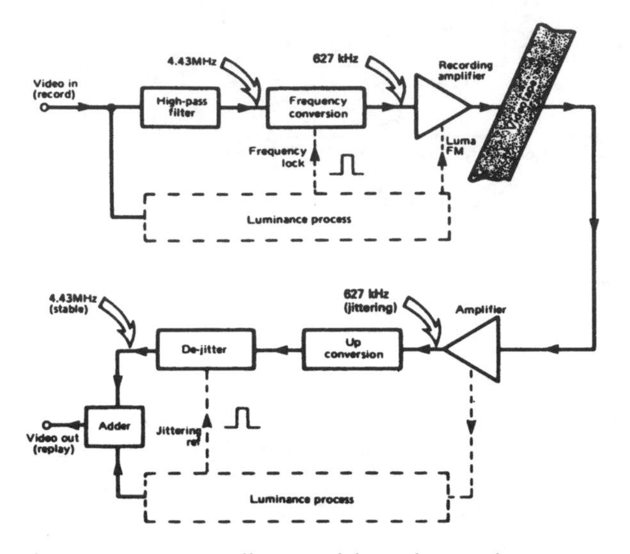

Figure 18.25 *Overall view of the colour-under system*

and adjusts the gain of the chrominance record amplifier to maintain it at a constant level; by these means the chrominance signal itself is held at correct amplitude. To prevent the recording of *confetti* or unlocked colours during monochrome transmissions or when the record down-conversion PLL is unlocked, a record colour-killer is provided.

On replay, too, these functions are required. The off-tape chrominance signal is vulnerable to level fluctuations and these are ironed out by the ACC loop. The replay colour-killer acts to prevent spurious colouring of replay pictures from monochrome videotapes, and shuts down the chrominance channel in the event of malfunction of the replay up-conversion processes.

Crosstalk compensation

In Chapter 17 we saw that early machines used a guard band between recorded tracks on the videotape to prevent the possibility of crosstalk between adjacent tracks. Later formats eliminate this guardband and achieve greater information density on the tape by laying down tracks with no intervening safety gap. To prevent crosstalk between tracks each head is given an azimuth offset; this ensures that readout from adjacent tracks is at a very low level and does not interfere with reproduction. As Figure 18.26 shows, however, the azimuth-offset idea is less effective at low frequencies, and the 0–1 MHz region occupied by the chroma signal will suffer unacceptable crosstalk between tracks unless steps are taken to remove the interfering signal.

We cannot prevent inter-track crosstalk taking place at colour-under frequencies, so methods of electronic cancellation of the unwanted signal have been devised. Each format uses the same basic idea but differs in the way that it is carried out. Let's look at a typical cancellation system, that used in the VHS format, to explain the principles involved.

The interlaced scanning system employed in broadcast TV calls for a half-line offset between the starting points of odd and even fields, as explained in Chapter 1. If the video tracks were laid down on the tape in this form, there would be a half-line offset between the recorded TV lines, just like that shown on the TV screen in Figure 17.14. However, it is not that simple, as the video tracks are laid on the tape at a slant angle, and the tape moves horizontally past the drum. The combined effects of these result in the individual lines of each TV field being laid exactly alongside each other, as shown in Figure 18.27. This ensures that when

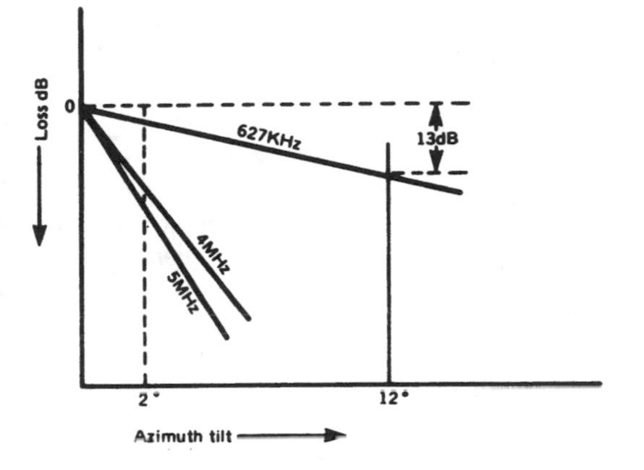

Figure 18.26 *The effects of azimuth tilting at different frequencies. LF signals have low immunity to crosstalk*

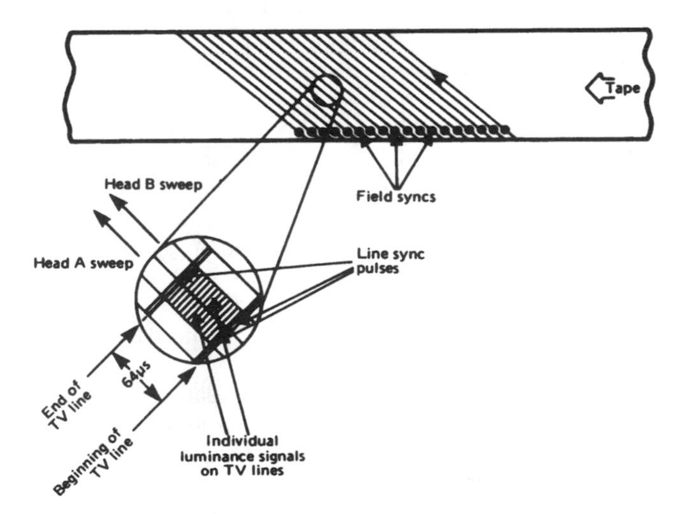

Figure 18.27 *The tape tracks are arranged so that individual TV lines lie physically adjacent on the tape*

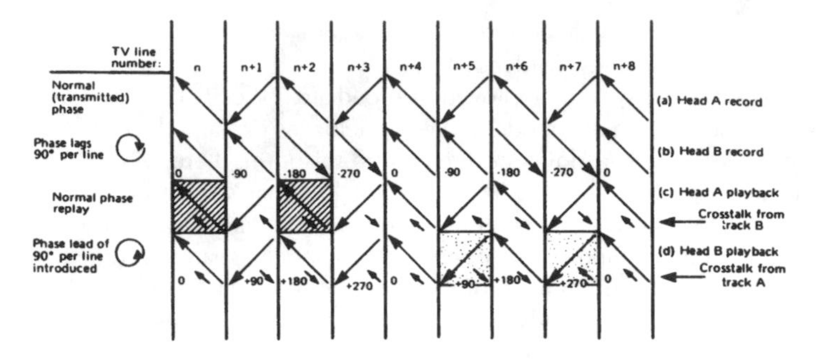

Figure 18.28 *Burst phasor diagram for the VHS colour-under system. Lines (c) and (d) can be seen to emerge with correct phasors as broadcast*

crosstalk occurs the interfering signal is coming from the 'same' line on the partnering field, and has the *same timing*, within that line, as the wanted signal. Thus there will be a fixed phase relationship between wanted and crosstalk signals, bearing in mind that the colour-under signal is based on a frequency (for the VHS system) of 626.9 kHz.

During record, head A is fed direct by the colour-under signal, and the PAL characteristic of this signal will result in the burst phases being laid down on the tape according to (a) in the chart of Figure 18.28. Thus TV line n has its burst at 135°, line $n+1$ at 225°, line $n+3$ 135° and so on. The chroma signal fed to head B is *delayed* by 90° on each successive TV line as shown in line (b) of the chart. Thus line n is 'normal', line $n+1$ has a 90° clockwise shift, line $n+2$ is rotated 180° clockwise, line $n+3$ 270° clockwise, line $n+4$ normal, and so on throughout the field period. During replay, head A will pick up a large signal from its own track and a smaller crosstalk signal from the adjacent tracks laid down by head B, with burst phases as shown in line (c) of the chart. To recover a 'normal' signal from head B during replay, we have to *advance* the phase of its output by 90° per TV line, to compensate for the 90° retard per line introduced during recording. Thus head B's corrected output will be line (b) on the chart, but phase advanced by 90° per TV line, as shown on chart line (d). The small arrows here show the crosstalk signals picked up from track A, *having undergone the same 90° per line phase advance*. Study of playback lines (c) and (d) of the chart will show that if we compare *any* playback TV line with its next-but-one neighbour, the main signal vectors add (i.e. they are pointing in the same direction), whereas the crosstalk vectors (small arrows) cancel. This is illustrated by the hatched boxes for head A playback (TV lines n and $n+2$) and the dotted boxes for head B playback (TV lines $n+5$ and $n+7$).

The reason for the carefully-contrived head B recording sequence is now clear. To eliminate the crosstalk signal it is only necessary to *add* each TV line to its next-but-one neighbour, which can be brought into time coincidence by a two-line (128 µs) delay, then added in a suitable matrix, as illustrated in Figure 18.29. The process is very similar to that which takes place in the familiar delay line circuit of a colour TV receiver, and the delay line is similar in construction (but with twice the path-length) to its TV counterpart.

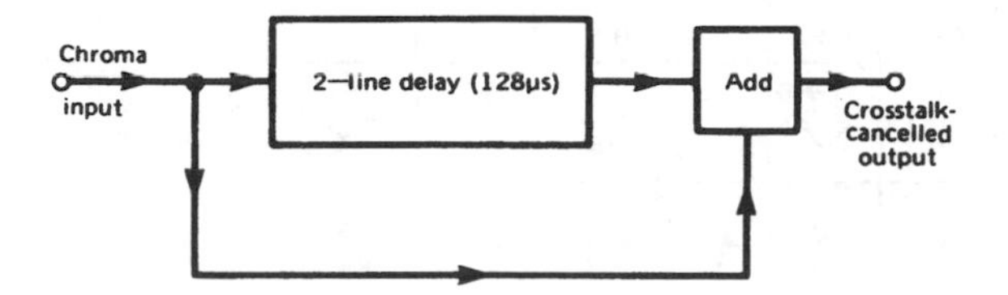

Figure 18.29 *Delay line matrix for chroma crosstalk cancellation*

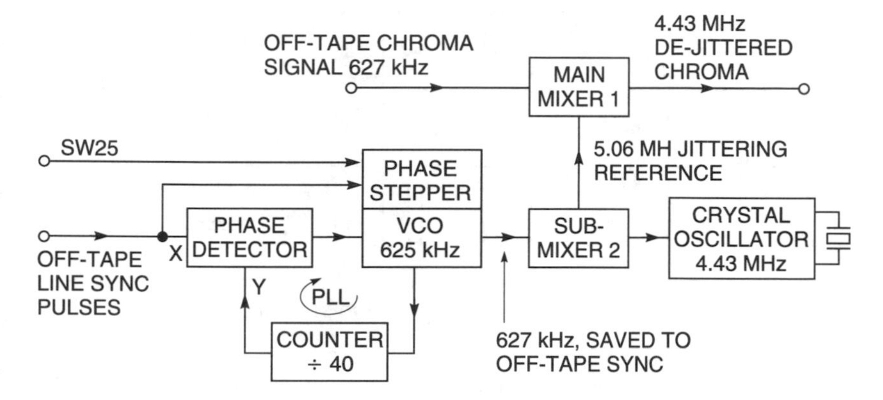

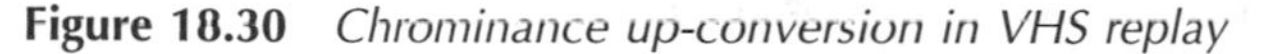

Figure 18.30 *Chrominance up-conversion in VHS replay*

We have described the crosstalk-compensation process in terms of burst phasors, whereas our real concern is eliminating crosstalk between the chrominance *signals* on adjacent tracks. Since the chroma signal is defined (in its encoded form) in amplitude and phase, the crosstalk compensation system works equally well on the subcarrier signal as on the burst itself.

The four-stage phase switching for head B is carried out in the phase-stepper block of Figure 18.30, which is an enlargement of previous Figure 18.24. The SW25 (head-switch) squarewave resets chroma record phase to normal (0° shift) for each head A scan, while the incoming line sync pulses govern the phase-stepping for the B head. During replay mixer 1 is used to up-convert the off-tape colour signal to 4.43 MHz, and the same phase-stepper now acts, via sub-mixer 2, to restore correct phase to the colour vectors as shown in lines (c) and (d) of Figure 18.28.

Video-8 format chrominance

The basic principle of the colour system of Video-8 format is the same as for VHS already described, in that the chroma phase recorded on tape is manipulated to ensure that crosstalk signals come off the tape in antiphase over a two-line period. The colour-under frequency for V8 is $(47 - \frac{1}{8})\ f_h$, which is 732 kHz. It is locally generated, but in order to implement de-jittering during playback, is locked to incoming line sync in a PLL incorporating a ÷375 stage. We start, then, with

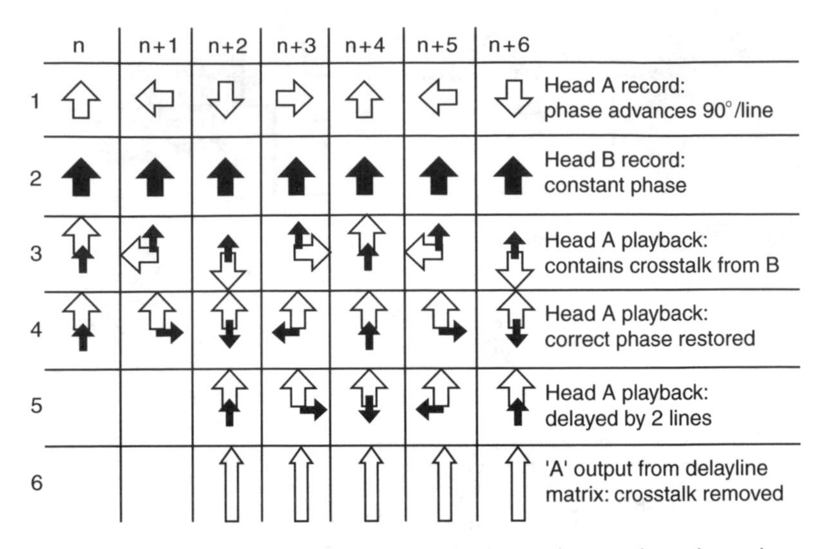

Figure 18.31 *The stages in removing crosstalk interference from the Video-8 chroma signal. The 'n' numbers refer to television scan lines*

$f_h \times 375 = 5.86$ MHz. This is divided by eight in a counter to render 732 kHz; the counter is under the influence of drum flip-flop and line-rate input pulses so that the phase of the colour-under carrier is advanced by 90° per line for head A's sweeps only; see row 1 of Figure 18.31. Head B's chroma record signal is not phase-changed, as row 2 of Figure 18.31 shows.

During replay the chroma signals are read out from tape according to row 3 in Figure 18.31: this shows the output from the Ch1 (a) head, and represents the main signal as recorded, together with a crosstalk contribution from adjacent B tracks, shown as small black arrows. To cancel the effect of the phase advance given to head A's chroma signal during record a corresponding phase-retard of 90° per line must be imparted to it during playback. This has the effect of restoring phase normality to the chroma *signal*, but introducing a 'twist' to the crosstalk component as shown in row 4 of the diagram. The process has not removed the crosstalk, but has paved the way for its cancellation.

Row 5 of Figure 18.31 shows the new timing of row 4 after its passage through a two-line delay system. Adding the delayed and non-delayed signals, so long as their timing is exactly right, gives addition of the wanted (in-phase) chroma signals, but complete cancellation of the crosstalk signals which are now in opposite phase for every line. The resultant 'clean' chroma signal is shown in the bottom row. Although we have only illustrated here the situation for a pure red signal recorded by head A, the same crosstalk-cancellation mechanism works for any colour, and also for head B's signals from which crosstalk from A tracks is removed.

The only time the system breaks down is when one line is markedly different in hue from its next-but-one neighbour. Under these circumstances of colour changes at horizontal edges in the picture the crosstalk-compensation system will introduce hue errors centred on the point of transition, the effect being worse than complete loss of crosstalk cancellation. To prevent this the chroma two-line delay circuit is governed by a *correlation detector* which looks for large disparities in hue over a two-line period. When they are detected the effect of the chroma delay line is cancelled.

Advanced techniques are applied elsewhere in the Video-8 colour circuitry. The burst signal, a crucial reference for both VCR and TV, must be kept in the best possible condition. To give it better immunity from tape-noise its amplitude is doubled during record, and restored to normal during playback. In addition to this *burst-emphasis*, the chroma signal itself is treated for noise-reduction in a frequency-conscious amplifier which gives increasing emphasis to the outer 'skirts' of the chrominance sidebands. Corresponding de-emphasis during playback gives a useful overall reduction in 'confetti' on the monitor screen. Indeed, these record techniques ensure that most chroma signals on tape, regardless of their amplitude in the 'real' signal, are recorded at high levels, clear of the 'noise floor.'

Chrominance definition of replayed pictures

We saw at the beginning of this chapter that the bandwidth of the chrominance signal is restricted during its passage through the recording system of a domestic VCR, resulting in a reduction of horizontal resolution (the ability to define vertical coloured edges) by about 50 per cent. The effect of the delay line matrices in the chroma crosstalk compensation system is to impair the *vertical* chrominance resolution, and the ability of the system to define horizontal coloured edges (vertical definition) is in fact reduced by a factor of eight overall. This has the effect of rendering roughly equal the vertical and horizontal chrominance definition of the displayed picture on replay, and is considered acceptable. There is little point in maintaining good vertical resolution when horizontal definition is restricted, though this is done in the broadcast signal, where vertical resolution is governed by the number of TV lines in the picture, whereas horizontal resolution depends on chroma channel bandwidth: 1.2 MHz or thereabouts in the UK system.

PAL encoding and crosstalk compensation

Although we have used the PAL chroma and burst vectors to illustrate the principle of the chrominance crosstalk compensation system, the PAL characteristic of the encoded chroma signal takes no part in the crosstalk compensation process. Thus the cancellation systems described above work equally well on any form of colour-encoding system. Because the PAL signal (and other line-alternating colour-encoding systems such as SECAM) has a two-line pattern, the crosstalk compensation is carried out on a two-line basis to retain the swinging burst (and chroma V) characteristics. With a simple encoding system such as NTSC, a one-line crosstalk cancellation system would be possible, conferring the advantage of better vertical chrominance resolution.

Signal processing chip

Throughout this chapter we have used block diagrams to illustrate the conversion and conditioning of the luminance and chrominance signals for their recording on tape. In practice now, two ICs are involved in providing all the processing described in this chapter: a head drive/preamplifier chip mounted in or close to the head drum; and a 'Y/C' chip on the main PC board. A typical chipset is shown in Figure 18.32, (a) for record and (b) for playback. Looking first at diagram (a),

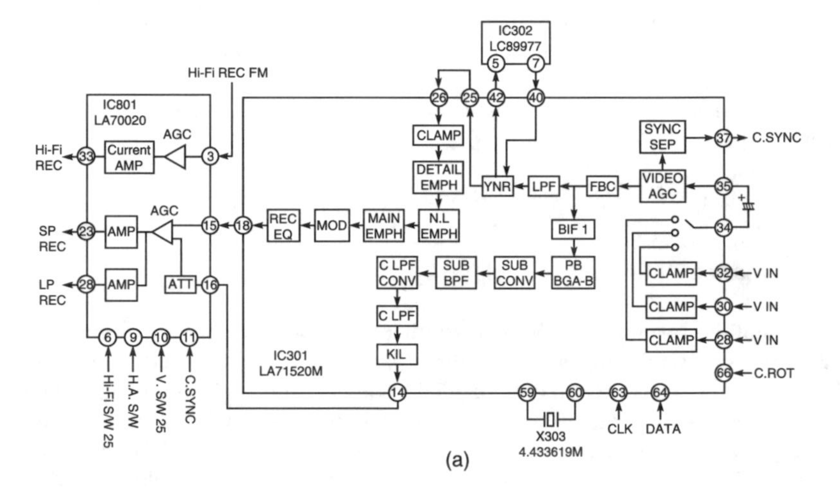

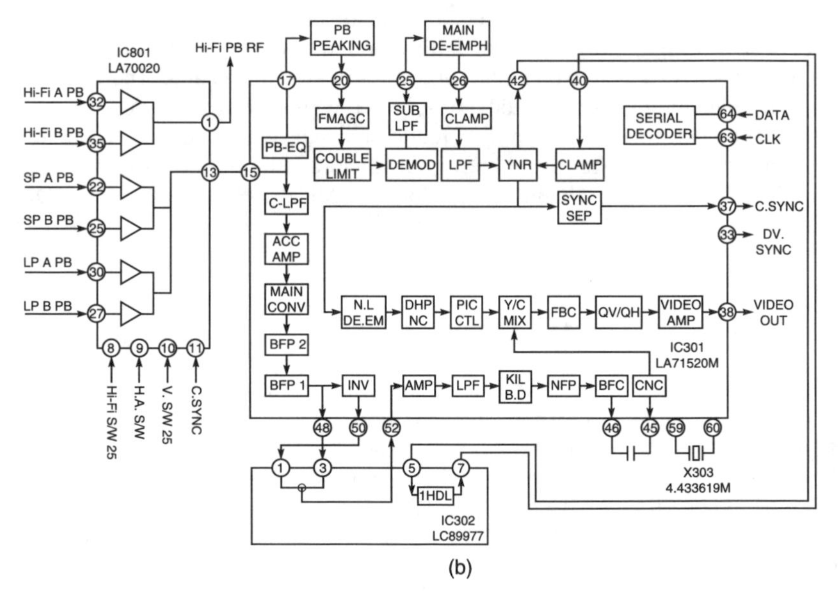

Figure 18.32 *Luminance processing, record mode at (a), playback mode at (b) (LG Electronics)*

there are three video inputs, selected by a switch at chip pin 34 under bus control. Luminance noise reduction uses an electronic 1-line signal delay in IC302 at main chip pins 40/42, after which the Y carrier signal passes through three stages of emphasis on its way to the modulator. Emerging at IC pin 18, the carrier enters IC801 on pin 15 for application to the heads. The chrominance signal undergoes the processes described earlier to pass from pin 14 of IC301 to pin 16 of IC801, where it meets the luminance carrier en route to the video heads. IC801 also handles the drive to the Hi-Fi audio heads, whose operation we shall meet in the next chapter.

Turning now to the same chipset working on replay, Figure 18.31(b), we can see on the left that this VCR is a dual-speed type with separate video head-pairs for SP and LP. The selected FM envelope signal passes into IC301 on pin 15 and splits two ways: up to the luminance limiter, demodulator and noise reduction stage (again using delay chip IC302); and down to the chroma up-converting stage, bandpass filters etc. Luminance and chrominance are reunited in the Y/C mix block to emerge from the chip as a composite video/chroma signal at IC pin 38. The colour phase rotation switching pulses enter IC301 at pin 66 (diagram a); the colour crystal is hooked to pins 59 and 60, while an I^2C bus (see Chapter 8) passes user- and system-control commands into the IC on pins 63 and 64. Two separate SW25 pulse trains enter head amp/switch IC801, one for video and a second for audio, as will be described in the next chapter.

19 *Signal processing: audio*

For several years the audio system used in domestic VCR formats depended on a longitudinal track laid along the extreme top edge of the ribbon. Figure 17.5 shows its position, and Figure 17.9 the placement of the stationary head which writes and reads the audio track. One of the virtues of home VCRs is their low tape consumption, but this depends on a slow tape speed, much slower for instance than an audio compact cassette. The result is a constriction on audio frequency response, particularly in LP modes where the rate of progress of the tape is halved.

The noise performance, too, of a longitudinal sound recording on video tape falls far short of other sound systems. This stems partly from the low tape speed (massive equalisation must be applied to the off-tape signal to maintain flat frequency response) but mainly from the narrowness of the tracks necessarily used. The standard longitudinal track width is 1 mm. Figure 19.10 highlights the shortcomings of VCR longitudinal track performance, and compares them with the results obtained from the new systems to be described now.

FM audio

The ingredients for success in recording TV pictures on tape are high writing speed and FM modulation as we saw in Chapters 15 and 16. Within the immovable constraints of existing formats it was obvious that for better sound performance these virtues must also be applied to the audio signal. The obvious solution was to record an FM-modulated sound carrier in or around the helical video tape tracks; but with the vision heads and tracks already chock-full of information (see Figures 17.31 and 18.19) the problem was where to squeeze in the audio recording with regard to frequency-spectrum space and the 'magnetic' capacity of the tape track. Some form of multiplex system is required. We have already met this in the description of MAC TV systems in Chapter 10, where a method of getting several streams of information through a single channel path was found in time-compression and time-division-multiplex. A similar technique is in fact used in the PCM sound systems to be discussed later in this chapter. All TDM setups require signal storage at each end of the link, however. Two alternative systems emerged for home VCR formats – *depth multiplex,* used in VHS models; and *frequency multiplex* in Video-8 format. The former is the most common, and will be covered first.

Depth-multiplex audio

This technique depends to some degree on the magnetic layer of the tape itself to discriminate between the video and audio signals. A separate pair of heads on the spinning drum is provided solely for the audio signal FM carriers. A typical layout of heads on the drum is shown in Figure 19.1. The two audio heads are mounted 180° apart and are arranged to 'lead' the video heads. The gaps cut in the audio heads have large azimuth angles: ±30°, sufficient to prevent crosstalk from adjacent tracks at the carrier frequencies (around 1–2 MHz) involved. The mounting height of the audio heads is set to place the hi-fi audio track in the centre-line of the corresponding video track. This track layout is shown in Figure 19.2, which also gives an idea of the relative audio track widths; the audio tracks are half the width of the vision tracks they accompany.

Figure 19.3 gives an idea of the depth-multiplex principle. Here the heads are moving towards the right across the tape, led by the audio head which has a relatively wide gap. The effect of this is to write into the tape a deep magnetic pattern penetrating some microns into its magnetic coating. Shortly following the audio head comes the regular video head, whose gap is in the region of 0.25 micron. Writing (for the most part) at higher frequencies, it creates shallower,

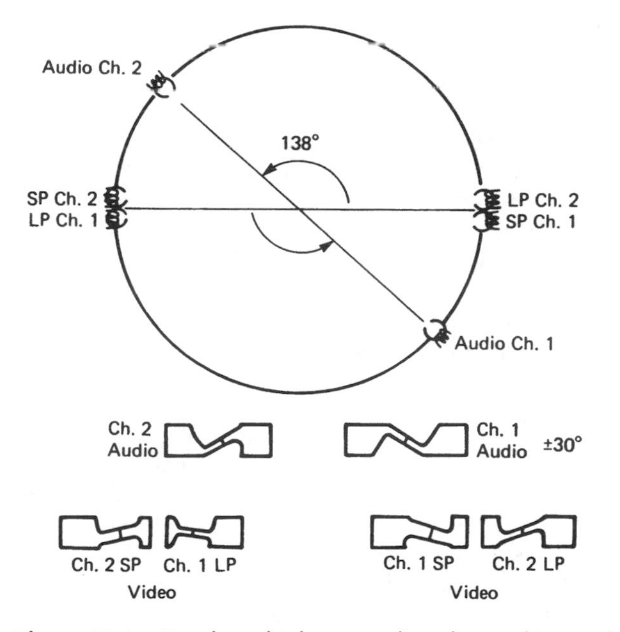

Figure 19.1 *Depth-multiplex recording drum. This one has the audio heads mounted 138° ahead of the double-gapped video heads*

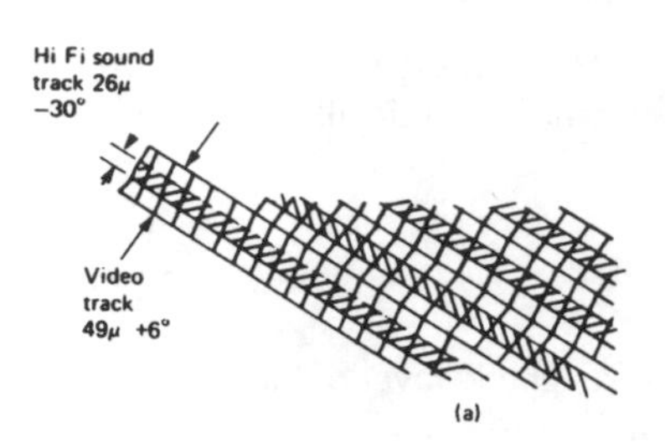

Figure 19.2 *Video/sound track relationship for VHS*

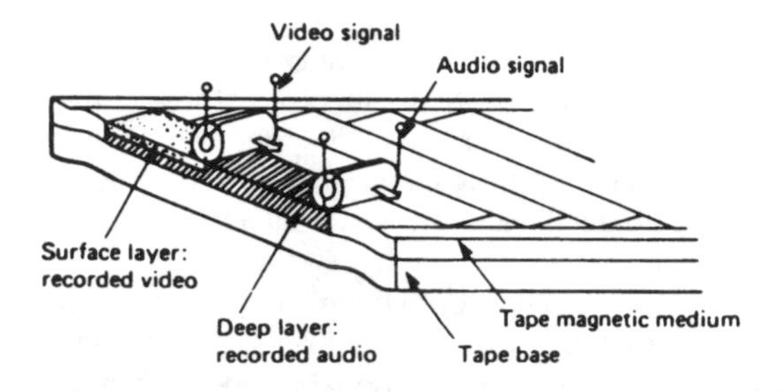

Figure 19.3 *Depth-multiplex principle: separate magnetic layers are written into the tape*

shorter magnetic patterns which penetrate less than 1 micron into the magnetic surface. Thus the recorded tape (top LHS of Figure 19.3) contains a 'two-storey' signal: a buried layer of long-wavelength audio patterns under a shallow top layer of video patterns.

During replay the same heads operate on the same tracks as before. The video head picks off its track with little impairment, and with minimum crosstalk from sound tracks because of the large disparity (36° for VHS-SP) between the azimuth angle of recorded track and replay head. The audio head during replay is handicapped by the barrier presented by the 'video layer' on tape, but the resulting 12 dB or so of attenuation – thanks to the use of bandpass filters and the noise-immunity of the FM carrier system – does not prevent noise-free reproduction of the baseband audio signal as long as the tracking is reasonably correct. For hi-fi sound the tracking performance is critical, especially in LP modes where narrow tracks tend to magnify any errors.

As with the video heads, switchover between the two rotating audio heads is carried out during the overlap period when both heads are momentarily in contact with the tape, one just leaving the wrap and one just entering. Because of the angular offset between video and audio heads on the drum (Figure 19.1) the head switchover point for the latter is set by a second, *delayed* head flip-flop squarewave triggered from the drum tacho-pulse. The FM carrier signals to and from the audio heads on the drum require their own rotary transformers, which may be concentric with those for the video signals under the drum; or mounted above as in the photo of Figure 19.4.

Figure 19.4 *Head drum mounting in a Panasonic hi-fi VHS machine. The rotary transformer for audio signals is above the drum*

The frequencies used for VHS stereo hi-fi in depth-multiplex systems are 1.4 MHz (L) and 1.8 MHz (R); the FM modulation sidebands extend for about 250 kHz on each side of the (unsuppressed) carriers. Each audio head deals with *both* FM carriers throughout its 20 ms sweep of the tape – during replay the carriers are separately intercepted by bandpass filters for processing in their own playback channels.

Frequency-multiplex audio

An alternative approach, used in Video-8 VCRs, to helical sound recording is to use the *video heads* to lay down on the tape an FM audio soundtrack.

The baseband frequency response is limited to about 15 kHz. As Figure 17.31 shows the audio FM carrier is based on 1.5 MHz and has a deviation-plus-sideband width of about 300 kHz. It is possible to insert this 'packet' between the outer skirts of the upper chrominance and lower luminance sidebands, permitting it to effectively become part of the video signal so far as the heads and tape are concerned. The FM audio signal carrier is added to the FM luminance signal in the recording amplifier, and laid on tape at a low level – some 18 dB below that of the luminance carrier. This suppression of sound carrier level helps prevent mutual interference between sound and vision channels, whose outer sidebands overlap to some degree.

The original AFM recording system made provision only for monaural sound. The advent of hi-fi stereo recording in the competing VHS format led to the adoption of a stereo AFM plan for Video-8 formats, using a second sound carrier at 1.7 MHz. Unlike VHS hi-fi, however, it is not possible here to simply use one carrier for each channel because compatibility with mono AFM equipment must be maintained. The stereo-AFM system is illustrated in Figure 19.5. Incoming L

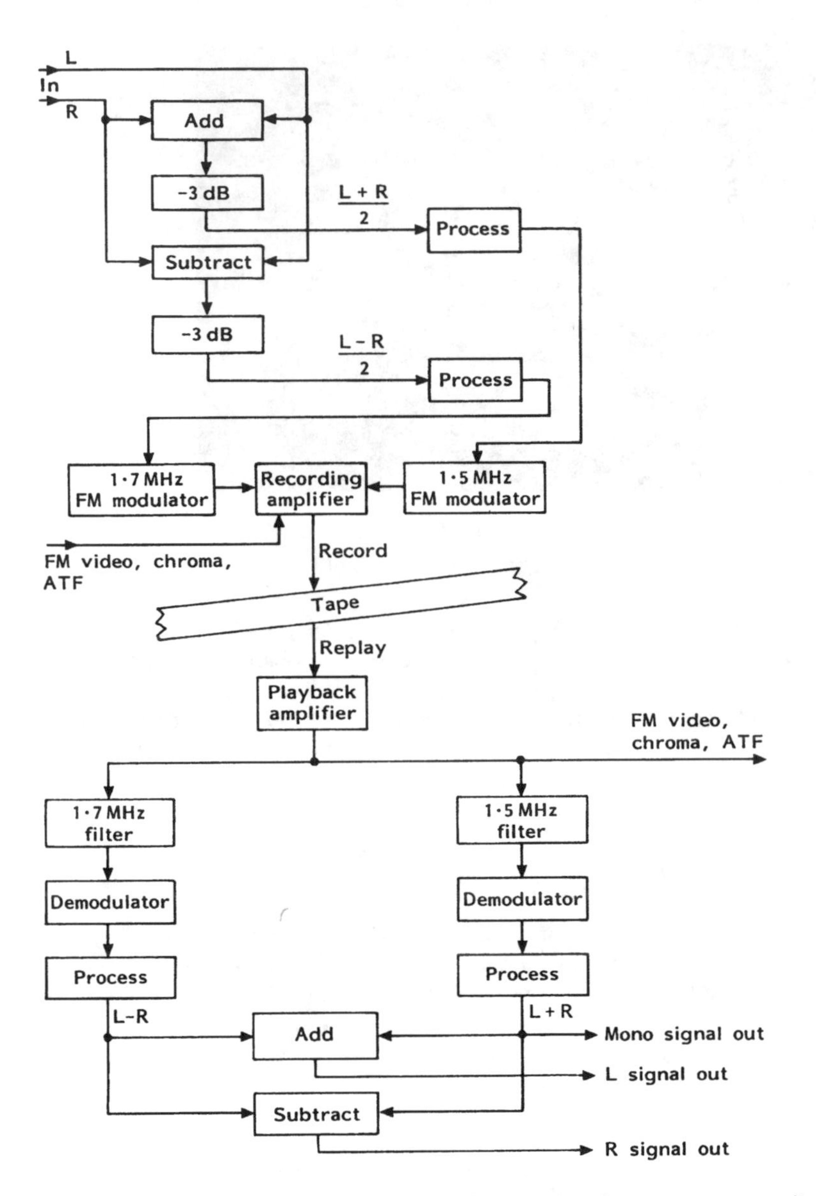

Figure 19.5 *Record and playback signal processing for AFM stereo audio in Video-8 format*

and R audio signals are brought together in an adder, and the result halved in an attenuator to derive an (L+R)/2 signal for FM modulation and recording on tape at 1.5 MHz. This forms the compatible mono signal. The L and R record signals are also routed to a matrix in which they are *subtracted* to produce an L – R signal, which is now halved to render an (L – R)/2 signal. It is this which is processed and frequency-modulated onto the 1.7 MHz carrier. During replay the recovered L + R and L – R signals are added and subtracted in separate matrices to derive L and R signals. Readers who are familiar with VHF-FM analogue stereo sound broadcast/reception techniques will recognise this 'stereo-difference' technique – in radio broadcasts the difference signal is conveyed in a subcarrier at +38 KHz.

Playback carrier routing

In replay mode the separation of the various off-tape signals in Figure 17.31 is carried out by four separate filters: a low-pass type with cutoff around 180 kHz for ATF tones; a low-bandpass filter centred on 732 kHz for colour-under signals; a narrow bandpass one tuned to 1.5 MHz for interception of the audio-FM carrier; and finally a high-pass (roll-on about 1.7 MHz) acceptor for the FM vision signal with its sidebands.

Audio electronics

In 'longitudinal' audio systems the only processing necessary for the audio signal itself is some recording equalisation and the addition of a bias signal as described in Chapter 16. For hi-fi recording a great deal more work must be done on the input signal to condition it for multiplex recording, and to reduce the overall system noise level to a point far lower than can be achieved with conventional sound recording systems on tape or disc. The techniques used are almost identical for depth-multiplex and frequency-multiplex systems. In both cases much of the audio processing circuitry is common to both record and replay.

Figure 19.6 shows, in generalised form, the arrangements for a single channel of hi-fi sound processing. In record the audio signal (left of diagram) is first passed through a high-cut filter to remove supersonic frequencies, then via switch S1 to the non-inverting input of an operational amplifier. The op-amp output (ignoring for a moment S3) takes two paths: one through a pre-emphasis 2 network to the input of a VCA (voltage controlled attenuator); and one via a weighting filter to a precision RMS-level detector. The output of this device reflects, from moment to moment, the true RMS (root-mean-square) amplitude value of the signal it sees. The RMS detector output is now applied as a control potential to the VCA. Thus the output from the VCA is large for high-level audio signals, small for low level signals. It is routed via S2 to the inverting input of the op-amp. The effect of this variable negative feedback is to 'compress' the dynamic range of the audio signal emerging at the op-amp output. The compression ratio is 2:1, expressed in decibels on the left-hand side of the diagram in Figure 19.7. The amplitude-compressed signal, since it will be conveyed by an FM carrier, has carefully-tailored pre-emphasis characteristic imparted by the emphasis-1, emphasis-2 and weighting filter sections in Figure 19.6.

Continuing with that diagram, the compressed and pre-emphasised audio signal now passes through S3 to a limiter whose purpose is to clip any signal excursions that may cause over-deviation in the FM modulator. Next follows an attenuator

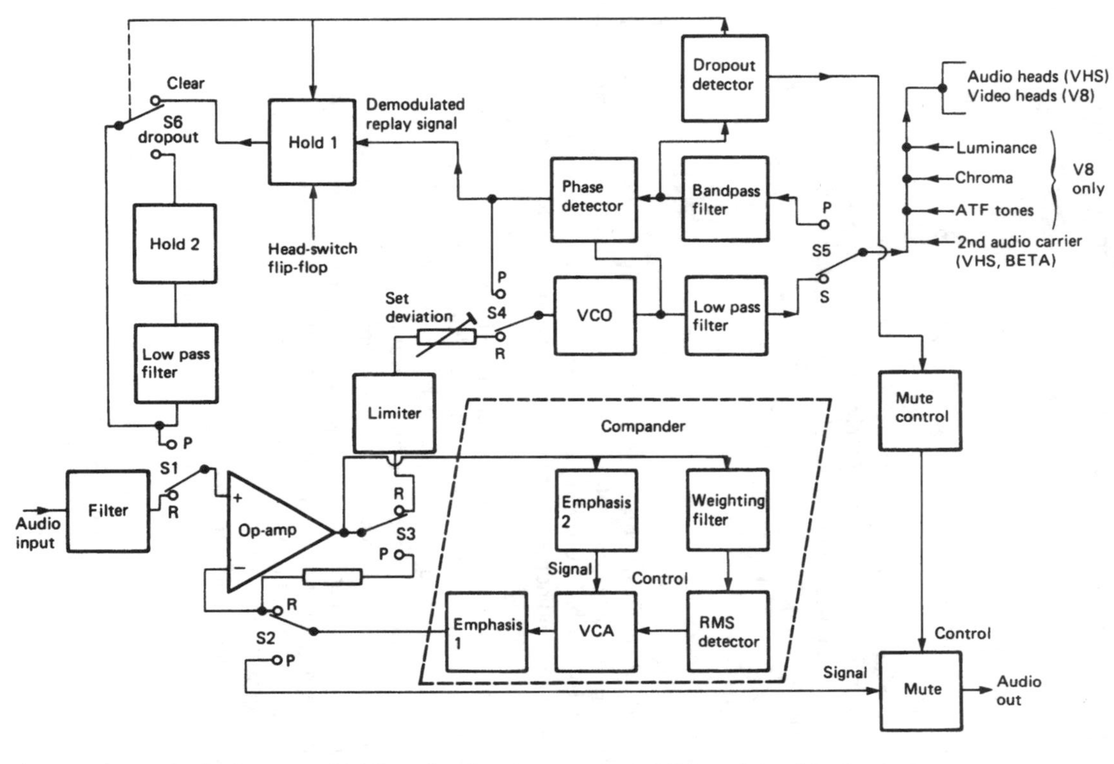

Figure 19.6 *Block diagram of hi-fi audio signal processing, fully explained in the text*

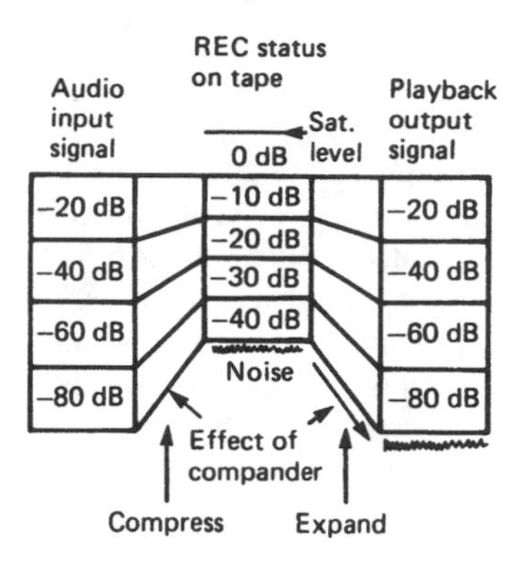

Figure 19.7 *Overall effect of companding on the audio signal. The dynamic range is greatly compressed for the 'on-tape' phase*

for pre-setting, via S4, of FM deviation in the VCO which performs frequency modulation. The FM carrier is conveyed by S5 to a low-pass filter whose output is added to ATF, chrominance and luminance-FM recording signals for passage to the video heads (V8 format); or passed directly to the rotary audio heads in VHS-format machines.

As is obvious from Figure 19.6 all the important processing blocks are used again during playback. Replay head input is now routed by S5 to a bandpass filter, centred on 1.5 MHz for Video-8, 1.4 MHz for VHS L-channel, 1.8 MHz for VHS R-channel etc. With its loop completed by S4 the VCO now forms part of a PLL detector whose demodulated output signal goes via S6 and S1 to the non-inverting input of the main op-amp, the gain of which is now set by the fixed feedback resistor from S3 to the *inverting* input.

Again the op-amp output signal is passed via the emphasis 2 filter to the VCA; again the latter is under the control, gain-wise, of the RMS detector whose input is conditioned in turn by the weighting filter. Once more the output of the VCA passes through the emphasis-1 circuit, now to form the audio output signal via S2. This time, however, the entire circuit within the dotted outline in Figure 19.6 is in the direct signal path rather than part of a negative feedback loop as it was during record. As a result its operation is reversed to 'mirror' its previous functions. Thus purely by the action of S2 what was the compressor/pre-emphasiser has become an expander/de-emphasiser, using all the same filters and components. Hence the expression *compander* and the symmetrical appearance of the diagram in Figure 19.7 whose right-hand side shows the effect of expansion of the signal's dynamic range during replay.

The most significant benefit of the companding circuit may not be obvious as yet. Plainly, it is advantageous to boost the low-level components of the audio signal so that they *significantly* deviate the FM carrier to swamp noise. The 'tape-noise' in an FM system is typically −45 dB, but since the lowest-level signals coming off tape are effectively 'attenuated' during playback, this −45 dB noise floor is depressed to −80 dB in practical systems. This ingenious combination of FM and companding techniques permits us to achieve an overall dynamic range and S/N ratio better than that inherent in the tape system itself as defined in Chapter 16 and Figure 16.7.

Dropouts can seriously upset the operation of a hi-fi audio system, and masking of these is even more important than for video. In Figure 19.6 a dropout detector (top right) monitors the off-tape audio FM carrier and throws switch S6 whenever it sees one, whereupon the 'hold 2' circuit and associated low-pass filter provide dropout compensation. Hold 1 circuit is primarily concerned with masking the dropout and disturbance associated with head-switching. If the dropout count becomes too high (e.g. severe mistracking) a mute circuit comes into operation; the audio output now reverts to the longitudinal sound track which is always provided as back-up in VHS, and to confer compatibility with other machines and tapes.

PCM audio

A brief outline of the method of quantising analogue signals was given in Chapter 12. All newly-developed hi-fi audio systems use digital encoding: examples are compact disc, the Nicam TV stereo plan, DAB radio broadcasting, digital TV and MAC TV sound transmissions. Although digital transmission and recording requires more bandwidth than other systems it has the advantages that the two-state signal is more robust than its analogue counterpart, and (so long as the quantising rate is high) is capable of superb S/N ratio and wider dynamic range.

The Video-8 format has provision for stereo PCM in addition to the monaural AFM sound facility described above. The most expensive V-8 camcorders and homebase VCRs are fitted up for PCM operation. It is impractical to give full details of the system in this book, and what follows is a basic outline of the technique used.

Figure 19.8 shows the functional blocks in the PCM sound processing. Initially the audio signal for record is amplitude-compressed in a compander similar to that in Figure 19.6. Next follows a quantisation process in which it is sampled at 2 f_h, (31.25 kHz) to 10-bit resolution. Since the tape system here cannot cope with 10-bit data a conversion is carried out to 8-bit: the process is a non-linear one, in which low-level signals are in effect given 10-bit (1024-level) descriptions, falling in three stages – as signal level increases – to 7-bit data (128-level) for the largest signal excursions. This has the effect (during playback) of concealing quantising noise in the loudest sound peaks, where they go virtually unnoticed; indeed the overall S/N ratio is subjectively equivalent to 90 dB.

Unless their transmission/recording media are very secure, digital data systems need error-correcting artifices to repair or conceal corruption of the data by noise and distortion. About 38.5 per cent *redundancy* is imparted to the PCM audio data by the addition to it of further data in the form of a cyclic redundancy check code (CRCC) – used during playback for corruption-test and 'first-aid' purposes. The data rate is high, and the effect of a tape dropout would ordinarily blow a hole in the information stream; to prevent this the data is 'scattered' on tape according to a cross-interleave code (CIC), part of the Video-8 format. The effect of a tape

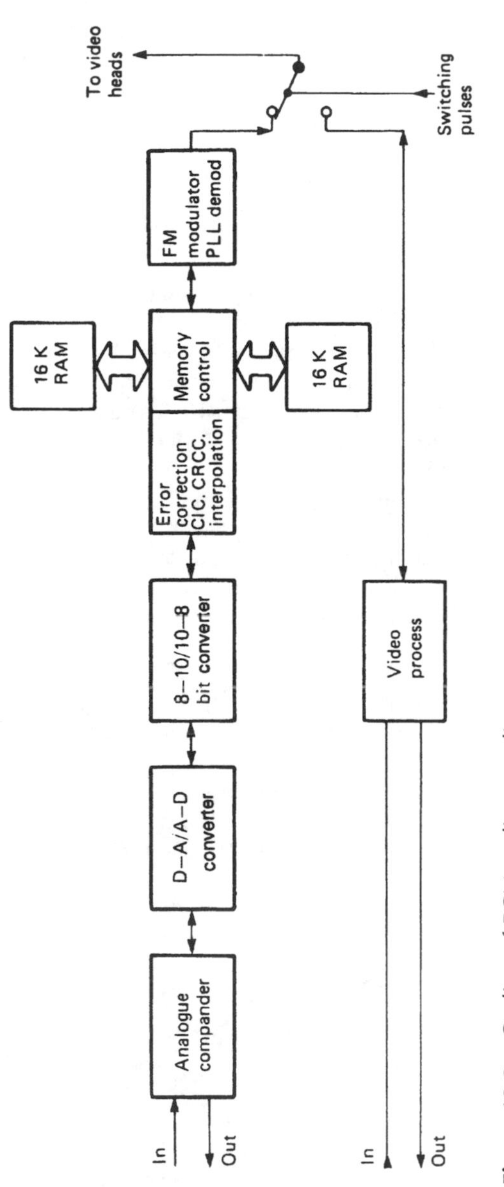

Figure 19.8 *Outline of PCM audio recording system*

dropout thus becomes distributed during replay, and the 'frayed edges' can be repaired by use of the CRCC and an additional parity check system.

The 8-bit data words are now temporarily stored in a pair of 16K RAM memories. Writing to memory is performed in real time. Readout from memory is much faster: all the data (which contains information on both stereo channels) is clocked out in less than 3 ms at 20 ms intervals. The effect of this 7:1 time-compression is to push up the data-rate to about 2 Mbits/second, but to confine the audio transmission period to a small time-slot, in very similar fashion to that of the MAC system in Figure 10.5. As we shall see in a moment, a separate place is found on the videotape for this data-burst. The two-state data signal is tone-modulated at 2.9 MHz for data 0 and 5.8 MHz for data 1; this is called frequency shift keying (FSK).

At this point the PCM signal is ready to go onto tape. The switches on the right of Figure 19.8 change over once per field period, in synchronism with TV field rate and RAM readout. This feeds data to the video heads alternately, during a period when each is scanning a 'forward extension' of the helical vision track on tape – this is illustrated on the right of Figure 19.9. The conventional video tracks (which also contain AFM audio information, duplicating the PCM sound track) are recorded over 180° of the head track, but the head/tape angle is such that they occupy about 5.4 mm of the 8 mm tape width. The extra 30° or so of tape wrap shown on the right of Figure 19.9 is devoted to PCM recording: while one head is writing PCM data, the other (diametrically opposite on the drum) is recording the last lines of the TV picture at the top of the *video* tracks on the right of the diagram.

During playback, head switching ensures that PCM data read off the tape is routed to the audio section for the appropriate 30° scan/3 ms. As in Figure 19.6 the record VCO FM modulator is now switched to perform PLL demodulation, producing binary data at its output. Continuing to the left in Figure 19.8, the data, still in time-compressed form, is read into the same pair of 16 K memories as was used during record.

Memory readout takes place in 'real time', expanding the data to give *continuous* data at a lower bit-rate – that at which the memories were loaded during record. Readout sequence is governed by the CIC (see above) in order to

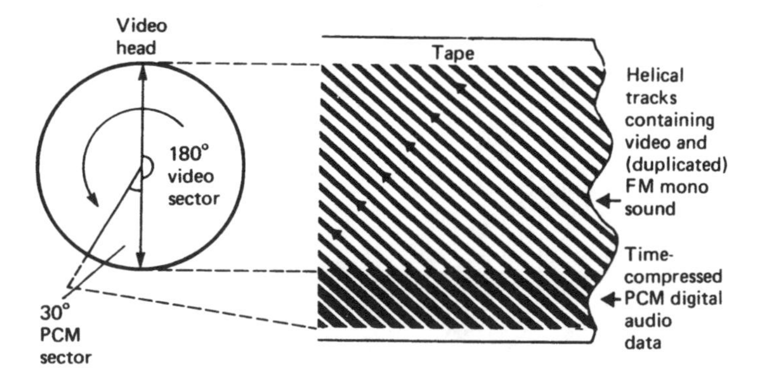

Figure 19.9 *The PCM data is recorded on a 'forward-extension' of the video tape tracks; an extra 30° of head rotation is reserved for this*

de-interleave the data, scattering and fragmenting errors in the process. In the memory control chip is also carried out error correction by means of the CRCC and parity checks mentioned earlier. For dropouts too severe to be repaired by these means, *interpolation* provides a 'patching' system which, in the face of sustained and continued corruption, devolves to a PCM mute action, switching the audio output line back to the AFM sound track. If this is also corrupted, silence will ensue!

The 8-bit 'reconditioned' data now passes to the 10-bit conversion stage to make ready for D-A conversion. As is common in these designs the record A-D converter is used for this, now switched to perform D-A operation. The analogue signal reconstituted at the D-A converter output is still in amplitude-compressed form, and is now expanded to full dynamic range in a logarithmic compander – the dotted-box section of Figure 19.6.

The relative performance of the audio systems described in this chapter is charted in Figure 19.10, where (a) shows the frequency responses, and (b) the S/N ratio capabilities.

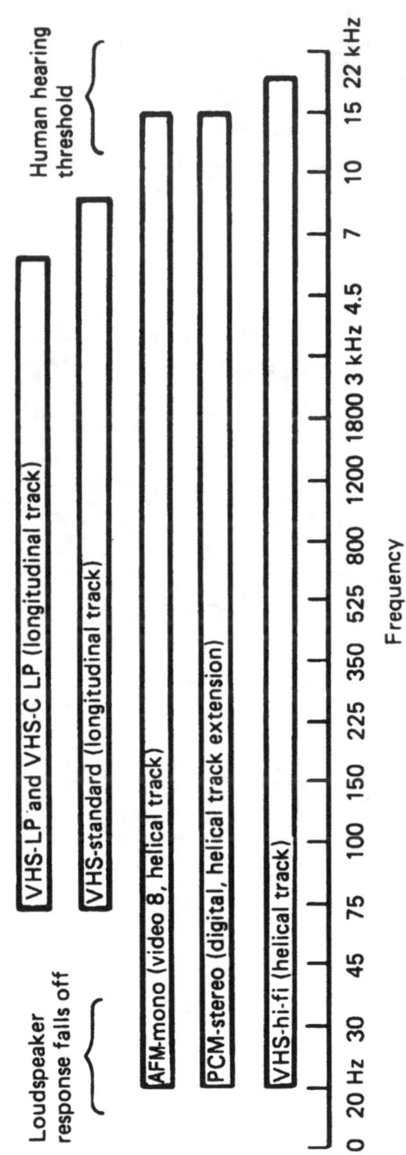

Loudspeaker response falls off
Human hearing threshold
VHS-LP and VHS-C LP (longitudinal track)
VHS-standard (longitudinal track)
AFM-mono (video 8, helical track)
PCM-stereo (digital, helical track extension)
VHS-hi-fi (helical track)
0 20 Hz 30 45 75 100 150 225 350 525 800 1200 1800 3 kHz 4.5 7 10 15 22 kHz
Frequency

(a)

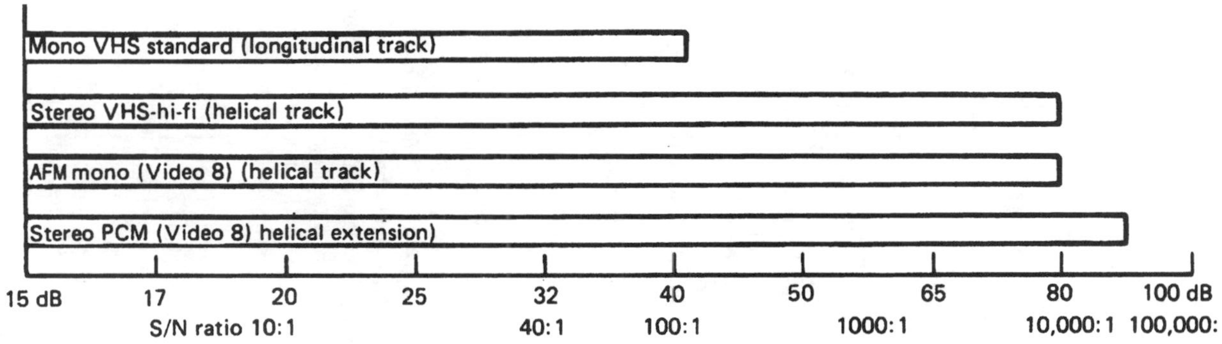

(b)

Figure 19.10 *Audio system performance of various formats. Bar-chart (a) illustratres frequency response capability; bar-chart (b) shows signal to noise ratio. The upper bar in (b) is shown in 'raw' state, i.e. without the sophisticated noise-reduction systems usually applied*

20 *Servo systems and motor drive*

In an audio tape recorder, we have a single drive system to rotate the capstan, and this pulls the recording tape past the sound head at a constant, fixed speed. Provided this speed is the same during record and playback, the programme will be correctly reproduced, and if the tape speed is arranged to conform to the standard (4.75 cm/second for audio cassette machines) we have the further advantage of interchangeability between tapes recorded and played back on different machines. The only requirements, then, of a drive system in an audio tape recorder are that it should run at a specific speed, and that short-term speed variations (wow and flutter) are kept at a low enough level to prevent noticeable changes of pitch in the reproduced sound. This simple drive system is possible because an audio waveform carries *all* the necessary information about the sound signal at any one instant, and this waveform is laid down along the tape as a single longitudinal stream of magnetic patterns.

The situation with TV signals is rather different. Because the TV image is a two-dimensional display, and only a single transmission channel is available, the image has to be built up by line-by-line scanning of the scene, with $312\frac{1}{2}$ lines making up one field, and taking 20 ms to do so, and two fields interlacing to form a complete frame, or picture. Thus for television signals it takes 40 ms to build up one picture, and we have introduced a second factor into the signal: time. The video signal, then, has built-in markers to indicate this second dimension of time, and the markers which signify the beginning of each new field every 20 ms are very important in video tape recording, as we shall see.

Another factor in the process of recording TV signals, which only has an audio-recording counterpart in DAT machines, is the idea of laying down parallel slanting tracks on the tape by means of a rotating head drum. If replay of these tracks is carried out in a random manner (even at the correct speed) there is no reason why the replay heads should follow the path taken by the heads during record, so that each replay head may well read between the lines, as it were, and reproduce a signal consisting of a mixture of the information from two adjacent tracks. We have already seen that adjacent tracks contain different video signals recorded at different azimuth angles, so that this form of mistracking will make a noisy mess of the reproduced signals.

To overcome these problems, a very precise control system, known as a servo, is required to govern the head drum and capstan drive system. In this chapter we shall examine the purpose and mode of operation for both head drum and capstan servos, and see how they govern the relationship between video head position and tape track placement.

Requirements during record

We have seen in Chapter 17 that two video heads are used in home VCRs, each of which lays down one complete field of $312\frac{1}{2}$ TV lines during its scan of the tape. At some point during the field period one video head has to 'hand over' to the other, and as this will cause a disturbance on the reproduced picture, the 'head interchange' or switching point is arranged to lie at the extreme bottom of the picture, just before the field sync pulse (Figure 17.17) so that the disturbance will be just off the screen in a correctly set-up monitor or TV set due to the normal slight vertical overscan. This means that during record we need to arrange for each head to start its scan of the tape about half a millisecond (corresponding to approximately eight TV lines) before the field sync pulse occurs in the signal being recorded; and hold this timing, or phase relationship, steady throughout the programme.

If this relationship were not present, the head-switching disturbance effect would move into the active picture area, and drift up or down the screen. The mechanics of the deck arrangement and the position of the 'entry guide' determine the point at which the moving tape starts its wrap of the head drum, as shown in Chapter 17. The electronic servo loop sees to it that each head passes this point and starts to write its field of information just before the field sync pulse occurs in the video waveform being recorded. To achieve this, we need some marker to tell the servo system the angular position of the head drum. There are many possible ways of generating this, but the most common is the use of a small permanent magnet fixed to the underside of the head drum, and arranged to induce a pulse in a nearby stationary pickup coil at each pass of the magnet, as shown in Figure 20.1.

Other methods of generating these pulses are possible. In some VCRs a drum-mounted 'flag' interrupts the light-path of an *optocoupler* once per revolution; or

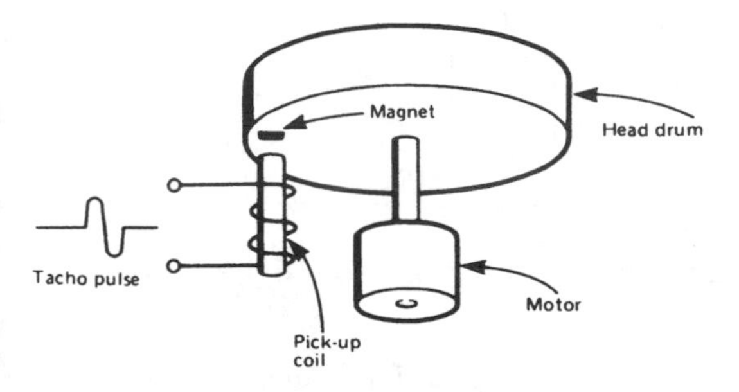

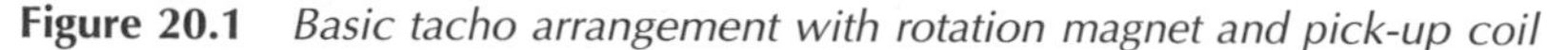
Figure 20.1 *Basic tacho arrangement with rotation magnet and pick-up coil*

Hall-effect devices are used. However it is made, each time this *tacho pulse* appears we know the head drum is at a certain angular position.

The tacho pulse, then, may be regarded as a phase reference for the head drum. If the tacho pulse is applied to a phase detector for comparison with the timing of the incoming field sync pulse, an output voltage will be produced (see page 278) proportional to the timing error between these two input signals. If this error voltage is applied to the head drum drive motor, the control loop will ensure a fixed relationship between the field sync pulse and the angular position of the head drum – yet another example of a PLL, but this time with a mechanical system within the loop. This simple servo system is illustrated in Figure 20.2. By delaying either the field sync pulse or the tacho pulse on its way to the phase detector, we can set up any desired phase relationship between incoming field sync and video head position, governed by the length of the delay introduced. It is arranged that the phase relationship is such that the video tracks are laid down with all the field syncs at the bottom of the tape, as in Figure 17.16. The fact that one 20 ms TV field is recorded by each head during one half-turn of the head drum precisely dictates the rotational speed of the drum – it comes out at 40 ms per revolution, or 1500 rpm.

What is required of the capstan servo during record? As we have seen, it takes no part in determining the timing of each one-field track, which is wholly the province of the head drum. The capstan speed governs the spacing of the video tracks, however, and as we know, they are laid down so that they just abut each other; this calls for a steady and even flow of tape past the head drum. The audio signal is recorded longitudinally on the tape in the normal way, so good sound and jitter-free pictures depend on a steady capstan speed. The actual capstan speed depends on the format, being 2.34 cm/second in VHS, for example. During record, then, the capstan needs only to be held at a constant speed, and in some early VCR designs no capstan servo was used. All current VCRs employ a capstan servo, however, which on record is slaved to a high-stability frequency reference known as a *clock*. This may be a stable crystal oscillator or the field sync from the signal being recorded.

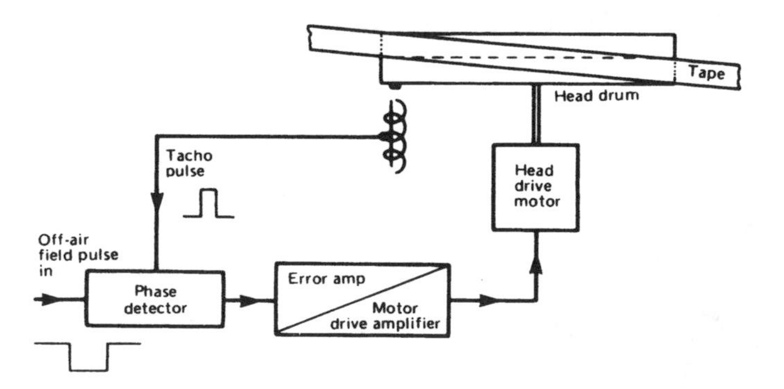

Figure 20.2 *Head drum servo loop during record. The head drum phase (angular timing) is locked to off-air field syncs*

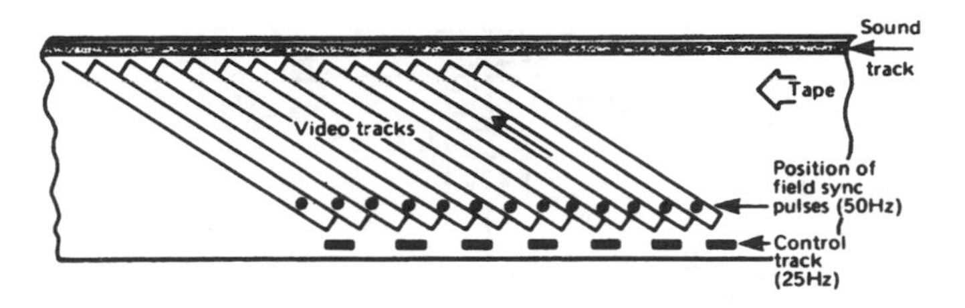

Figure 20.3 *Configuration of tape track pattern as a result of the action of the servo loop in Figure 20.2*

The control track

When the video tape being recorded is played back, we will need a reference signal to define the physical position of each recorded video track on the tape. The situation is akin to that of cine film, where sprocket holes are provided to ensure that each frame is exactly in position in the film gate before the projector's shutter is opened. Our video tape 'sprocket-holes' take the form of a control track recorded along the edge of the video tape by a separate stationary head just like the sound head. The control track consists of a 25 Hz squarewave pulse train. One pulse, then, is recorded for each two fields and the track pattern is shown in Figure 20.3 which also shows the position of the sound track. We shall see later how the control track is used during playback.

Servo operation during playback

Let's assume we load into the VCR a good pre-recorded tape, and initiate the replay mode. To get a stable and noise-free picture we need the same close control over the capstan and head drum drive as takes place during recording, but the control references are somewhat different. We'll take the operation of the head drum servo first. The head-changeover point, which needs to correspond with that used during recording, is controlled during replay by the head drum tacho pulse, which triggers the head-switching bistable via a variable time delay. The delay is adjusted so that head switching takes place at the correct instant, just into each tape track.

The speed of the head drum during replay determines the line and field frequencies of the off-tape video signal. Thus the drum needs to spin at precisely 1500 rpm as during record, and this is achieved by slaving it to a stable frequency reference – again a crystal oscillator is used. Phase-locking takes place in the same phase detector as was used during record; the positional feedback input to this is again the tacho pulse from the drum pickup coil, but the reference input to the phase detector now becomes the local high-stability frequency source – a crystal oscillator as shown in Figure 20.4.

During replay we have to ensure that each video head scans down the centre of its intended track, and this is achieved by adjusting the lateral position of the tape around the head drum to line up the head scans with the recorded tape tracks, and

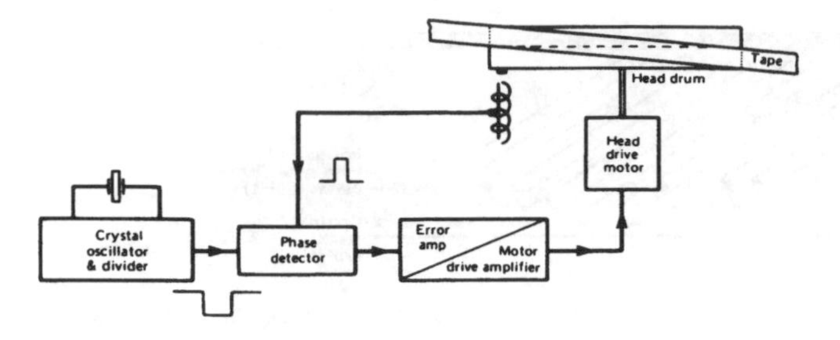

Figure 20.4 *Head servo on replay, locked to stable local reference*

hold them constantly in alignment with each other. This calls for another servo operation, and a moment's thought will show that this 'lateral lining-up' procedure can be carried out by adjusting tape position (via the capstan servo) or head drum position (via the drum servo). In either case we need a reference signal related to the position of the video tracks on the tape, and this is where the control track, laid down during record, comes into use. We'll describe the capstan-tracking system, as used in most machines.

In addition to its duty of pulling the tape through the machine at a steady speed equal to that used on record, the capstan is now required to position the tape relative to the head drum to achieve correct tracking. Although the video tape tracks are very narrow, they are at such a shallow angle (say 5°) to the tape path that a relatively large lateral movement is required to 'move' the tape across from one track to the next. Let's see how this drum/tape-position phasing is effected.

During playback the control track head reads the 50 Hz pulse train off the tape to provide a 'track position' reference signal for the capstan servo. This is compared in a phase detector with the same stable local signal (50 Hz mains or the frequency-divided output of a crystal oscillator) to which the head drum is slaved, and any error is fed to the capstan motor to pull it into lock. Thus a fixed relationship is set up between drum position and tape track placement, and the phasing, or timing, of this relationship can be adjusted by including a variable delay in the path of the control track signal to the phase detector. This variable delay is set by the tracking control, and it is adjusted on playback for minimum noise in the picture, signifying optimum tracking. A diagram of this system appears in Figure 20.5, while the effect of adjusting the tracking control is conveyed in Figure 20.6. Some VCRs carry out the replay tracking function via the head drum servo.

Automatic tracking systems

So long as the physical position of the sync head is correct, with reference to the entry-point of the tape onto the drum in the recording machine, its tapes will replay in any similar machine, provided that the latter is also in good mechanical alignment. The tracking control only operates during replay, and is only required to take up tolerances due to mechanical wear, tape tension variations etc. It is a user control, and maladjustment often impairs picture performance.

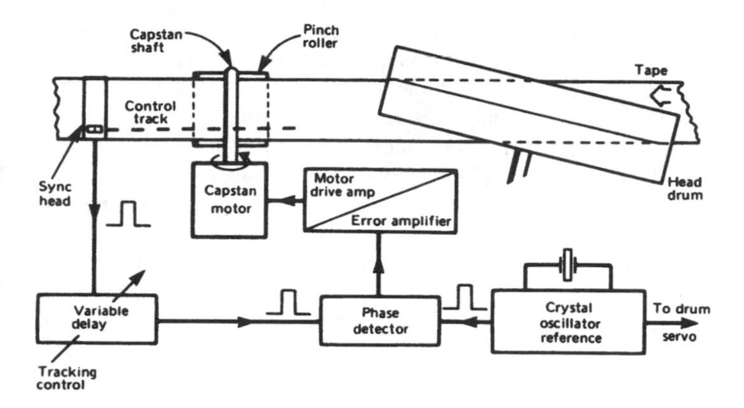

Figure 20.5 *The capstan servo during replay, locked to control track pulses*

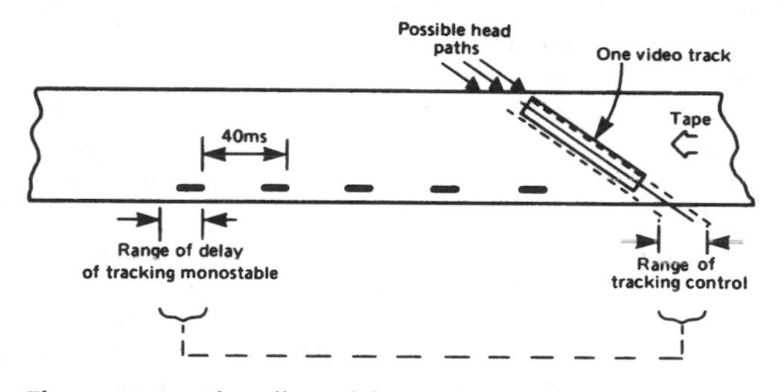

Figure 20.6 *The effect of the tracking control on replay. It 'phases' the lateral tape position to line up video tracks with the path of the vision head*

Various forms of automatic tracking system have been developed to overcome this problem: we have examined the ATF arrangements of the Video-8 format. With this no manual tracking control is necessary or provided since 'tolerance' errors, both long term and short term, are automatically taken up. Any errors due to tape-path anomalies between record and playback machines are similarly removed. In VCRs using longitudinal control tracks a form of automatic tracking control is possible. The off-tape FM video carrier's *amplitude* is independent of signal content, depending only on the efficiency of tape-head transfer. This means that tracking accuracy can be measured by monitoring the replay FM 'envelope' amplitude. Auto-tracking systems operate by varying the phase delay (bottom left-hand corner of Figure 20.5) to maximise replay head output. This can be done on a continuous or a 'stab and store' basis.

Servo basics

We have seen what servo systems do in a VCR and why they are necessary. Time now to find out how they work! All servos are closed-loop devices, where the 'end result' is fed back to modify the 'processing'. Perhaps the simplest example of a closed-loop system is an electric immersion heater, where the water temperature is monitored by a thermostat which switches the heating element on and off to maintain a reasonably constant temperature regardless (within the capabilities of the heating element) of the demand being made on the hot water supply. A more scientific example of a closed-loop system is found in a TV receiver where incoming line-synchronising pulses are used to hold in step the horizontal scanning lines of the picture. A sample of output frequency and phase is fed back to the flywheel sync circuit and its timing compared with that of the incoming sync pulses. Any discrepancy gives rise to an error signal, and this, after filtering, controls the frequency of a VCO. The local frequency and phase are pulled into lock with the controlling, or reference pulses, and the oscillator is then said to be slaved to the reference. In discussing colour-under systems and colour crosstalk compensation processes we met the closed loop (PLL) system several times, and the phase discriminator and VCO at the heart of it are discussed on pages 277–8.

Figure 20.7 shows the basic servo loop. In this case the motor is a permanent-magnet DC type, so that its speed is proportional (for a given mechanical load) to armature current. On its lower shaft is a disc with a magnet attached so that one pulse is induced in the tachogenerator coil for each revolution of the shaft. This pulse, at 25 Hz, is timed and shaped in the pulse amplifier and then fed to the phase detector. A second input to the phase detector forms the reference, and in our example it is derived from the field sync pulse of a video signal via a sync separator, divide-by-two counter and monostable delay, whose function will be explained shortly. Assume the DC coupled motor drive amplifier is arranged to have a standing 5 V DC at its output with zero input voltage, and that the motor is designed to rotate at about 1500 rpm with such an input voltage. This rotational speed will give rise to tacho pulses at 25 Hz, and if these are coincident in time with the reference pulses, the phase detector will provide zero output and the motor will continue to rotate at 1500 rpm. Now, if the mechanical load on the motor is increased, its speed tends to fall, and the phase detector will start to see a

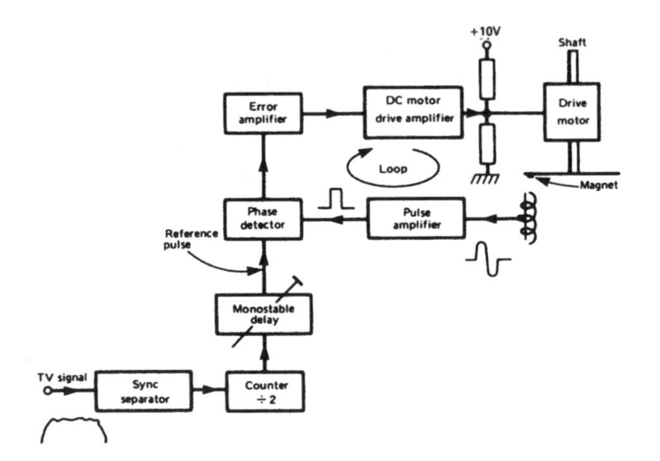

Figure 20.7 *Basic servo loop. The reference signal is incoming field sync pulses*

time-lapse between the arrival of a reference pulse and that of the tacho pulse, giving rise to an error voltage into the DC amplifier. This error voltage then drives more current through the motor and so increases its speed to restore the status quo. When reference and tacho pulses again become coincident in time, steady-state conditions are restored in the servo loop. Should a speeding-up of the motor occur, the circuit operates in the opposite way to correct this.

Loop transient response and damping

The design of a servo system is critical if its performance is to be good, and VCR servo systems are very demanding in this respect. The performance of a servo is traditionally analysed in terms of a step input function, in which a step, or instantaneous change, in control voltage is applied and the output function (in our case, capstan or head drum phase position) analysed to see how closely it corresponds to the step input.

This is illustrated in Figure 20.8, where the step input is shown along with three possible responses at the servo output. Plainly it is not possible to achieve an instantaneous change in servo output where a mechanical system is involved – the inertia of the moving parts will cause an inevitable time lag before the new phase condition is established, and the designer's aim is to do this in the shortest possible time without spurious mechanical effects. Three factors are involved – damping, gain and time-constant – and each of these are present in mechanical and electrical form. An ideal servo system has each of these factors optimised, and the electrical design of the motor drive amplifier takes into account the mechanical characteristics of the motor and shaft coupling systems, and inertia of the rotating parts. Electrical analogies of mechanical functions can be built into a circuit in such a way that deficiencies are catered for – for instance, positive feedback via a capacitor will reinforce the gain of a motor drive amplifier only when the *rate of change* of drive amplifier output is high; the same effect can be realised by negative feedback via an integrator. This may be used to overcome the inertia effect of a heavy flywheel. Very often two time-constants are used to control the response time of the servo loop: one to quickly achieve correct speed (frequency) and a second to establish exact positioning (phase).

Let's now look at the servo responses illustrated in Figure 20.8. The step input is shown in curve (a). Curve (b) illustrates the effect of excessive damping, in which

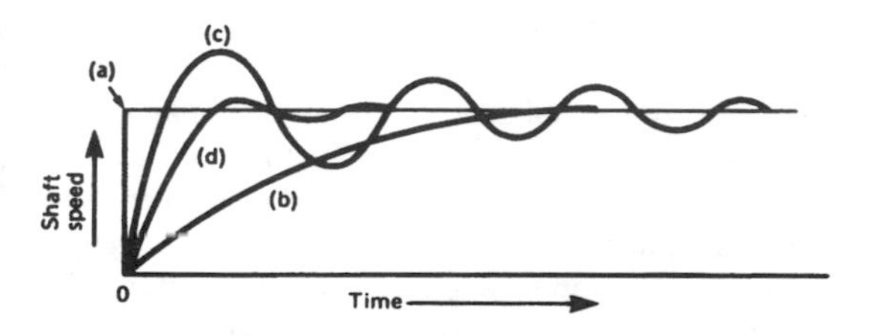

Figure 20.8 *Loop transient responses compared. Waveform (a) is the step input; the other waveforms show possible reactions at the drive shaft*

the output takes a long time to follow the input signal resulting in sluggish operation; curve (c) shows insufficient damping, causing the servo output to *hunt* or oscillate about the correct point for a period before settling down. Without any damping, continuous oscillation would take place! The ideal situation is shown in curve (d), that for *critical damping*. This is the one which designers normally aim for as offering the best compromise between response speed and hunting effect. A single slight overshoot and one small undershoot is followed by stable operation at the new level. This effect is easily demonstrated in most machines by making a sudden and well-defined adjustment of the tracking control to simulate a step input function to the servo control. The picture will be seen to move sideways in one or two damped cycles of oscillation before settling down.

There is great variation in the design of motor drive amplifiers, depending on the mechanical system and intended conditions of operation of the machine. Much depends on motor design; brushless multipole DC drive motors with purpose-designed IC control systems give good servo performance in current VCR designs.

Pulse-counting servo

Figure 20.7 showed a basic 'analogue' servo system (as used in the earliest designs of VCR) to illustrate the principles involved. In fact practical systems achieve the same end with more precision and versatility by using a pulse-counting technique as shown in simplified form in Figure 20.9. A clock oscillator (generally tied to the machine's 4.43 MHz colour subcarrier crystal) runs at a fixed frequency to operate the system. Its pulses have no effect on counter 1 until a reference pulse arrives there to *enable* it, whereupon it starts to count clock pulses, accumulating the count in a register. As soon as a sample pulse appears the contents of the register are 'frozen', and at that moment in time the stored count is proportional to elapsed time between the appearance of the reference and sample pulses. We now need to compare this stored value with a reference, or time marker, and this appears in the form of counted-down clock pulses from counter 2. The accumulated count here (depending only on elapsed time) is compared with the memory contents in a comparator whose 'difference' output is converted to an analogue value in a D-A converter consisting of a triggered (set/reset) flip-flop and an R/C integrator. The error potential thus produced forms the drive voltage to control motor speed.

What's been described is basically a measurer of time and thus suitable for phase control of a drive motor. In a practical servo system a second control loop is required to quickly establish correct running speed of capstan and head-drum, even before phase control becomes relevant. For this *speed loop* the requirements are slightly different: it's necessary to get a sample of shaft speed for comparison with a frequency reference like a crystal, and produce an error voltage which will quickly 'pull' the motor into the correct rotational speed. The sample comes from a frequency generator inside the motor, typically a printed 'coil' over which a shaft-mounted magnet rotates. It's called an FG (frequency generator) and after clipping to form a square wave is applied to what amounts to a differentiator, see

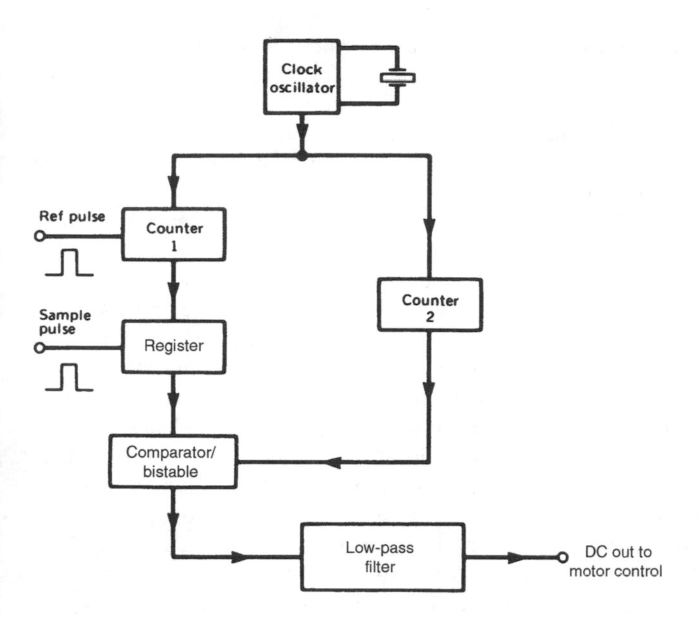

Figure 20.9 *Simplified block diagram of one form of digital servo*

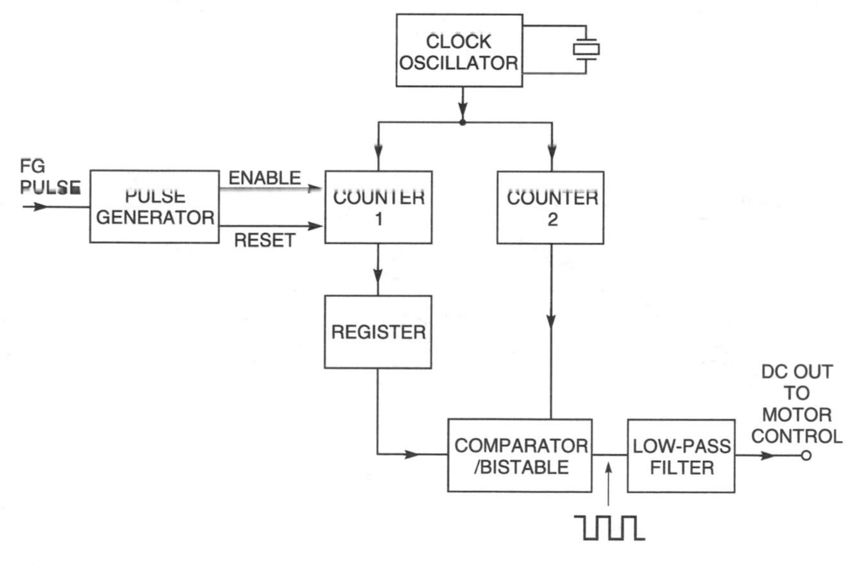

Figure 20.10 *Pulse-counting servo system for motor speed control*

Figure 20.10, which produces a pulse at each rising and falling edge of the FG signal. The rising-edge pulse *enables* counter 1 so that it can begin to fill a register; counting stops and the register's contents are frozen upon the arrival of the FG's falling-edge pulse from the differentiator. Now the count held in the register is proportional to the period of one half-cycle of the FG waveform. Meanwhile counter 2 has been steadily incrementing the reference clock pulses and presenting the count to a comparator where it is compared, on each FG cycle, with that which has accumulated in counter 1's register. 'The difference' value produced represents the motor's deviation from correct speed, and is applied as an error voltage to the motor to pull it quickly into the correct rotational speed; this error voltage is produced in the same way as described above for the phase-control loop. In fact Figure 20.10 forms an accurate and fast-acting frequency-to-voltage converter.

VCR servo chips combine the phase- and speed control loops shown in the two diagrams. They work at different rates, typically 15 kHz for speed and 2 kHz for phase, and the DC control voltages from their integrating filters are combined on their way to the motors. In picture-search and still modes the drum speed control loop counter operates at different points, programmed by the control system, to maintain correct line frequency in the off-tape video signal in the face of effectively-different scanning speeds: see Figure 17.18 and accompanying text. Likewise, the capstan speed control loop's count-division ratio is varied during record for SP/LP operation, and during replay for picture-search functions. In fact the capstan motor rotates 'backwards' in reverse-search modes, and in either direction of search/cue/review its speed/phase is tied by the servo system to a multiple of off-tape control-track pulses in order to lock the noise-bars (due to track-crossing, see Figure 20.11) stationary on the screen. In most VCR deck designs the capstan motor has other functions: tape threading, cassette-loading and/or reel drive, each of which is governed by the system-control rather than the servo circuit, and is mechanically 'switched' on the deck.

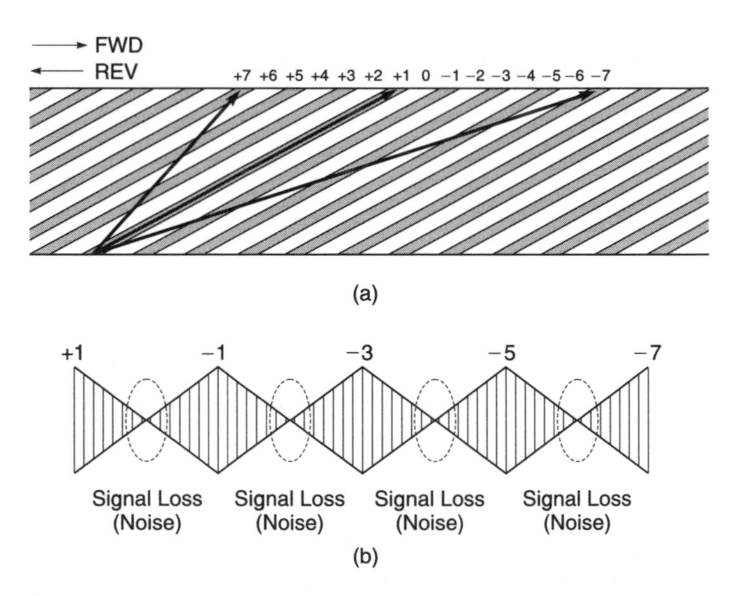

Figure 20.11 *The effect of track-crossing by the video head in trick-speed playback modes. The effect is noise-bars on the picture*

Still-frame requirements

Chapter 17 related how various machines solve the problem of noise-free freeze-frame reproduction by means of four-head drums or extra-wide replay heads to overcome the change in effective tape track angle between moving and stationary tape. When the tape transport is stopped at a random point, track scanning by the vision heads may be such that a mistracking bar is seen on the still frame.

It is necessary to position the tape so that this noise bar is 'shuttled' out of the picture to give satisfactory reproduction. This is carried out by the capstan servo which 'inches' the tape forward until the noise bar moves out of the picture into the field blanking period, when capstan drive ceases. The reference for the inching system is off-tape video, usually a sample of FM luminance carrier. The presence of the noise bar in the field blanking period would tend to obliterate the field sync pulse during replay, and even it were available, the half-line 'jump' in its timing between successive fields would cause picture judder on still frame in machines where this system is used. For these reasons, an 'artificial' vertical sync pulse generator is used in still-frame mode, and its pulse timing is governed by a preset, adjusted for minimum vertical judder of frozen pictures.

An extension of the inching idea outlined above makes possible slow-motion replay in suitably equipped machines. Let's assume we need to replay at one-third normal speed. It will be necessary to position the tape accurately on the head drum to achieve good still-frame reproduction as described, and hold it in position for three sweeps of the video heads, i.e. three revolutions of the drum. After this the tape is quickly moved forward by the capstan to line up the next track accurately for the next three revolutions of the head-drum, and so on. The same process can be carried out with the capstan going backwards to give a slow-motion sequence in reverse. The precision required during these processes is of a high order and the circuits to carry out the operation use a mixture of analogue and digital techniques, often with specially designed ICs.

Video-8/Hi-8 servo

As we saw in Chapter 17 and Figures 17.28 and 17.29, Video-8 format machines do not have a control track as such: video head tracking is achieved by auto-following 'tones' laid down in the video tracks themselves during record mode. The end-result (Figure 17.29) is an error voltage for application to the sorts of motor we shall look at next. The tracking tones are also used during still-frame reproduction from 8 mm tapes, whereby a single tone is kept 'in sight' as far as possible during each head-sweep of the (now stationary) tape ribbon.

Motors

Some VCRs use a simple permanent-magnet DC motor for capstan drive, coupled to the capstan shaft by a drive belt. Motor speed is proportional to armature current, so that the error voltage from the servo has merely to govern the

Figure 20.12 *Direct-drive motor for capstan drive in a VCR. On the left is the stator with three drive coil pairs, PG pick-up head (one o'clock) and drive IC at top; on the right the printed-magnet rotor whose shaft forms the capstan proper*

voltage/current applied to the motor. This approach has the advantages of cheapness and simplicity.

Most capstan motors, and all drum motors, are *direct-drive* types, in which the capstan shaft is actually the motor spindle in the one case, and the drum is directly mounted on the spindle in the other. In many current VCR designs the head-drive motor is mounted atop the drum because the main PC board is fixed to the underside of the deck. In a direct-drive motor the rotor takes the form of a flat or cup-shaped disc magnet, into which are 'printed' a number of alternate N and S poles. The stator is formed by several coils (typically six) in a radial pattern, and fed sequentially by current from an IC multiphase driver circuit built onto the stator board. Current-pulse timings are governed by Hall-effect magnetic sensors, and speed/torque by stator coil current, in turn controlled by the voltage applied to the speed-control pin of the IC. Figure 20.12 shows the construction of a direct-drive motor for a full-size VCR; those used in camcorders use the same construction and working principles, but are built on a tiny scale.

Tension servo

During record or playback, it is important that the tension on the tape is held at a constant level, regardless of the amount of tape on each spool and the friction encountered by the tape on its journey through the deck. Insufficient tension in the tape during its wrap round the head drum leads to poor head-tape contact and results in poor signal transfer and noisy pictures. Too great a tension causes head

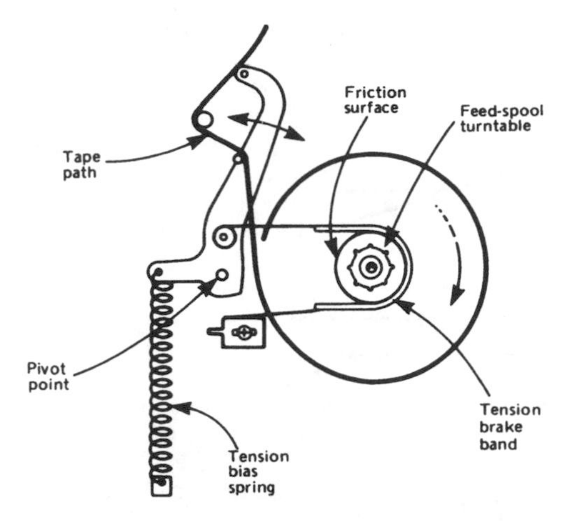

Figure 20.13 *Simple brake-band tension regulator*

and tape wear and tape stretching, and can result in poor compatibility with other machines due to 'stretch-distortion' of the magnetic pattern of the tape. When replaying a 'foreign' tape on a machine with excessive back tension, the registration between luminance and chrominance signals is lost, and the colour will be 'printed' to the left of the luminance image in the reproduced picture.

To prevent these problems, VCRs incorporate a tension servo to maintain a constant 'stretch factor' in the tape. In most machines this takes the form of a passive mechanical negative feedback system, in which a tension arm bears against the tape as it leaves the feed spool in the cassette. The arm is linked via a lever to a brake band wrapped around the supply spool turntable, so that low tension causes the brake band to bind against the supply turntable and vice-versa, maintaining reasonably constant tension in the tape. The arrangement is shown in Figure 20.13.

In machines where each tape spool is driven from its own motor, a more precise control of tension is possible. Reverse current in the supply-spool motor sets the back-tension in the tape. Typically the tape emerging from the cassette passes over a spring-loaded feeler which carries a shutter; this is in the light path of an optocoupler (LED-phototransistor combination) so that conduction in the phototransistor is proportional to tape tension. The phototransistor current is amplified and used to control the current in the *supply* spool motor, which is arranged to run 'backwards' at this time. Because of the steady pull on the tape from the capstan, the supply motor will be pulled in the opposite direction to which it is trying to run, and the motor current will then determine tape tension under the control of the optocoupler sensor. At the beginning of a cassette, when all the tape is on the supply reel, the reverse current in the supply spool motor may be of the order of 250 mA. As the tape is wound on to the take-up reel this current drops to maintain correct tension, as illustrated in Figure 20.14, which also shows the constant forward current in the take-up spool motor, representing the latter's steady pull on the tape away from the capstan and pinch roller.

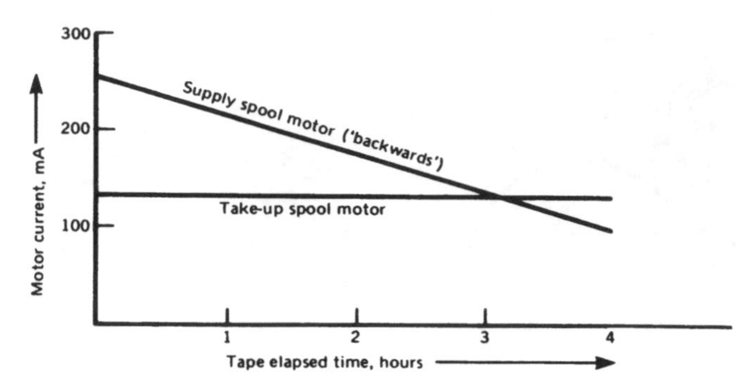

Figure 20.14 *Electronic tension servo: motor-control currents vary with the amount of tape on each spool*

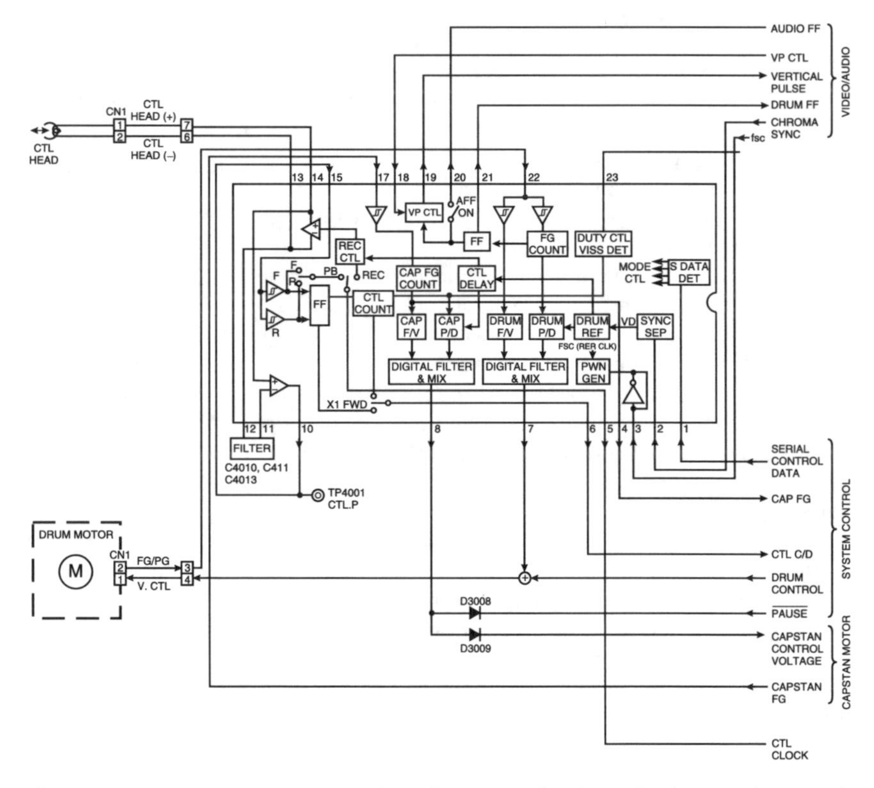

Figure 20.15 *VCR servo system based on a single chip which provides speed and phase control for both capstan and drum motors (JVC)*

Servo IC

In practice the entire servo system for both drum and capstan motors is fabricated inside an IC, often the one which carries out system control. Figure 20.15 outlines

Figure 20.16 *Combined FG and PG pulse train from the drum motor in Figure 20.15; the PG 'needle' pulses ride on top of the FG squarewave*

in block diagram form the arrangement of a dedicated servo chip used in a JVC VHS deck, whose main connections and functions we shall now briefly examine. Pin 1 carries all the user and system-control commands into the chip, governing such modes as 'trick'-play, SP/LP and 'emergency' functions, in a serial-data control line. TV sync pulses for drum motor phase control during record come in on pin 2, while no.3 takes a sample of 4.43 MHz subcarrier frequency from the chrominance section to act as a stable clock/timing reference for motor speed control. Pin 17 is the entry point for FG feedback from the capstan motor for its speed control; no. 22 does the same for drum motor feedback which comes as a combined FG/PG waveform in this design, as shown in Figure 20.16. The two pulse streams are separated by an amplitude discriminator inside the chip. Control track pulses from the CTL head enter, during playback, at IC pins 13 and 14 to control the tape track positioning relative to head sweep trajectory. 'Auxiliary' outputs from the device are at pins 4 and 21, respectively capstan FG and drum flip-flop (SW25) en route to system control and elsewhere; 'delayed' SW25 squarewave to the audio hi-fi section from pin 20 for head-switching; and a vertical drive pulse out of chip pin 19 to time the artificial field sync generator during still-frame reproduction.

The main functional outputs from the IC are at pin 8, whose voltage governs the capstan motor speed and phase; and pin 7, speed/phase control for the drum motor. Both of these lines are overridden – grounded in fact – by the system control chip when it's required to stop either of these motors. In the case of pause mode, capstan motor control is switched by diodes from the servo to the syscon chip in order to perform the 'inching' function described earlier in this chapter. During record the CTL head is driven with pulses via IC pins 13 and 14; their duty-cycle changes at the beginning of each new recording to provide a tape-indexing mark as described on page 332. During replay the control track marker is detected on the line emerging from IC pin 5. Control track pulses loop out of the chip via pins 11 and 12 for filtering, and 10/15 for provision of a test point, used in fault diagnosis and setting-up.

In respect of its internal workings the IC of Figure 20.15 follows the principles described earlier in this chapter; the blocks below centre of the diagram marked CAP F/D and P/D (frequency and phase detector respectively) and DRUM F/D and P/D correspond to our diagrams 20.9 and 20.10. In our look at this IC diagram we have made several references to system control, which will be the subject of the next chapter.

21 *System control for VCRs*

Several times in this book, we have compared aspects of audio and video tape recording and the machines for doing so, and in each case we have seen how much more sophisticated is the video system. Simple audio tape recorders have only a single motor; no electronic circuit or motor design will make it rotate in opposite directions at the same time! All that is necessary, then, is a simple mechanical interlock to prevent more than one transport button being depressed at a time. In high-quality audio machines with fast spooling and more than one motor, a simple form of system control is provided. Once again, then, in the area of system control, the VCR has a multiplicity of requirements which do not exist in its audio counterpart. These arise from: the threading in and out of the tape from the cassette; the high rotational speed of the head drum; the fragility of the vision heads; the use of several drive motors whose efforts could conflict; the high speed and torque of the fast-wind and rewind processes; the need to control the machine by light-touch sensors or a remote keypad; and several minor factors.

Need for system control

The system control (we'll call it syscon for short) section of a VCR may be regarded as a 'policing' system for the mechanical functions of the machine. It has two basic interrelated functions, preventing user 'abuse' and programming the machine's mechanical operations, while protecting VCR and tape in the event of machine malfunction or tape faults. The syscon receives inputs from the user's mode keys (command functions) and also from monitoring points around the tape deck (protection functions). These are correlated in the syscon and so long as no conflict exists between command and deck or tape status, the mode required is permitted. If at any time during any mode, a relevant protection input is received (e.g. drum motor stops during play or tape-end sensor is activated during rewind), the syscon will invoke a new mode – 'stop'! The syscon recognises on each input line only two states, on (1) and off (0), and presents its outputs to solenoids, motors etc. in the same way.

Syscon inputs

Let's examine the inputs to the syscon in turn. The command inputs are the user control keys – play, record, rewind, fast foward, fast rewind, pause, stop and forward/reverse picture search. The tape counter also provides a command function (stop) if it reaches 0000 or a pre-programmed number ('go-to' facility) with its memory key depressed. The protection inputs are many and varied, and we'll look at each of them, and the effect they have.

End sensors. The tape-end sensors are activated by markers 150 mm or so from each end of the tape: an LED shines through a transparent leader tape to activate a photo-cell.

The tape-end signal thus derived invokes stop mode and then prevents any further tape transport in the 'unsafe' direction. Many machines also incorporate a protection circuit which monitors the current in the tape LED; if an open-circuit here went undetected no tape-end signal could be given, and damage might result.

Rotation detector. We have seen that the head drum generates tacho pulses as it rotates, and their cessation would mean that the head drum is stalled. This is monitored by the syscon, along with tacho pulses from the take-up spool carrier. If either fail, stop is entered after a short time. During freeze-frame or picture pause modes both tape spools will stop, so the spool rotation detector output is overridden in these modes. To prevent tape and head wear, a timer is started which enters stop after a few minutes' scanning time over a single tape track.

Cassette-in. All modes are inhibited when there is no cassette in the machine. A lever and microswitch detect the presence of a cassette, being deflected by its plastic envelope as the cassette carriage is driven home into the machine.

Recording tab. As with audio cassettes, video types have a removable safety tab to prevent accidental erasure or recording-over the programme it contains; where the tab has been removed, a lever and microswitch detect the fact, and signal the syscon to prevent record mode being entered.

Threading completion. Sometimes known as the after-load (AL) switch, this closes when the tape is fully threaded around the head drum. Until threading is complete, the syscon will inhibit capstan drive to prevent forward motion of the tape. *Mode-switches* give more comprehensive information.

Dew sensor. If the ambient temperature in which the VCR operates changes suddenly, usually due to bringing it indoors from location work, there is a risk of vapour condensation, particularly on the head drum. If the machine were allowed to operate in this condition the tape would stick to the head drum with disastrous results! To detect condensation, then, many machines are fitted with a dew sensor on the stationary part of the head-drum assembly; it consists of a pair of metallic fingers or a high-resistance 'element'. The presence of moisture increases the resistance of the sensor, and this is monitored by the syscon which acts to enter stop and give a warning. Some machines incorporate a low-wattage head heating element to drive off condensation. These features are important in portable equipment, and are present in many 'stationary' home VCRs.

Power interruption. When the machine is running in any normal mode, the syscon is in a 'latched' state, so that interruption of mains power will unlatch the circuit. The result, in most machines, is that stop will be entered on restoration of power.

Tape duration measurement. Sophisticated machines have a 'time-elapsed indicator' as a front-panel readout, and to operate this it is necessary to know what the cassette type is (2 hour, 3 hour etc.) and how much tape is left on each spool. Some cassettes have a code embossed into the cassette body to indicate running time, and this operates sensors in the VCR; some sophisticated machines of various formats compute spool speeds relative to each other against the fixed tape speed to arrive at a figure for cassette type in hours'-worth of tape. Once the tape running time is known, it remains to determine the amount of tape on each spool. This may be carried out via the tension servo system in some designs, or by an extension of the spool-speed-versus-tape-speed computation described above.

Syscon outputs

Motor stop. The capstan and drum motor can be enabled or disabled by on-off signals from the syscon. These usually operate transistor or diode switches in the motor drive amplifiers.

Loading motor drive. Most VCRs have a *loading motor*, which via a series of cams and levers, drives the tape-threading guides, pinch-roller engagement, reel brakes, back-tension lever, front-loading mechanism, half-load arm etc. It is driven in both directions (via a power-drive IC) by the syscon IC.

Reel brakes. In some VCR designs, tape reel brakes are operated mechanically from the loading motor cam. Other designs use magnetically applied brakes under the control of the syscon. During the threading phase the spools need to be unbraked to enable a loop of tape to be drawn out of the cassette. During unthreading, spool drive needs to be applied to take up the tape loop, pulling in the slack as it becomes available.

User Indicators. The syscon drives panel-mounted LED, fluorescent, or on-screen displays to indicate machine and tape status. The number and sophistication of these depends very much on make, model (and price!) of the VCR.

Programming muting. To prevent unlocked pictures and mistracking effects from being seen as the machine runs up to speed (*lock-up time*) the VCR's video – and sometimes audio – sections are muted, either for a fixed time-period after play mode is entered, or until the servo systems are locked up. In the latter case, the mute will operate in the absence of off-tape tracking pulses (e.g. blank or faulty tape), and in the event of servo malfunction. During 'trick-speed' operation, such as pause, shuttle search and slow or reverse motion, the servos will not be operating normally, and the syscon must override the auto-mute circuit.

Signal routing. The signal-processing circuit functions are switched by the syscon between record and playback functions. The syscon is also responsible for tuning and timer operations in many VCRs, and often auto-set-up (see later) in modern designs, where these and other functions are governed via a serial data command line like the I^2C system we met in Chapter 8.

Early syscon

We have seen that the inputs and outputs from the syscon are generally in binary form, or logic levels 0 or 1, and that the primary function is one of decision making. This means that a syscon can be made entirely of logic gates, invertors and counters, though interfacing circuits are required to enable their outputs to drive motors, solenoids and the like. The command-and-co-ordinate role of a VCR syscon is ideal for a microprocessor, and an early design is shown in block diagram form in Figure 21.1. Inputs, both command and protection/feedback, enter the chip on the left, and outputs leave on the right. By way of interfacing, IC801 in the diagram is supported by eight further chips and a multiplicity of transistors and diodes.

The control keys are (electrically) arranged in a matrix – pressing a key links two input pins on the processor, and the command required is interpreted within the chip. The other input lines have already been described; here UL is UnLoad, a microswitch which closes at completion of tape unthreading to permit stop or eject modes to be entered. Timer CTL comes from a separate tuner/timer chip, and has the same effect as manually keying record so far as IC801 is concerned.

Looking now at the output lines on the right of the chip, 33 is a signal-mute line, while pin 30 switches down most of the power-supply section during off (in fact 'standby') mode. Pin 58 receives the SW25 pulse to show that the head-drum is rotating OK, while no. 55 takes pulses from the take-up reel rotation-sensor opto-coupler. Pin 54 is the IC's source of clock pulses. Pins 25, 24, 26, 27 and 28 drive deck-mounted solenoids (via transistor power-switches) in this early design. Apart from pin 38 (loading motor control) all the other output ports drive simple LED function indicators here.

While this 'basic' microprocessor syscon was an improvement over the very first manual/mechanical control system, its relative inflexibility and need for lots of interfacing circuitry and mechanics was a limiting factor.

Multifunction control chip

Current designs of control microprocessors and their peripheral circuits offer greater (external) simplicity, versatility and lower cost. Figure 21.2 illustrates this, with a syscon device incorporating all the servo electronics for the Hi-Fi VHS deck of which it is a part. In this diagram the only interfacing devices visible are IC502, an 8-pin loading motor driver, and transistors Q501/2, end-sensor pulse invertors. Most of the input and output functions are self-explanatory in the light of the foregoing text: we shall examine those which merit further explanation now. Pins 74–77 are normally held high by their pull-up resistors unless grounded by the deck-mounted mode switch or *rotary encoder*. Taking the place of the AL and UL switches of older designs, it gives the control chip information on the position of the deck mechanics, and its messages are mainly reflected in the outputs (loading +/–), of IC pins 79 and 78 whenever the deck mode changes. These may be commanded by the user or auto-invoked by the syscon itself, for

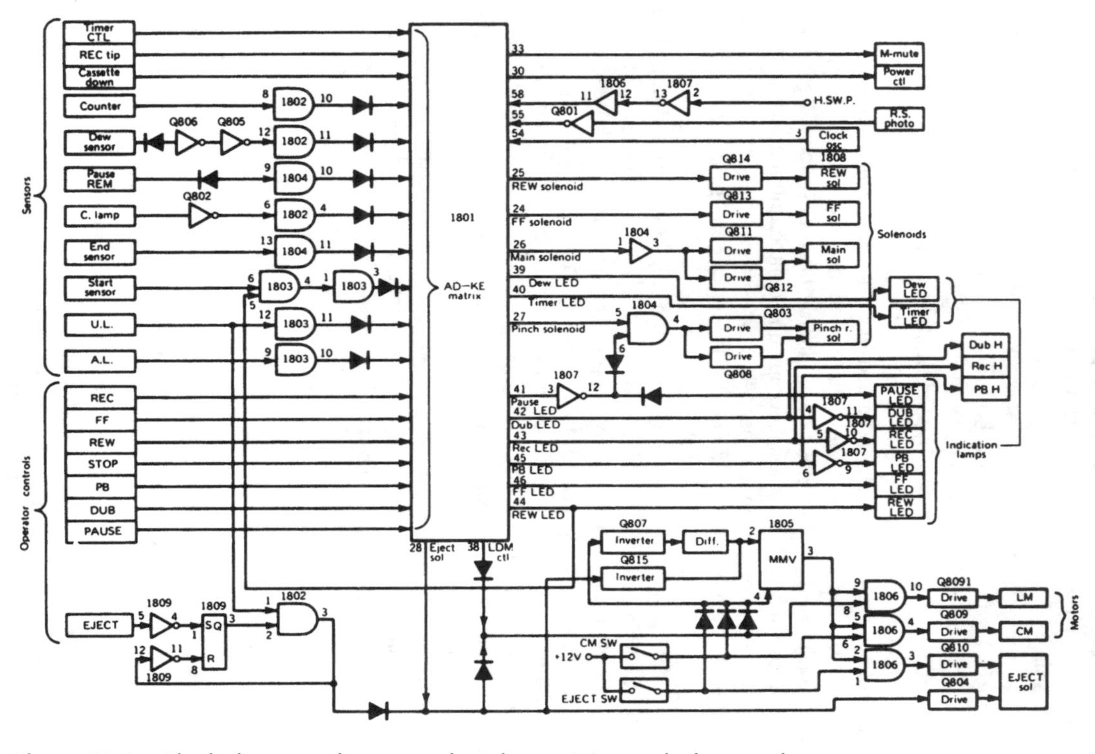

Figure 21.1 *Block diagram of a syscon for 'electronic' tape deck control.*

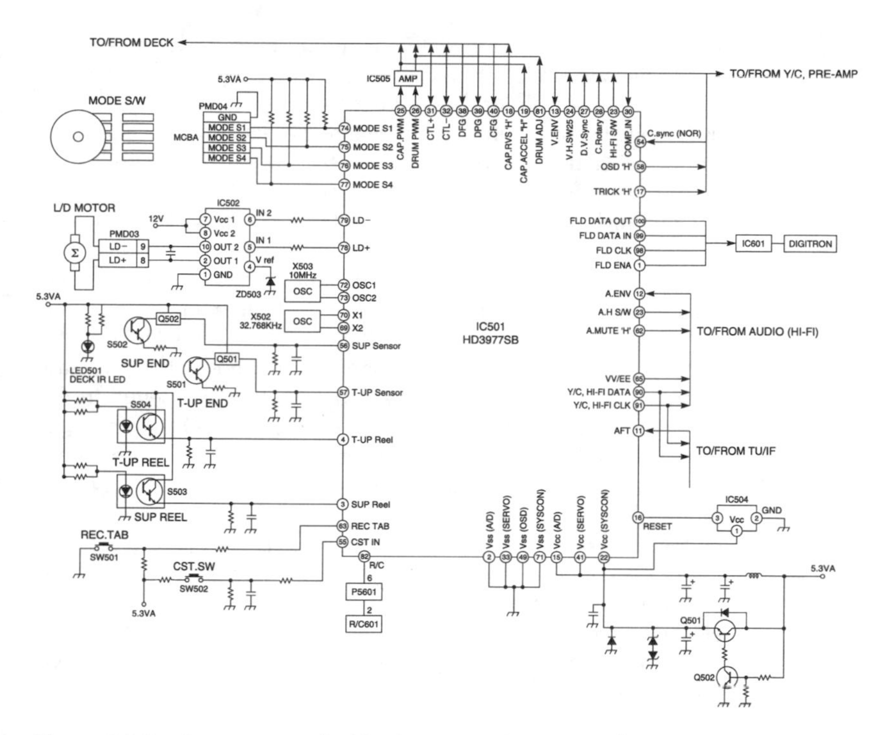

Figure 21.2 *System control chip, incorporating servo electronics (LG Electronics)*

instance at tape-end as indicated by a 'high' at pin 56, supply side tape-end detector. The reel-rotation sensors at IC pins 3 and 4 consist of optocouplers through whose gaps pass the castellated lower skirts of the two spool-drive turntables to give low-rate pulse trains. Pin 82 is the input port for serial control data from the remote control sensor, a module containing an infra-red phototransistor as described on page 120. Its output pulse train is decoded and actioned within IC501 here. Separately-filtered Vcc supplies for syscon and servo functions enter the IC at pins 22 and 41 respectively, while a reset pulse is applied to pin 16 shortly after switch-on to initialise the chip. In its turn the control IC resets the deck – via the loading motor – to stop mode if for any reason it is elsewhere at switch-on.

Examining now the output ports on the right of the IC diagram, pins 90 and 91 connect the I^2C bus to audio and tuner/IF sections of the VCR, conveying control data in serial-pulse form, while pin 11 samples the tuning point to provide AFT action to maintain accurate tuning: the resulting 'offset' value is passed back to the tuner via the I^2C bus, along with commands for the sound and vision sections of the VCR. Pin 65 switches the signal electronics between record and play modes. Pin 62 goes high to mute audio; 23 switches between longitudinal and Hi-Fi sound tracks during replay, and no. 12 takes a sample of off-tape Hi-Fi carrier envelope signal for sound-switching and ident purposes. The next four pins up in the diagram, nos 1, 98–100, are concerned with the FLD (florescent display) panel on the front of the machine: they govern the 'digitron' indicator via a separate front-panel-mounted decoder/driver chip IC601. IC501 output pins 17 and 58 go high for 'trick-play' and on-screen display respectively.

The ports drawn along the top of the chip in Figure 21.2 are mainly concerned with the servo functions, and have counterparts in the diagram of Figure 20.15, except nos 18, which goes high to reverse the capstan motor; and 19, whose 'boost' function when high quickly changes the rotational speed of the capstan motor during mode changes. Finally, at the top right hand corner of the chip diagram are some ports concerned with the Y/C section of the VCR's electronics. Pin 13 takes off-tape luminance envelope signal to govern auto-tracking; 24 and 23 provide the head-switching squarewaves for luminance and Hi-Fi audio respectively; 28 the colour-under phase-rotational signal whose effect we saw in Chapter 18; and 27 the artificial vertical sync pulse which is inserted in the video output waveform to stabilise still-frame pictures. Input pin 30 takes composite video during record, and leads to a sync separator within IC501 whose output provides the phasing reference for the head drum when magnetic tracks are being laid down on tape.

Not shown on this diagram, but important nevertheless is pin 11, connected to a ladder-network of operating-key switched resistors on the front control panel. The 'step' input thus provided is A-D converted inside the chip into deck-mode and channel-select commands. Also present inside IC501 here is a complete OSD (on-screen display) caption and symbol generator whose output is added to the video signal which loops through the chip via its pins 43 and 45.

Key scanning

The resistive switched-ladder method of interfacing user control keys to the syscon processor chip is a good one, economic of IC pins and capable of supporting a large number of functions. An alternative technique, also used in infra-red remote control handsets, is sequential scanning of the control keyboard. A keyswitch matrix is set up as shown in Figure 21.3, and is fed at the top with 'strobed' scanning pulses of perhaps 1 ms duration in sequence, so that SP1 is energised for 1 ms, then SP2, SP3 and SP4. The output lines are monitored by the micro for returning pulses, so that if line OPB 'lights up' during SP3, key 10 has been pressed; if line OPD comes on during SP4, key 16 has been selected, and so on. The returning pulse from an OP line is decoded within the micro and stored in its internal memory as a command to be processed. This key-scanning system uses a total of eight lines to handle 16 keyboard commands; the number of keyboard lines can be further reduced by more complex scanning systems. Some VCR designs, notably those in camcorders, have a scanning system embracing not only control keys but such things as deck sensors.

Multiplexing

In the sorts of machine we've described it's not hard to imagine that the number of command and feedback inputs, and control outputs from the syscon chip will run

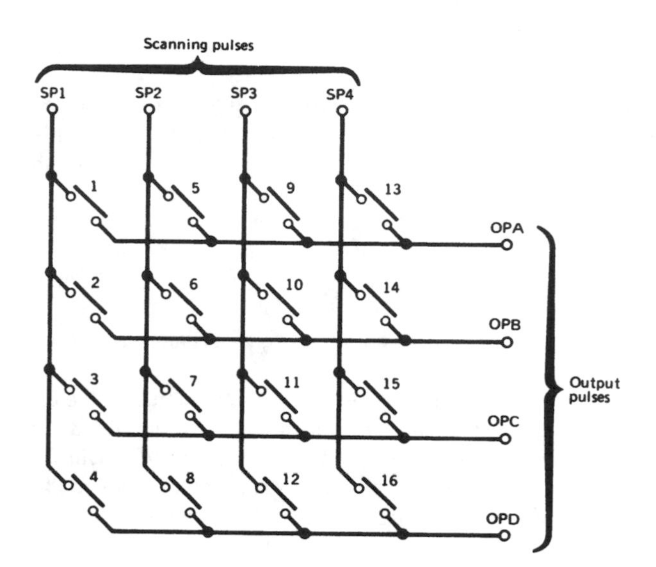

Figure 21.3 *Key-scanning matrix for economy in wiring and IC pins. Operation is explained in the text*

to over 100. It is impractical to provide more than 100 or so pins on this sort of IC, so pin-sharing systems are used on input and output ports. In early designs it was done by time-division multiplexing, in which outboard *expander* chips work on a sequential basis, switching several data lines to and from the micro ports in turn when requested by the control or 'chip-select/enable' line from the micro's CPU. During the 'waiting' period the expander chip stores data in its own internal memory.

Later control-system designs solve the problem by the use of a serial-bus control system like I^{2}C, covered above and in Chapter 8. Even with the benefit of that, the chip featured in Figure 21.2 sports 100 pins, every one of which is used for some purpose!

Real-time tape counter

The simplest tape counter systems merely count the revolutions of the take-up (TU) spool. This is a very non-linear measuring system because the tape length (and hence running period) between counts depends entirely on the amount of tape which has accumulated on the TU spool during playback and the resulting diameter of the reel. Real-time tape counters provide a reasonably accurate idea of the elapsed time at any point in hours, minutes and seconds. They depend on counting control-track pulses at the CTL head in record and playback modes, reckoning them at the rate of 25 per second and displaying the accumulated count on an LCD, fluorescent or on-monitor-screen readout.

Index search

In Chapter 20 we saw that the control track pulses are written onto the tape primarily for the purpose of servo phase-lock – aligning the paths of the video heads with the centres of the tracks over which they are sweeping. In fact the servo system only uses the leading edge of each pulse induced in the CTL head by the control track. This opens the way to the use of the CTL pulse for a third purpose, that of writing an index mark on the tape. Normally the mark/space ratio of a CTL pulse is 60:40. To make an index mark the ratio is briefly altered to 28:72, which makes no difference to the operation of the servo circuit, but can easily be picked up by an integrator or time-counter to give a 'flag' signal whenever a mark is detected.

At the start of each new recording – and at other points programmed by the user if required – an index mark is automatically put on the tape. During replay the user can invoke index search in either direction, whereupon the VCR will fast-transport the tape as required, reverting to play mode as soon as an index mark is encountered. Developments of this system are *intro-search*, in which the machine will play for a few seconds at each index mark before fast-transporting to the next; and multi-index search, whereby the user can program the VCR to stop and play at the *nth* index mark in either direction. Hence the need for the half-loading mechanics described in Chapter 17. Video-8 formats have similar facilities for both indexing and real-time tape counting, but necessarily rely on track-counts from the video heads.

Remote control

For remote control of VCR functions an infra-red light link is used between a hand-held commander unit and the machine. For economy of battery power consumption within the handset the encoder IC there is virtually dormant until a key is pressed, whereupon it comes to life, starts up its clock oscillator (governed by a crystal or ceramic resonator) and scans the keyboard matrix to ascertain which key has been pressed. From that a command code is generated by reference to a ROM look-up table. The encoder chip prefaces each command by a *framing code* (which identifies the commander handset type and primes the receiving-end decoder) and then sends a binary pulse train in the form of a string of *words*, each unique to the mode requested. With only a single infra-red link, the data is necessarily sent in serial form! On receipt inside the VCR, a *shift register* converts the serial data to parallel form ('static' data on separate lines) for decoding within the control microprocessor.

Record timer

Many remote control units have a timer-setting facility in which the times, dates and channels of the TV programmes required to be recorded can be key-stroked in and displayed on an LCD panel on the hand unit. When they have been checked and verified by the user, the data, which has been stored in a 'long' register within the handset, is transmitted to the VCR in a long string of binary words, prefaced by the usual framing code. In a fraction of a second all the data is conveyed and stored in memory in the VCR control section; for confirmation purposes it's decoded and displayed on the front control panel of the machine or on the TV/monitor screen. When the control section detects a coincidence between stored and real-time data it switches the machine out of standby, selects the specified channel and starts the recording process. At the end of the set period it powers-down the VCR and re-enters the waiting phase until the next time-coincidence is detected, and so on.

An extension of this technique is used in VideoPlus remote control handsets. Every possible combination of date, time, programme length and transmission channel is given a unique code consisting of up to seven numbers. These are published in the programme guides of newspapers and in listing magazines, and keyed in by the user rather like a telephone number. Within the VCR they are stored in RAM as binary data, decoded by reference to a ROM look-up table and then used to select the broadcast channel and invoke the recording as described above: again the time/date/channel settings can be displayed for confirmation by the user. The advantage of this VideoPlus system, designed and patented by Gemstar, is its extreme simplicity of operation for the user.

PDC

A problem can arise with VideoPlus if the programme schedules are disrupted, for example by extended sports or news coverage. To cater for this, and in effect turn the timer-record process into a 'closed-loop' one, a PDC (Programme Delivery Control) protocol has been agreed between broadcasters and receiver-chip manufacturers.

When a PDC-equipped VCR is set to timer-record mode, it is primed to look for specific codes carried in the teletext datastream. We saw in Chapter 9 how 'packets' are used there, and that Packet 8/30 carries PDC codes. System control processors like that shown in Figure 21.2 incorporate a PDC decoder attuned to this particular slice of data, transmitted once per second on TV line 16. When the appropriate codes for the required programme (even be it at the 'wrong' time *or channel*) come up, the control system springs to life to record what's been requested, terminating the session when instructed over the air. PDC only works reliably when the received signal is good enough to afford clear text decoding, *and* when the broadcaster transmits the correct code and uses the correct protocol.

Auto set-up

The presence in teletext packet 8/30 of broadcaster and channel ident, and real time data paves the way, with a suitably-equipped decoder/control system in the TV or VCR, to fully automatic setting up of tuning and clock time. The process is initiated – manually or automatically – during installation, and starts with an auto-tuning-sweep of all available broadcast bands. Whenever a transmission is found the sweep stops while its packet 8/30 data is downloaded into temporary memory, together with data on carrier frequency and received signal strength/quality. When all available transmissions have been logged in this way a sorting process takes place, in which the broadcast channels are identified and (where there's a choice) the best and strongest selected. The tuning point for each of these is binary-coded and stored in non-volatile memory in the order of BBC1, BBC2, ITV, CH4, CH5 etc. so that they are correctly selected by user-programme keys 1–5 etc. Real-time and -date information is also read out of the text data for setting of the front-panel or on-screen clock and calendar. This auto-set operation is similar to what takes place in a DTV receiver, and relieves the user of the need to do anything but provide an adequate aerial, and hook up to the TV, also tune in the latter if the link is an RF (aerial-lead) one.

Basic microprocessor operation

A microprocessor is an LSI (large-scale integration) integrated circuit, and functions in similar manner to the CPU (central processing unit) of a computer. In normal computer applications it is supported by one or more separate memory chips, in which data is stored and retrieved by the CPU as required.

The sorts of microprocessors found in VCR machines, however, need very little memory capacity, and it is possible to design a complete simple computer system into a single package, perhaps better termed as in 'integrated microcomputer'. It contains all the elements of a small general-purpose computer on a single chip. Thus (Figure 21.4) we find a CPU (central processing unit), ROM (read-only memory) RAM (random-access memory) *clock*, and I/O (input/output) *ports*. The CPU is the heart of the device, controlling the interchange of information between the various sections of the chip and the I/O ports. The ROM contains control programmes covering all combinations of input signals to the system, and these instruct the CPU when the latter calls for them in turn. The RAM is used as a temporary 'storeroom' for data which needs to be held pending the completion of mechanical or data-processing operations. Input and output ports accept requests and dispense commands to and from the microprocessor. The clock is a steady source of timing pulses which 'steps' data sequentially through and around the microprocessor. Address, data and control bus systems are the internal highways for data interchange between the sections of the chip; data is loaded into these on a sequential basis, preceded by an electronic 'label' to indicate its routing and destination.

Under the control of its internal and pre-programmed 'action memory' the micro can cope with any combination of circumstances, and come up with the most logical answer. Taking the example of the user keying in fast-forward during a rewind operation, the micro-syscon will enter stop, wait for that to happen, then programme the deck for fast-forward operation. In other designs, fast-forward will be entered immediately, with reel motor drive and braking controlled in such a way that the changeover is effected in the shortest possible time consistent with

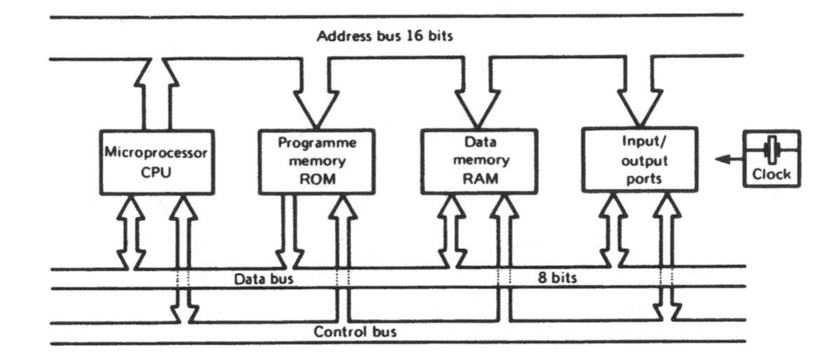

Figure 21.4 *Internal block diagram of a microprocessor IC*

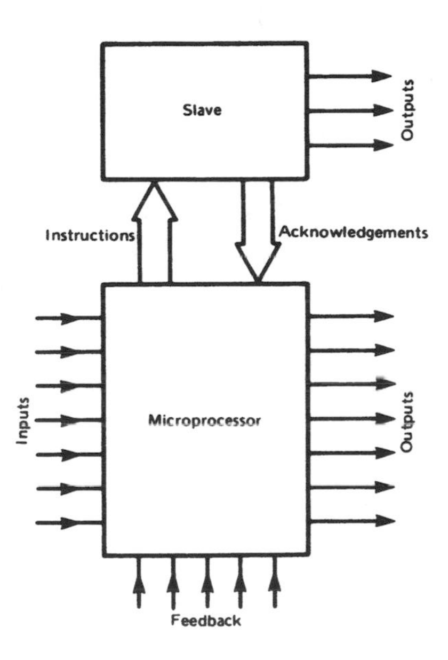

Figure 21.5 *The versatility of the microprocessor can be increased by the addition of a 'slave' chip*

not stretching or 'looping' the tape. If more memory or processing capacity than can be accommodated in one microprocessor is required, then an additional slave chip can be tacked on, in the manner of Figure 21.5.

Regarding memory, the only external requirement is an EEPROM (Electrically Erasable Programmable Read-Only Memory) to hold such data as programme and output-modulator tuning points, video-headswitch point, model features and identity, and user-programmed settings and features from the installation/set-up menus. The EEPROM is connected to the serial-data control (e.g. I^2C) bus, along which it is accessed by its own address for transfer of data to or from its internal store.

Summary

In this chapter we have examined the role and operation of the VCR's system control section, and explored the interfaces by which it communicates with the user, the tape deck and the signal circuits. In practice syscon ICs are very reliable, and most problems which crop up are external to the chip itself, with 'user' and 'software' troubles near the top of the list.

22 *The complete VCR*

The circuits and processes discussed earlier in this book form the basics of a home VCR and are, for the most part, unique (in the domestic environment) to videotape machines. In this chapter we shall deal with the peripheral circuits, i.e. those which take no part in the basic video recording process but are essential to 'service' and power the machine and provide operating convenience and flexibility. All the circuits to be discussed have counterparts in other domestic equipment such as clock-radios, TV games, audio tape recorders and TV receivers. Figure 22.1 shows a block diagram of a VCR.

Mixer/booster

To avoid regular plugging and unplugging operations, the VCR is permanently connected 'in series' with the aerial lead to the TV receiver. When the machine is off, or recording a programme other than the one being viewed, it is important that normal TV reception is not affected by the machine's presence, so a *loop through* facility is provided in the aerial booster, a small RF amplifier which is permanently powered. Its modest gain cancels the losses incurred in the extra RF plugs, sockets and internal splitting of the RF signal within the VCR.

UHF tuner and IF amplifier

Most recordings made on a home VCR come via broadcast transmissions, so the machine needs a tuner and receiver built in to select and demodulate broadcast programmes. At the time of writing, all terrestrial UK programmes are radiated in the UHF band, and the tuner and IF arrangements are identical to those provided in TV sets. An effective AGC circuit is provided to ensure a constant signal level to the recording section, and AFC (Automatic Frequency Control, sometimes known as AFT, Auto-Fine-Tuning) feedback maintains correct RF tuning. Some VCRs have auto-set routines, in which they tune themselves, working from a program

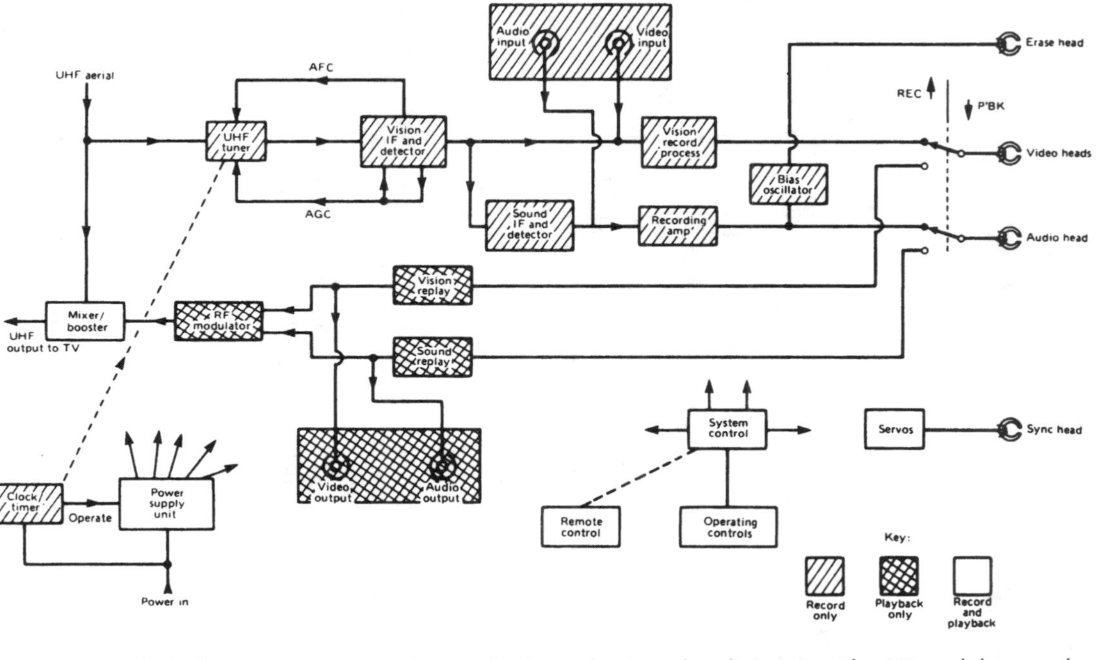

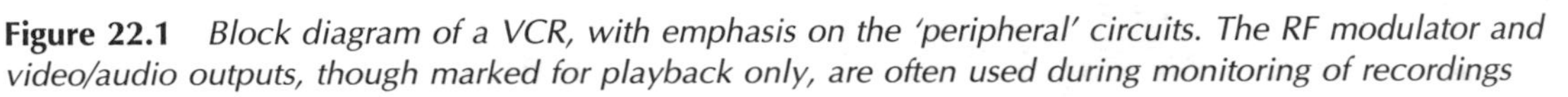
Figure 22.1 *Block diagram of a VCR, with emphasis on the 'peripheral' circuits. The RF modulator and video/audio outputs, though marked for playback only, are often used during monitoring of recordings*

in the control section, and set their own clock from teletext. Also found in some VCRs are built-in satellite tuners and/or surround-sound decoders.

Sound

An intercarrier IF amplifier and detector follows the vision detector to provide a sound signal for recording on a tape. After the detector, the sound recording processes follow audio cassette practice, with the audio signal being added to an HF bias source before recording on a longitudinal track on one edge of the tape as shown in Figure 17.5. The slow speed of the tape and narrow audio tape track do not make for ideal recording conditions, and noise reduction systems are commonly used to improve performance. Foremost among these is the Dolby system, which involves a form of non-linear and frequency-dependent pre-emphasis during the recording process, and complementary de-emphasis in replay.

Stereo machines use one of the hi-fi systems, VHS hi-fi, AFM-stereo or PCM, described in Chapter 19. Virtually all hi-fi stereo VCRs have built-in Nicam decoders of the type described in Chapter 11. Most pre-recorded cassettes offered for sale or rent also incorporate good stereo and surround sound tracks.

The concept of stereo-with-TV, be it from broadcast or video tape, is a difficult one. The whole reason-for-being of a stereo setup is to recreate a wide, vibrant and 'living' sound stage in the listening area. Even with a very large screen by current standards the picture size (and position, very often) has little correlation with the sound field in most domestic situations. Stereo TV sets are now commonplace, and while they are much better than early monaural models, the loudspeaker spacing is necessarily closer, drawing in and tightening up the sound stage to match the small picture 'window'. This is not ideal for concerts and recitals, whose main feature is the sound itself! Ideally, link a hi-fi VCR to hi-fi amplifiers and loudspeakers.

Bias oscillator

To provide a suitable AC recording bias for the 'longitudinal' audio signal, and to generate an erasing signal to wipe out video tracks, a power oscillator is used. Usually consisting of a discrete transistor oscillator built round a feedback/driver transformer, it operates at about 60 kHz, and is used only in record mode. Besides the audio head it powers a full-width erase head to wipe all signals, sound, picture, and control track, off the tape on its journey towards the head drum during recording. As described in earlier chapters, no recording bias signal is required for the video heads as this function is performed for luminance and chrominance by the constant-level FM luminance recording signal.

Clock-timer

Depending on vintage and price, VCRs vary in their timing and programming facilities. All machines have an on-screen or front-panel clock display, fluorescent digital in mains-powered static machines and LCD (reflective) digital in battery-operated portables to conserve power. The clock is arranged to show real time as a permanent display, counted down from a local accurate reference – usually a quartz crystal, sometimes 50 Hz mains frequency. A digital memory in the clock/timer chip is capable of storing the clock display data for pre-set time, and this stored data is continually compared with the real-time display. When the two correspond exactly, a coincidence-detector opens several gates to invoke record mode in just the same way as a clock-radio, or clock-cassette/radio.

In the first machines, that was the limit of operation of the clock/timer, except that in some models the clock display doubled as tape counter indicator when this function is selected by the user. Modern designs of VCR have comprehensive timer facilities, typically offering a choice of eight or more pre-programmed recording 'events' of any desired length, on any channel over many consecutive unattended days. This is achieved by an extension of the above technique, with the clock detector now looking for a day-coincidence as well as time-display coincidence, then gating on the broadcast channel held in its memory as well as record mode for the tape deck. These are then held on for the requested time. The timer circuit is based on a microprocessor (see Chapter 21) which is capable of informing the operator if he keys in impossible requests like 'start at 1930 hrs, stop at 1900 hrs, or overlapped recording instructions on different channels!

Florescent display

Most VCRs have a clock/status display on the front panel, as do DVD players, some set-top receiver boxes etc. Their operating principle is similar to picture tubes in that they have a filament and a florescent screen. Here, though, all the required symbols are 'printed' in phosphor on the faceplate, and the hot cathode is in the form of a bar running the entire length of the panel. The electrons released from it are attracted towards the florescent screen segments when not inhibited by a negative potential on the nearby switching electrode, which latter is controlled by a drive IC, very often a purpose-designed microprocessor which may also look after other 'front-panel' duties: user-key and remote-control decoding, timer functions etc, where the main micro does not carry them out. The switching electrodes work on a 'strobe' basis, whereby two of them ('grid' and 'segment') must be simultaneously pulsed on to light up an icon or one of the segments which make up an alphanumeric symbol. An idea of this is given in Figure 22.2. Here the grid areas (labelled along the top) are sequentially 'primed' by pulses at connection pins 4–11. Within each grid are up to 12 segments, pulsed likewise via pins 21–32. To light up the REC symbol, then, its segment pin (shared with segments in other grids) must be pulsed on during the periods while G1 (pin 4) is energised. To get a number 2 at the right of the panel, segments a,b,d,e and g must be pulsed on during the strobe period of grid 8. And so on – the drive IC carries out decoding and matrixing. Filament voltage is applied to outer pins 1 and 35. The accelerating potential can be as low as 25 V because of the very close spacing of 'anode' (screen) and cathode electrodes.

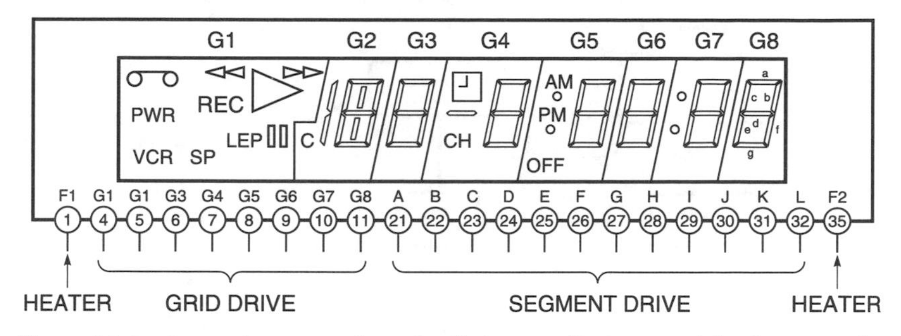

Figure 22.2 *Internal construction of a florescent display panel for front-panel readout from a VCR*

RF modulator

When the machine is playing back, a TV signal at baseband is produced, and this cannot be applied to a TV receiver whose only signal input facility is via its aerial socket. To cater for this, an RF modulator is provided within the VCR, working as a tiny TV transmitter. Baseband audio and video signals are applied to its input for modulation on to a UHF carrier, whose frequency is chosen to fall into a convenient gap in the broadcast spectrum. A rear-access preset control swings this frequency to avoid beat effects with other RF signals in the vicinity. Some VCR modulators are programmable for any channel in the broadcast band. Modulator characteristics are according to CCIR system I, the UK standard, and a diagram of the modulator is given in Figure 22.3. Output level is set to be 1–3 mV (the optimum for a TV receiver), and this modulated RF carrier signal is added to the booster output for application to the VCR's RF output socket.

During record, regardless of the signal source which may be off-air, TV camera, cable, another VCR or whatever, the signal being recorded on tape is applied to the RF modulator so that it may be monitored on the TV set if desired. This is called the E-E (electronics to electronics) mode to distinguish it from off-tape playback mode. The E-E signal is taken off from the record electronics as late as possible, and several of the record and playback circuits (such as the luminance/chrominance adding stage) are usually included in the loop.

Set channel facility

During initial setting up, when neither the TV nor the VCR are correctly tuned, it is difficult to establish the correct tuning point for the TV when adjusting it to the output channel of the VCR's RF modulator, unless a pre-recorded tape is available to provide a playback signal. To assist with this, most machines have a simple video pattern generator incorporated, which modulates the RF carrier with a 'marker' to enable the TV to be tuned. Some VCRs generate a plain coloured raster or a pattern/text for this purpose.

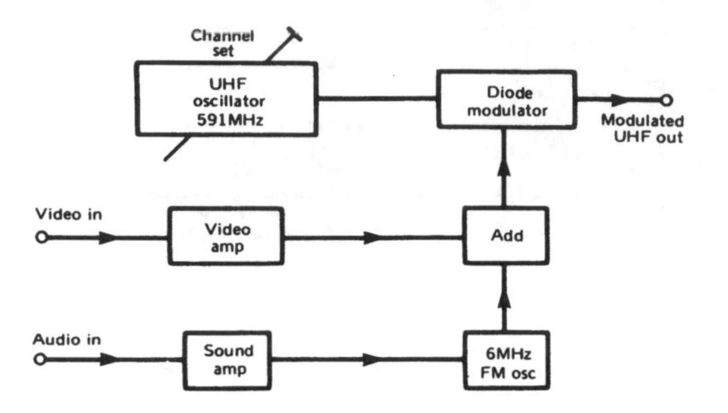

Figure 22.3 *Basics of UHF RF modulator*

Audio/video in/out connectors

Many of the signals required to be recorded are baseband for video and audio, typically coming from a TV camera or from another VCR for copying/editing. Sockets are provided in the form of (e.g.) phono for sound, BNC or phono for vision, and a 21-pin SCART socket for both.

Baseband outputs are also provided in similar connectors to those mentioned above, so that connection to audio systems, TV *monitors* and other equipment is possible. The standard for baseband video signals is universal at 1 V peak-to-peak (sync tip to peak white), and for sound a level of 0.5 V rms.

Record-playback switching

In almost every department of the VCR, from sound to servos, from tuner to syscon, function switching is required between record and playback modes. The record/playback switching is carried out by electronic switches in the form of diodes, transistors and ICs, toggled by supply lines which appear during the relevant mode. Thus we find '+12 V record' or '+9 V play' lines distributed to all operational blocks of the VCR. While the source of these lines is the PSU, they are enabled by the user's operating keys, usually via the syscon.

Power supplies

A home VCR consists, as far as its electrical section is concerned, of a lot of transistor/IC circuits, perhaps a solenoid, and between two and four low-voltage DC motors. Most VCR circuits operate at supply-line voltages between +5 and +20 V, with some critical applications requiring close stabilisation and decoupling. In addition to these, the varicap tuning system calls for a highly-stabilised 33 V supply, and the florescent display panel one of about −30 V. Older models

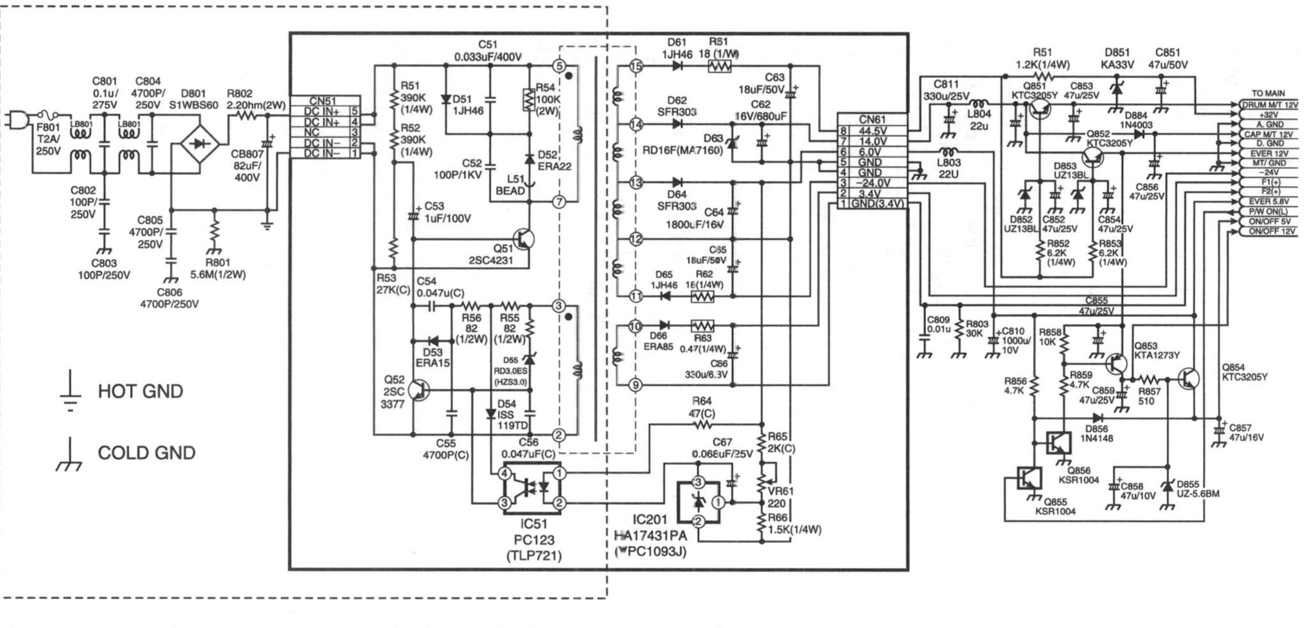

Figure 22.4 *Mains power converter for home-deck VCR (Daewoo Electronics)*

also require a potential of −25 V or so for reprogramming of the EEPROM data store.

While some (mainly older-model) VCRs use a power supply based on a 50 Hz mains transformer and large reservoir capacitors, the requirements are best and most efficiently served by a switch mode power converter, an example of which is given in Figure 22.4. Mains power enters on the left, passing through EMC filters on its way to full-wave rectifier D801, whose action charges reservoir C807 to about 325 V. This voltage is switched, rapidly and cyclically, by chopper transistor Q51 to the primary winding (pins 5–7) of converter transformer T51. Initial conduction in Q51 is triggered by the charging of kick-start capacitor C53, after which oscillation is maintained by T51 feedback winding 3/2. The energy passing through T51 is governed by the switching duty-cycle of Q51, which depends in turn on the degree of conduction in regulator transistor Q52. A large influence on Q52 conduction is the feedback coming from optocoupler IC51 which samples the level of the 6 V line from converter transformer secondary winding 12–13. Thus is a stabiliser feedback loop set up, with fine voltage adjustment preset by resistor VR61. Mains isolation is provided by the gaps inside T51 and the optocoupler, as shown by the heavy dotted line in the diagram.

On the secondary side of T51 there are five rectifier/capacitor sets, all conducting during the 'off' period of Q51. D61/C63 provide (via a further stabiliser) the UHF tuning voltage of +33, while D62/C62's 14 V line is used to drive the deck motors. D64/C64 generate power to operate the control microprocessor and other chips; D66/C66 provide a 3.4 V filament/heater drive for the florescent display panel, whose – 24 V accelerating potential comes from D65 and C65. Outboard of this supply panel there is a series of regulators and stabilisers, and a pair of transistor switches, controlled by the syscon, to provide 'on/off' 5 V and 12 V lines, only present when the VCR is switched out of standby mode. The VCR of which this power supply is a part draws 17 W from the mains when in use, much less in standby mode.

Clock back-up

Although most types of EEPROM can retain stored data (for ten years or more) without the need for a sustaining voltage, it's necessary to keep the clock oscillator – but not its display – ticking over during power cuts and while the VCR is temporarily disconnected from the mains. It saves having to reset the time, and retains stored timer-recording data. Clock back-up supply comes from a small battery or electrically-large capacitor which becomes isolated from the main +5 V or +3.3 V supply by a diode when mains power is lost, and can sustain the oscillator and timer-programme memory for a period varying from a few minutes to several hours, depending on the design of the machine.

23 *Analogue camcorders and home video*

Camcorders (camera-recorders) are the most advanced form of 'domestic' technology. Combining the elements of camera, encoder, VCR electronics, deck- and lens-mechanics and computer control, they draw on the systems, technologies and techniques described in most of the chapters of this book. Here we shall draw together these many threads and see (necessarily in broad terms) how they go together to make up the most expensive – in terms of pounds sterling per kilogram – bunch of electronics and mechanics available on the shop shelves.

Optical section

The lens assembly is a complex ensemble of precision glassware, with adjustment for focus and zoom by sliding members within the barrel. Both focus and zoom rings are driven by miniature electric motors; where manual (hands-on lens ring) adjustment is provided for these, the motors drive through slipping clutches. Zoom motor control is provided for in a body-mounted rocker switch, sometimes offering two zoom speeds. The auto-focus system consists of a servo loop with TTL (through-the-lens) picture-sharpness sensors. It's controlled by a microprocessor, and many models offer a choice of picture zones for auto-focus operation.

Also inside the lens assembly is a multi-bladed iris, whose operation is similar to that of a moving-coil meter. It too is part of a servo loop, this time controlled by the level of luminance signal coming from the image-sensor. The higher the light level in the televised scene the smaller the iris opening, with a consequential improvement in depth of focus field. There is also an AGC system in the video amplifier: between these two control loops the signal level – and hence contrast in the reproduced picture – is held constant over a huge range of ambient scene brightness. In very low light situations, however, noise (grain, snow, confetti) intrudes on the picture.

Interposed between the lens and the image sensor are two optical filters. An infra-red cut filter prevents most infra-red radiation reaching the sensor so that heat and similar energy sources have little effect on the picture, while a crystal filter takes out the finest detail in the incoming scene to prevent 'beat' and patterning effects due to the dot-matrix structure of the sensor IC and its colour filter. It also minimises cross-colour effects due to (e.g.) PAL signal bandsharing.

The image sensor itself consists of a mosaic of up to 500 000 photodiodes, each of which acquires and stores a charge proportional to the intensity of light falling upon it. As described on page 2 the charges are transferred by horizontal and vertical CCDs (charge-coupled devices) to the image-sensor's output terminal, whence they emerge (after processing in a sample-and-hold stage) as a serial video signal. Thus the scanning process is controlled in a series of small discrete steps, one pixel at a time, by the CCD driver stage in box A of Figure 23.1. Its timing is governed by the pulse generator, whose 'heartbeat' is provided by an SSG (sync and subcarrier generator) working from a precision crystal reference. As well as timing the scan and subcarrier frequencies and the colour burst and blanking intervals to broadcast TV specification, the SSG also synchronises such functions as internal caption and symbol generation for viewfinder indications and timecode generation, the latter used in precision editing. The colour components of the image are captured by a coloured translucent dot-matrix overlay bonded to the face of the pick-up sensor chip; the principle of operation is broadly similar to the striped-faceplate camera tube described in Chapter 3. Here each coloured dot-filter is aligned with its own silicon photodiode element below.

All camcorders offer a *fast-shutter* feature, in which a very short exposure is taken in each 20 ms field period. It prevents blurring of the image when freeze-frame mode is selected during replay of fast-moving images recorded on tape. It's achieved by sweeping all the charges out of the photodiode imager at a late stage in the field period, then allowing a relatively short time-slot for integration of the image before the charges are transferred into the 'bucket-brigade' CCD.

Image stabilisation

A problem with very small camcorders *(palmcorders)* is shake and wobble in the reproduced image due to the natural tremor in the human hand which holds it. It's especially troublesome at extreme zoom settings. To mitigate this, two very different technologies have been developed. The first, brought to the market by Sony, involves a vari-angle refractive prism, consisting of two silicon-oil filled plate-glass panels linked by bellows, mounted in the lens assembly. The prism's angle, relative to the pick-up sensor's image plane, is varied by drive coils under the control of a piezo-electric sensor of pitch and yaw of the camcorder's body.

The second anti-shake system is electronic in operation. It depends on the use of an A-D converter like that described in Chapter 12 in conjunction with a large DRAM digital memory in which one whole field of picture information is temporarily stored. The image for recording on tape is read from the centre section of the memory bank, using approximately 85% of its contents, area-wise. The pixel data corresponding to the outer periphery of the picture is selectively used by the EIS (electronic image stabiliser) processor with reference to four motion detection zones in the picture and an algorithm which distinguishes between camera shake and natural movement in the picture.

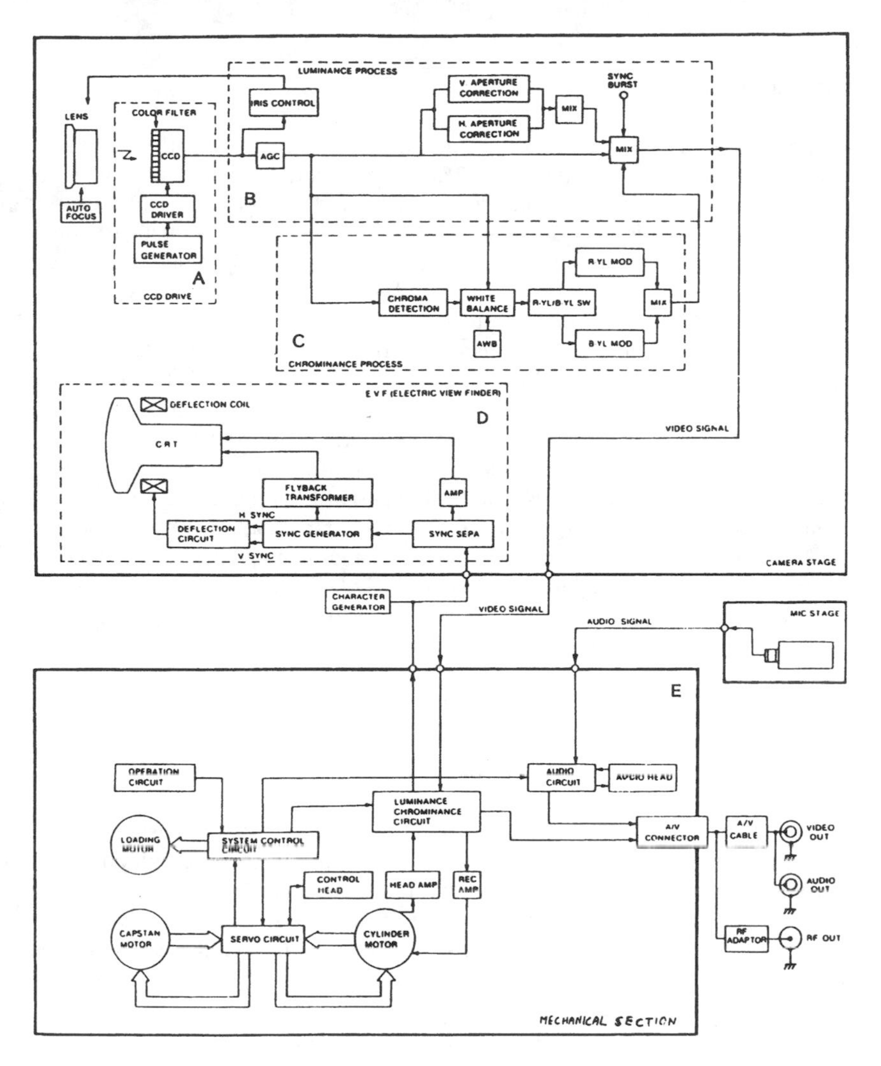

Figure 23.1 *Simplified block diagram of the electronic sections of a camcorder. The lettered blocks are described individually in the text*

Digital picture zoom

In addition to the optical zoom facility provided on all camcorders, those fitted with a digital field store are able to offer a further 'zoom' feature in which only the memory-data corresponding to the central section of the picture is read out, at normal field and line scanning rates. The effect on screen is of fewer but larger pixels rather than real image magnification, so it counts more as an 'effect' than as a true extension to the zoom range. Many other effects are possible when a field store is used in a picture-processing system, as we saw in Chapter 12, and will further explore towards the end of this chapter.

Luminance processing

Returning to the block diagram of Figure 23.1, and looking now at the processes shown in section B, the iris- and AGC-regulated luminance signal undergoes *aperture correction*, in which the sharp edges and outlines of picture objects are enhanced by adding overshoot and 'undershoot' before and after the transition in the video waveform as shown in Figure 18.12 and 18.13. While this process does not actually increase definition, it subjectively sharpens the reproduced image as viewed on screen. In fact these contour enhancers are used on replay for the same effect; some camcorders are fitted with an 'edit' switch to soften the action of the enhancer when the camcorder tape is to be copied onto another tape in a machine incorporating its own record-signal sharpener: the effect of two such circuits acting in series can lead to spurious effects and instability.

Another function of the camera luminance processor, not shown in Figure 23.1, is that of gamma correction, described at the end of Chapter 2. It's achieved by a voltage-controlled attenuator (VCA) circuit in which the control voltage is derived from the input signal itself. The way in which the luminance gamut is stretched is shown in Figure 23.2. The luminance signal has blanking, sync pulses and colour bursts added at the mix box on the right of block B in the diagram. At this point, too, the chroma subcarrier signal is superimposed, and we shall next examine its derivation.

Chrominance processing

Box C in Figure 24.1 is deceptively simple! In practice it incorporates a great deal of IC-based processing circuitry, which in advanced models is carried out (along with the luminance signal processing) in the digital realm.

The first step is to derive R and B primary-colour signals from the CCD image sensor output. It's done by a pair of sample-and-hold gates driven by a pulse output from the CCD driver circuit. The pulses are timed to coincide with the repetition frequency of complementary colour filter dots on the sensor faceplate.

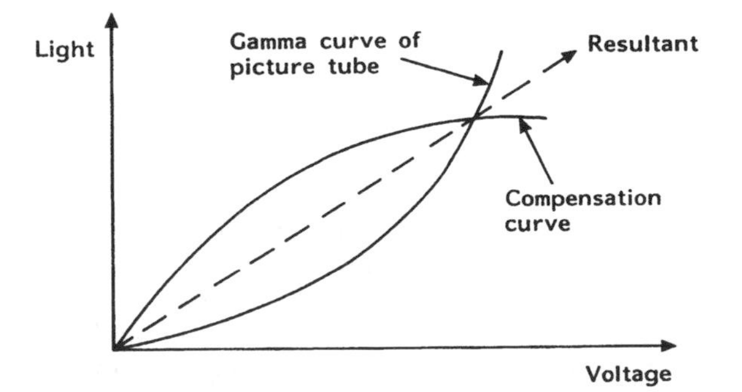

Figure 23.2 *Gamma-correction curves for light output compensation in a cathode-ray picture tube*

The samples thus obtained are fed to a switched matrix whose outputs form R and B signals with relatively low definition due to the 'coarse' nature of the optical colour filtering system. This is not a problem because the U and V signals derived from them have limited bandwidth anyway, as we saw in Chapter 3.

At this point the white-balance correction is applied. Different light sources (daylight, sunlight, artificial illumination etc.) have widely different hues. Human eyes can 'neutralise' these when viewing the actual scene but not, strangely enough, when they are reproduced on screen. All must be rounded up or down then, colour-temperature-wise, to the standard Illuminant D (see Chapter 2) for television purposes. Some camcorders have pre-selectable settings for white balance, typically correcting for indoor/outdoor conditions. All have an automatic facility for white balance (AWB) in which the incoming light is sampled, either through the lens or via a separate light-sensor on the camera body, and the gains of R and B amplifiers set accordingly. If the scene being viewed is a white card – or simply the ambient light via a translucent white lens-cap – the colour-balance is correct when both the R – Y and B – Y signals are at zero. The gain settings for R and B amplifiers can be stored in memory at each sampling and used thereafter until different lighting conditions obtain, when once again the sample-and-store process is manually invoked. Some camcorders continually auto-adjust colour balance on the basis of the sampled hue of the picture, but this can lead to desaturation and other spurious effects on certain types of scene.

Only when the colour balance is spot-on will white objects in the televised scene appear truly white in the TV picture, and only then will other hues be correctly reproduced. The most critical ones in this respect are flesh-tones, one of the few 'real' references for the viewer of a TV picture.

Now the B–Y and R–Y colour difference signals are derived in a matrix involving a bandwidth limited luminance (called Y_L) signal from block A in Figure 23.1. They are weighted to become U and V components and encoded into a PAL (or whatever) chrominance signal as described at length in Chapters 3 and 4 of this book. The chroma subcarrier joins the Y signal in the mix box of section B of the diagram. In high-band (S-VHS, Hi-8) camcorders the Y and C components are kept separate throughout to avoid the defects (primarily cross-colour, see the end of Chapter 4) of band-sharing inherent in PAL (Secam, NTSC) broadcast transmissions. In our camcorder example, however, the video signal is a composite one, now ready to leave the camera section. It takes three paths: to the tape-recording electronics, to the video output socket (for monitoring) and to the viewfinder, which we'll examine next.

EVF

The electronic viewfinder (EVF) can take several forms, but most often is based on a miniature black and white picture tube with a very small deflection angle and a high-definition screen, supported by a tiny bunch of electronics in the form of a single IC and a few peripheral components. These latter include a small line output section – transistor and flyback transformer – to provide scanning current in the line deflection yoke, and operating voltage for the electrodes of the mini-CRT, as shown in section D of Figure 23.1. The whole ensemble, is barely bigger than half a sausage even though it contains all the elements of a monochrome monitor, the operation of which is discussed variously in Chapters 1 and 8 of this book. The tiny screen is fronted by a slide- or twist-to-focus magnifying lens built into an eyecap.

Peripheral to the EVF, and drawn below it in our diagram, is the character- and symbol-generator which superimposes data on the EVF screen. The readouts thus provided can give the operator a wide variety of indications and prompts: battery charge state, light level, record/pause status, tape counter, focus zone, date/time (this can if desired, be superimposed on the recorded picture, too) white balance status, etc. etc. These are gathered by sensors associated with the control system and composed by a (usually bus-controlled) caption generator IC in synchronism with the field and line scan timings.

Alternative EVF systems include colour LCD screens of the sort described in Chapter 7, some types of which are on a side-mounted flip-out panel, and others built onto the camcorder's backplate, in either case removing the need for an eyecup magnifier system. They have the advantage of giving the operator some idea (though not as accurate a one as a conventional colour monitor) of the colour-balance of the pictures being recorded, and can – at a pinch – be viewed by more than one person at a time. On the debit side is their relative lack of picture definition; and their vulnerability to being 'washed out' by sunlight. Some camcorders offer the best of both EVF systems with both a monochrome tube and a colour LCD screen.

Audio section

All camcorders have an integral microphone, which may be a mono or – with hi-fi camcorders – a stereo type. Many have external microphone sockets for greater versatility and an opportunity to eliminate the motor- and handling noise which is almost inevitable in quiet situations where the ALC (automatic level control) drives up the gain of the audio amplifier. The action of the ALC circuit is necessary to enable the camcorder to cope with the huge range of sound levels it may encounter, from the rustle of leaves to the roar of a jet aircraft. All these have to be brought within the dynamic range of the tape system, discussed in Chapter 16. Where a hi-fi sound recording system is provided in the more expensive models, the dynamic range for sound recording is much greater and the ALC action can be less harsh. The camera sound is passed out of the camcorder during record along with the composite video signal for monitoring if required: very often a headphone socket is also provided for audio monitoring 'on the hoof'.

Sound recording and playback techniques were discussed in Chapter 19, and the processing circuits of camcorders conform to them. In VHS camcorders using small head-drums (Chapter 17) the provision of the extra heads for depth-multiplex recording makes for a complex and expensive drum assembly on the little tape deck! Video-8 and Hi-8 formats use AFM or (top-grade models only) PCM sound systems, again like those covered in Chapter 19.

Camcorder tape deck

Section E of Figure 23.1 depicts the essential parts of the tape deck, whose main distinction, compared with a homebase deck, is its extreme miniaturisation. Small head drums are used, with the multi-head and long-wrap techniques described in Chapter 17. The servos are digital types developed from the basic concept illustrated in Figure 20.9. The capstan and cylinder drive motors are physically very flat with direct drive (Figure 20.12) and no electrical moving parts, and are designed to react fast enough to counter the effect of camera movement.

The electronic (sound and vision) processing circuits associated with the tape deck are the same, in essence, as those already covered in the relevant chapters of this book; again the main difference is in the physical size of the components and circuit boards. In camcorders wide use is made of sub-miniature and surface-mounted components, mounted on tiny double-sided or multi-layer printed circuit boards, some of which are flexible, and most of which are peculiar in shape to match the profile of the case and the mechanical tape deck. An idea of the internal construction of the camcorder is given by the photo of Figure 23.3.

Control system

In a camcorder the control microprocessor has many more interdependent functions to control and co-ordinate than in a homedeck VCR, although the basis of its operation is similar to those already described for TV sets and VCRs. Although the system control heart in Figure 23.1 is shown as being based on – and confined to – the mechanical deck section, its electronic tentacles extend over every section of the camcorder, from the lens drive system to the AV output switching and power supply section. Control is effected by one or more two-wire serial bus communication systems, which also interface with subsidiary microprocessors such as that which controls auto-focusing.

An EEPROM is linked to, and governed by, the control microprocessor in similar style to that described in connection with the TV control system outlined in Chapter 8. Here its purpose is not to store broadcast-reception tuning points; it holds data on set-up levels and reference-points throughout the camcorder, set by the user, the service engineer and the manufacturer. It contains such diverse data as the head-switch point, the white-balance setting, audio record level and PAL encoding parameters. No longer does a factory operator or a service engineer twiddle pre-sets in the camera section, PAL encoder or luminance circuits: the data for these is written into the EEPROM via the remote control handset or from an external computer, whence they are conveyed during camcorder operation to the appropriate circuit section and there (where relevant) translated into control voltages or currents by D-A converters within the processing ICs on the spot. This simplifies and speeds up factory production and eases the job of the service technician – so long as he has the necessary hardware, software and data to do the job! Allied to this concept are diagnostic readouts, generated by the

Figure 23.3 *Inside a Sony camcorder: (a) optical block and picture processing sections; (b) circuit boards unfolded to show layout and flexible inter-board connectors*

control microprocessor, to relay to factory and service personnel fault data, typically: deck shutdown was invoked because of lack of rotation of the take-up spool/incorrect frequency of drum tacho pulses/excessive tape-loading period – or whatever.

Some camcorders have a serial data link to the control microprocessor from an edit-control port, using an operating code similar to that of the remote control system we examined in Chapters 8 and 21. Examples of these are the Sony-designed LANC system and the 5-pin link system incorporated in Panasonic camcorders and homedecks. They permit control of the camcorder by an editing system of the sort we shall look at shortly here.

Timecode generation

Also designed for use in post-production editing, some of the more expensive camcorders incorporate a *timecode generator*, in which every TV frame recorded on tape is given its own unique electronic 'label' in terms of elapsed hours, minutes and frames. These labels can be detected by suitable editing machines and used to control a copying tape deck for precise and frame-accurate editing. There are two timecode systems in common use in analogue home camcorders. Video-8 and Hi-8 types use RCTC (re-writable consumer time-code) in which eight-bit digital data bytes are stored on a video tape-track extension between the video and PCM (where incorporated, see Chapter 19) portions. This is shown in Figure 23.4. VHS-based camcorders incorporating timecode use a similar system called VITC (vertical interval time-code) in which the data are carried in the unused lines of the field blanking interval – the area which carries teletext data in broadcast TV transmissions. The advantage of RCTC is that it can be written (in a suitable camcorder or video deck) onto an existing tape recording, whereas adding VITC to a recording involves copying it onto a second tape in the VITC-equipped VCR.

VHS-C camcorders

The Video-8 cassette is a small one, intended specifically for use in camcorders; it permits the design of very small 'palmcorders'. The VHS system, on the other hand, is based on a much larger cassette originally designed for home-deck VCRs; its use in a camcorder makes for a large, bulky and heavy unit. To overcome this problem the VHS-C (compact) cassette made its appearance. It contains enough tape for a maximum shooting time of 1 hour (SP) or 2 hours (LP) and forms a much smaller package. Even so, it retains compatibility with the VHS – or S-VHS – format, and can be operated in a full-size deck; in most cases an adapter, whose body-shell conforms to VHS dimensions and mechanics, is required to facilitate this. More on this and allied subjects will be found in Chapter 25.

An internal view of a camcorder is shown in Figure 23.3.

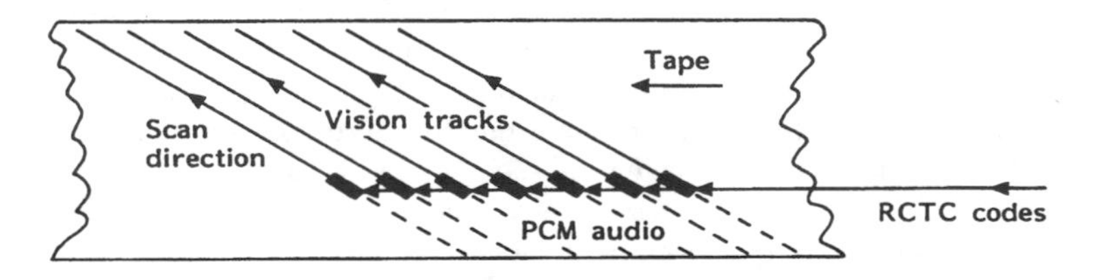

Figure 23.4 *In Video-8 format the RCTC timecodes are written onto the tape at the start points of the video tracks. Not all V-8 VCRs use the PCM sound tracks shown dotted here*

Home movie post-production

It is seldom that the tape which comes straight from the camcorder provides a complete and satisfactory 'programme' by itself. As in the days of amateur cine cameras it's usually necessary to cut out the mistakes and the unwanted footage, and to reassemble the rest in a different order to that in which it was shot, while inserting titles and perhaps adding special effects.

Dubbing and editing

Audio dubbing is carried out by operating the machine in replay so far as the vision signals are concerned, but with the sound section of the machine in record mode, and the audio input jacks 'live'. Apart from the necessary electronic switching facilities, the machine needs only an additional audio erase head (positioned on the tape path just before the audio rec/play head) to achieve this, and in dub mode the replay picture is monitored while adding a new sound track to the recording. Hi-Fi audio tracks cannot be dubbed because it's not possible to manipulate them without disturbing the video tracks under which they are buried on tape. In a Hi-Fi VCR, then, audio-dub function is confined to the 'low-Fi' longitudinal track, whose output on replay can be mixed with the original Hi-Fi sound if required.

With cine film the picture editing process is a physical one, involving scissors and glue. For half a dozen good reasons videotape cannot be cut and spliced, so the technique is changed to that of selectively copying sections of the original footage onto a second tape in the required order, and thus building up a *master copy*, on which the soundtrack, too, can be composed and controlled by the editor. The problem here (except with digital tape recording systems) is that each time the picture and sound are copied from tape to tape their quality is compromised, especially with low-band domestic formats. It's important, then, to limit the number of 'generations' involved in the editing process, and to get the best-possible quality of picture and sound at the outset.

The basic editing process consists of setting the source VCR (generally in budget-limited home operation, the camcorder itself) to play, the copy VCR (which may be the family's general-purpose machine) to record, then to operate the latter's record-pause control to hold up the copying action as each new scene or sequence is found and set to play in the source machine. This is called *assemble editing*. Ordinarily this would lead to 'glitches' and instability in the played-back picture at the transition points due to discontinuities in the vision/sound recording and the control track. There are a number of reasons for this. During record and playback the servos, as we have seen, are locked up to incoming field syncs and off-tape control tracks respectively. When a change is made in input signal the servos have to run up to speed and lock-in to the new video source, and so the signal recorded on the tape goes haywire during this period. On replay the discontinuity in the control track also causes a 'hiccup' in servo operation, and these effects result in a very ragged transition between the two pre-recorded sequences. Further problems arise from the fact that the full-width erase head is placed about 8 cm before the video head drum on the tape path, giving rise to three or four seconds crosstalk from the original tape tracks before newly-erased tape reaches the recording heads.

'Clever' edit

To overcome these problems and give a reasonably clean transition at edit points, the technique of *back space edit* was introduced. In its most common form, Figure 23.5, it involves the VCR machine back-spacing (or rewinding) 20 to 25 frames (about a second) when record pause is selected, then stopping. At the end of the pause period the machine will re-start in playback mode so far as its vision systems are concerned, but with the capstan servo slaved to the new incoming video signal. When the servo has locked up, the vision circuits switch automatically to record mode during a convenient field-blanking interval so that the last frame of the old material is followed by a new and fully synchronised field of the new programme with no lack of continuity in control track or video tape tracks; very little disturbance will be seen on the replayed programme over the edit point. A certain amount of chroma crosstalk can occur briefly at the changeover point in some designs, and much depends on the sophistication of the VCR's electronics – this is another area in which the microprocessor has come to the fore.

The alternative and less-used *insert-edit* system consists of the 'dropping-in' of a sequence into a pre-recorded programme, examples being titles, credits and commercials. Here the original control track is retained to give 'glitch-free' transitions at the start and end of the insert. Some home-decks and camcorders are equipped with ports for *synchro-edit,* in which the pause and run functions of both VCRs can be controlled by a single key at one end or the other.

Manual editing is a long and wearisome business, especially when the required sequences are scattered widely along the source tape and the programme being assembled is a long one. It can be automated to a greater or lesser degreee by an *edit controller,* which may be incorporated into the source or copy machines;

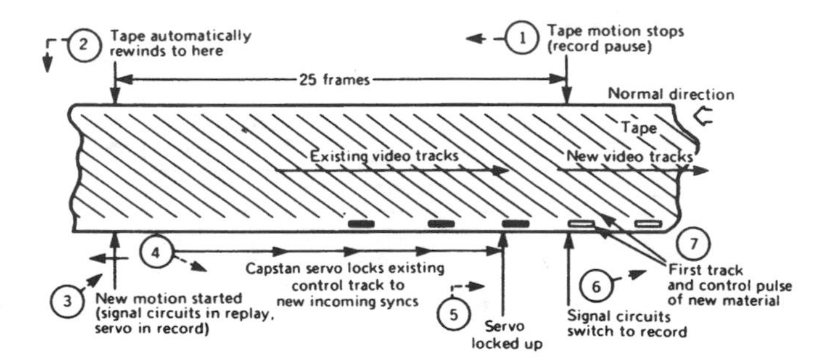

Figure 23.5 *Assemble edit: the sequence of events is indicated by the ringed numbers, and the result is a clean switch from old to new programme material*

provided as a stand-alone unit; or take the form of a home computer working in conjunction with suitable software and interfacing hardware. All work on the same basic principle of storing the (human) editor's cut-point instructions (known as an EDL, edit decision list) in a memory and manipulating the transport controls of source and copy VCRs to assemble the master copy as required. In the simplest systems the cut-point references are defined by an electronic frame-counting process. This becomes less accurate as time (and frames!) passes, and more sophisticated – and expensive – systems make use of the timecode schemes mentioned earlier in this chapter.

One of the simplest auto-edit outfits is based on the random assemble (RAS) concept, introduced by JVC. Here the edit memory is built into the camcorder, and programmed by the user keying in the required start and finish points of each sequence required. A hard-wire link is provided for connection to a suitably-equipped homedeck, and when set to go in this mode the camcorder finds and runs the selected sections of tape, controlling the pause function of the copy deck meanwhile.

Stand-alone and computer-based systems work on the same principle, and have provision for storage of up to 200 'edit events'. For them to work there has to be some form of access to the camcorder's (or source machine's) control system so that the deck can be manipulated. Control of the copy VCR can be achieved by a hard-wire link (e.g. LANC) or by an infra-red wand whose emitted signal mimics that of the deck's own infra-red remote control unit. Some edit-control systems are capable of operating by frame-counting, VITC or RCTC, and of controlling the slave VCRs by LANC, Panasonic, pause-jack or infra-red means, offering great versatility in operation.

Computer-based edit systems, as well as governing the automatic-assembly modes described above, can also generate titles, captions and effects and (with the help of a *genlock*) synchronise them with the video picture and superimpose them upon it. The basic setup for harnessing a computer to home video production is shown in Figure 23.6. Packages are available for both the popular types of home computer, and some of them can offer a wide range of facilities including frame-grabbing, animation, digital data-rate compression and audio processing.

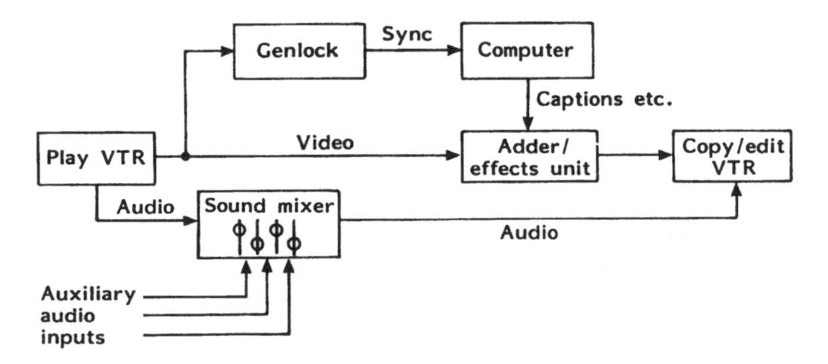

Figure 23.6 *Use of a genlock, computer and sound mixer to enhance an edited video tape*

Simpler units available for home editing include enhancers, which artificially sharpen up images in the way shown in Figure 18.13; colour correctors, which decode the composite signal to Y and RGB to facilitate control of each colour individually for hue manipulation, then recode to (e.g.) PAL standard; caption generators containing a keyboard and picture-synchronised character-generation IC system; video inverters for viewing photographic negatives and for effects; and sound mixers in which a variety of audio sources can be blended, faded and mixed for recording on the master tape.

Digital effects units

By converting complete video fields into digital form and storing them in a large memory as described in Chapter 12, it becomes possible to achieve the effect of genlock, whereby two autonomously-generated pictures can be mixed, superimposed or cross-faded. One of the video signals is A – D converted and written into field memory in seven-bit or eight-bit form, to be read out again, one or two fields later, at a rate governed by the synchronising pulses of the second picture source as illustrated in Figure 23.7. In addition to the basic 'genlock' function, many other effects become possible when a one- or two-field picture memory is available: 'wipes', mosaic effects, picture zoom, still-frame and strobe images, picture-in-picture (Figure 23.8), and chroma- and luma-key, in which selected types of background in one picture can be replaced by the corresponding parts of a second picture. Thus, for instance, a person shot against a plain blue background can be made, on the master-tape or the monitor screen, to appear on a beach, in a city street or on the railway track with a fast-moving train approaching!

Timebase correction

As we have seen earlier in this book, the necessity for videotape replay to depend on a mechanical process imparts to the reproduced video signal a certain amount of timing jitter. In multi-generation video copies (the result of editing processes) this jitter can be sufficient to upset the line synchronisation of the TV or monitor used for display; the result is horizontal instability of vertical edges in the picture, and often a bending or wobbling at the top of the image, perhaps with hue errors

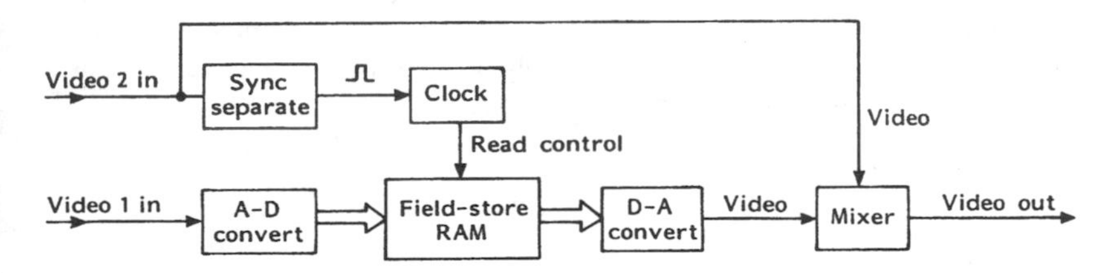

Figure 23.7 *Digital frame synchroniser for superimposition of TV images from two autonomous sources*

Figure 23.8 *Picture-in-picture is one of the many effects afforded by a digital picture field store*

there as a result of poor synchronisation of the colour subcarrier regenerator in the chroma decoder. This problem can be overcome by the use of a timebase corrector (TBC) whose principle of operation is the same as that shown in Figure 23.7. The jittering video signal is A-D converted and the data produced written into memory one field at a time. Memory readout is governed by a stable local crystal oscillator which also times the newly-generated outgoing sync pulses. Indeed the sort of effects unit shown in Figure 23.7 has a TBC function so long as the video-2 input (master) has no timing jitter superimposed on it. Some top-class camcorders and home VCRs/edit decks have digital TBCs built in – they are switched into operation during the replay mode.

In the next chapter we shall examine the digital equivalents of the camcorders and editing systems described here. Because a digital camcorder has analogue outputs and timecode facility it can be used for editing as described above; similarly an analogue camcorder or VCR lends itself to the computer-based editing systems to be described in the second half of the next chapter so long as a suitable A-D capture card is fitted to the computer.

24 *Digital tape formats and computer editing*

As with digital TV broadcasting, digital video recording had to wait for the development of exotic LSI chips and advanced bit-reduction techniques before it became economically viable for the home- and enthusiast market. It was in 1995 that the first DVC (Digital Video Cassette) camcorder went on sale in the UK. The DVC format had been agreed and accepted by all the world's major manufacturers, so here at last was a 'common' system to end the 'format wars'. Well, almost! Within another four years two more digital cassette systems had been introduced, both intended to bridge the gap between existing analogue technologies and the new world of 'digi-tape': they were the Sony Digital-8 format and JVC's D-VHS system, each of which will receive attention later in this chapter. No sooner had digital camcorders been introduced than their prices began to drop sharply, driven down by competition between manufacturers and – at a more local level – between retailers. Within five years the average prices of DVC camcorders had fallen to the point where they were comparable to those of high-band analogue machines with similar features, gravely threatening the established formats in the home video-movie sphere.

The advantages of a digital videorecording system are several. The format itself affords higher picture resolution, better in practice than a colour TV broadcast. The digital signal can be transferred to another (digital) tape without any loss of quality, and edited and manipulated likewise, wholly in the digital realm. Both video and audio signals can be 'dubbed' because they occupy different areas on tape. The colour-bleed and -smear which arises from low chroma bandwidth in analogue tape systems disappears, as do the Y/C bandsharing artifacts due to PAL and other colour-encoding systems. Because the picture is built up 'from scratch' as it were, in the playback electronic system, there is no potential for noise or interference on picture or sound, though of course when the digital bitstream from tape gets too corrupt the sound disappears and the picture becomes progressively 'blocky'; intermittently 'frozen' wholly or in segments; and then disappears altogether, just like DTV broadcast pictures when the signal quality deteriorates.

DVC format

Although the DVC format has four standards, embracing high-definition and DVB (digital video broadcast system) variants, we shall concentrate here on DVC-SD (standard definition), which has been adopted for home- and 'hobby' camcorder use. It uses a cassette much smaller than an audio or Video-8 type, capable of storing one hour of high-quality sound and vision in SP mode, or $1\frac{1}{2}$ hours in LP mode on 6.35 mm-wide tape, with excellent sound quality and 500-line picture definition, a good match for large-screen home-TV sets. We shall look more closely at the tape, cassette and system specifications later.

Signal processing: record

Chapter 12 showed how analogue TV signals are digitised, and described the MPEG system of bit-rate reduction. The process used for DVC is very similar to that used for DTV broadcast: it uses the same MPEG 'toolkit' and has largely the same building blocks and techniques as those shown in Chapter 12. The main difference is the degree of data compression: for DVC the ratio is about 5:1. This arises from the need to 'trick-play', freeze and edit the pictures. Spacial redundancy (intraframe compression) is applied, but not temporal redundancy.

Figure 24.1 outlines the initial processes in the DVC record system. Sampling rates are 13.5 MHz for luminance and 6.75 MHz for chroma, the latter dealt with line-sequentially for Cb and Cr. 8-bit quantisation is used on Y, Cb and Cr signals, making an initial data rate of 216 Mbit/sec: it is reduced to about 25 Mbit/sec for recording by use of spacial redundancy coding, DCT technique and variable-length coding. As with DTV systems the bit-reduction process is based on blocks of 8 × 8 pixels – indeed these blocks can sometimes be *seen* during replay if the data becomes corrupt, e.g. by mistracking.

The record and replay system for DVC is shown in block diagram form in Figure 24.2. Following A-D conversion in record mode the data is 'shuffled' by bringing together macroblock samples from widely-separated picture areas and

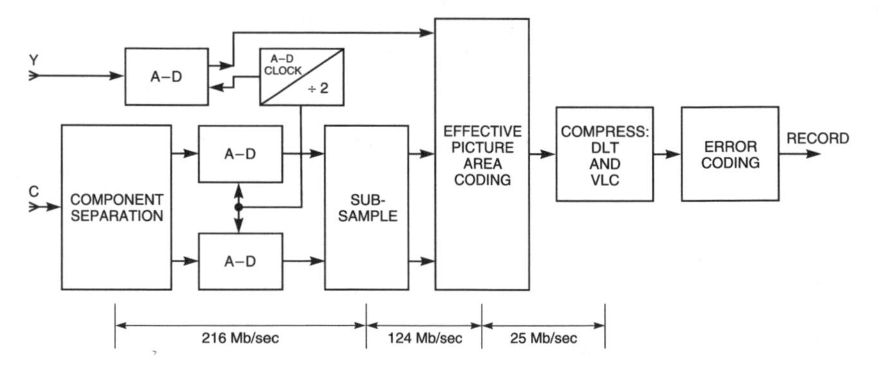

Figure 24.1 *Bit-reduction and coding in the record stages of a DVC tape system*

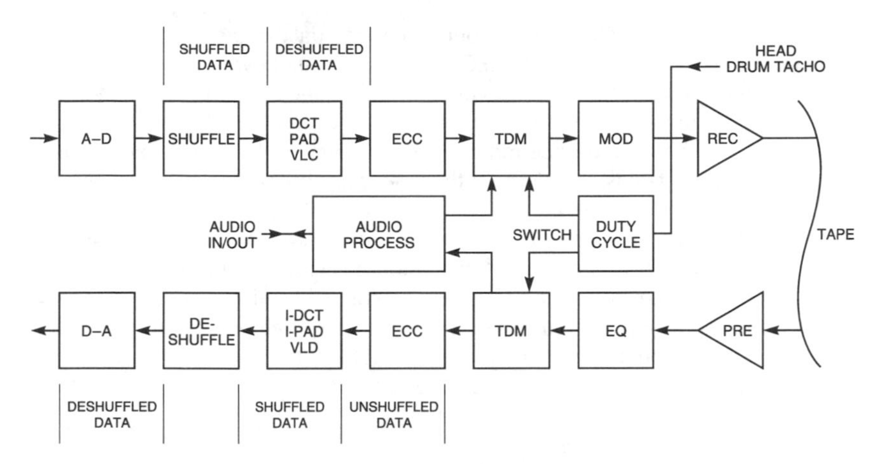

Figure 24.2 *DVC data processing system throughout*

assembling them in what appears to be a random sequence to reduce statistical distortion due to compression ratio bias, and to give best-possible 'trick' and still-frame reproduction during playback. Later in the chain the data is deshuffled once more to regain correct order for recording on tape: LHS of picture at the tape-entry side and RHS picture data at the exit side, top of picture first on the lower edge of the tape ribbon.

Next comes the data-compression stage, and then ECC (error correction coding), a form of Reed-Solomon protection, primarily to combat the effects of tape-mistracking and dropout. The next block in the diagram is the TDM (time-division multiplex) switch: the video data only occupies about one-half of the duration of a head scan, and for part of the remainder the audio datastream is recorded on tape. The selected data is modulated and fed via a drive amplifier to the recording heads.

DVC audio

The user can select either 16-bit or 12-bit sound mode, the former containing high-quality stereo signals, and the latter offering four separate channels and suitable for dubbing. The sound data is 'chopped', stored and released to the modulator in time with the TDM switch changeover, itself synchronised by the head-drum PG/tacho pulse.

Signal processing: replay

The first stages in DVC replay are signal amplification and equalisation, the latter not to give best bandwidth and noise performance as in analogue tape systems, but to optimise the demodulator's discrimination between 0 and 1 symbols. Thus furbished, the off-tape signals are gated by the TDM switch to the audio or video stages in synchronism with their readout. The video data is checked and – where possible – errors are repaired in the ECC stage; any data beyond repair is deleted and replaced by good data from the same position in the previous scanning line, a process akin to the action of a dropout compensator in an analogue VCR.

Now the data is shuffled once more, using the same look-up table as was used during record, for its passage through the bit-expansion process, where 'inverse' DCT and decompression are carried out. After deshuffling the picture-data is built up, complete frames at a time, in a large (5 Mbit) SRAM memory store: it is read from here into the D-A converter whose output provides the reconstituted video signal. The analogue audio signal is reconstituted likewise, using the inverse processes to those applied during the record phase. Both video and audio can be conveyed from the machine in digital form (IEE1392/Firewire/i-Link) into other digital equipment – more details in Chapter 28.

DVC track configuration

DVC head-drums have a diameter of 21.7 mm, and rotate at 9000 rpm to give a writing speed of about 10 m/sec. Two video heads are fitted, with azimuth angles of ±20°: this amounts to a 40° offset between them, with a correspondingly low level of crosstalk pick-up from adjacent tracks. The heads are very narrow to give a track pitch of just ten microns in SP, 6.67 microns in LP mode, corresponding to tape speeds of 18.8 and 12.6 mm/sec respectively. The tape's helical wrap embraces 186° of the drum's periphery; the signal envelope is clipped at the 180° point by the head-switch.

A complete PAL picture frame (with corresponding sound signals) takes 12 tracks on DVC tape as shown in Figure 24.3. Each 32.89 mm-long track consists of four segments, written sequentially by one head as it scans across the tape at an angle of 9.167°. First comes the ITI segment, containing reference data for track height/positioning and to provide a tracking signal for audio and video dubbing and inserts. The ITI data is used purely for 'housekeeping' purposes by the VCR

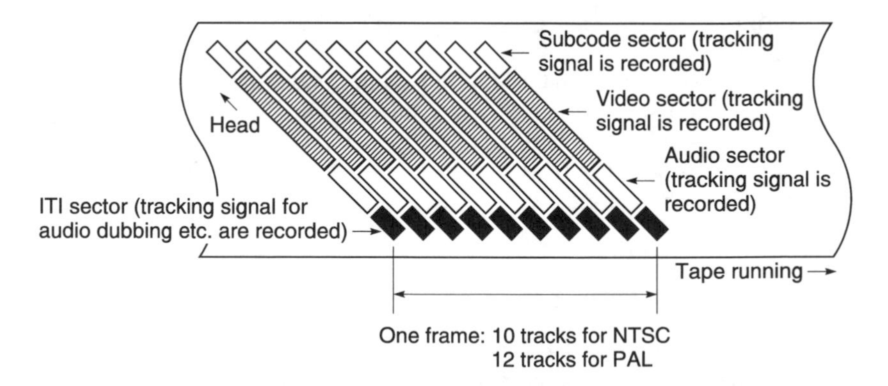

Figure 24.3 *Track layout on DVC tape: the physically-separate video and audio sectors permit dubbing of either*

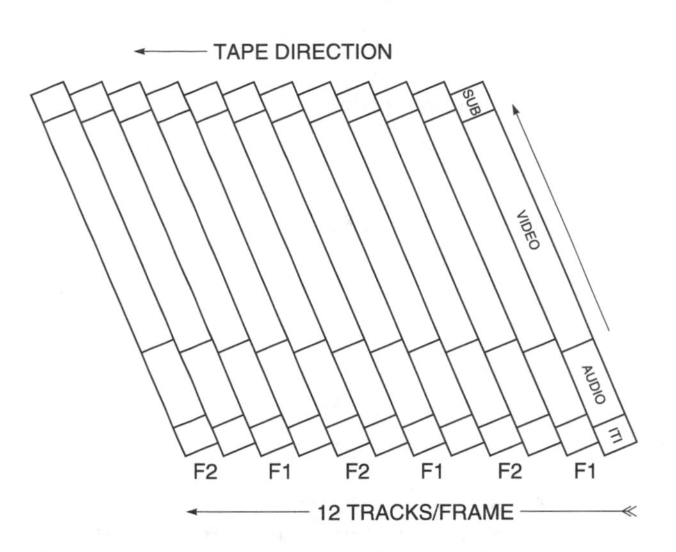

Figure 24.4 *Auto-tracking for DVC. Two frequencies F1 and F2 are used as shown at the bottom*

itself during playback. The next two segments are the audio and video data we have already met; in the video sector are recorded additional data such as date and time of recording, camera-settings, widescreen mode flag, etc. These can be displayed during replay if required. The final sector in the tape track shown in Figure 24.3 contains *subcode* data, primarily a timecode which can be used to identify every frame, useful for editing. Also in the subcode are cue flags for location of each new shot, and for marking still-frame/photo shots when the camera is set to that mode.

DVC tracking system

As with the Video-8 format described on page 259, automatic head-tracking during replay is effected by the use of 'tones' recorded along the tracks. In DVC the system is simpler than for Video-8, as Figure 24.4 shows. Here two tone-frequencies are used, 465 kHz and 697 kHz, alternating with no recorded tone in a four-track sequence. When scanning a track with no pilot tone the off-tape F1 and F2 crosstalk signals are made equal by the tracking-servo system, signifying a track-centre trajectory of the head; this condition corresponds to maximum pilot-tone amplitude during replay of an F1 or F2 track.

System parameters

The specifications for DVC-SD format are set out in Table 24.1 for the electrical and coding sections, and in Table 24.2 for the mechanical and tape-track characteristics.

Table 24.1 *Data and coding specifications for DVC*

	Error correction code	Reed-Solomon code	
	Channel coding	Pre-coding S-INRZI (PR4) Modulation: 24–25 code F0:0, F1:465 kHz, F2:697:5 kHz	
	Recording clock (Fclk)	41.85 MHz	
	Data transfer rate	Video: 124 Mbps – 24.9 Mbps (1/5 compress) Audio: 48K × 16b × 2 ch – 1.55 Mbps Total: 41.85 Mbps (rec.)	
	Video signal recording system	Digital component recording 525/60	 625/50
	Effective pixels	Y: (H) 720 × (V) 480 C: (H) 180 × (V) 480	Y: (H) 720 × (V) 576 C: (H) 360 × (V) 576
Video	Sampling frequency	4:1:1 Y: 13.5 MHz, C:3:375 MHz	4:2:0 (Line sequential) Y: 13.5 MHz, C: 6.75 MHz
	Digital video compression system	DCT variable length coding	
	Audio signal recording system	PCM digital recording	
	Channel	2 ch	4 ch
Audio	Sampling frequency	48/44.1/32 kHz	32kHz
	Equalisation bits	16 bits linear	12 bits non-linear

DVC tape and cassette

A new type of tape was developed for DVC recording, having five layers as shown in Figure 24.5(a). It is 6.35 mm wide and seven microns thick. The third layer is the magnetic one, having two high coercivity evaporated metal coats for good reliability, high output and low noise. The back coating has very low surface friction, while the overcoat hard carbon layer is durable, flexible and 'tough' in terms of abrasion resistance.

Mini-DV cassettes come in 30 min and 60 min types, DV-M30 and DV-M60 respectively. They are capable of 50% more running time in the LP mode, whose picture is just as good, but whose tracking tolerance is 'tighter', with a greater risk of data errors. The mini-DV cassette package measures 66 × 48 × 12.2 mm, and is illustrated in Figure 24.5b. ID terminals 1–4 may merely connect to internal resistors to indicate to the deck's syscon such details as tape type, capacity and grade; where the cassette contains a memory chip, they function as serial data ports for read and write. In this case the non-volatile memory chip typically has

Table 24.2 *Mechanical and tape-track specifications for DVC*

Item		*525/60 system*	*625/50 system*
	Standard	DV system (SD)	
	Recording system	Rotary 2 heads azimuth recording	
	Drum rotation speed (rps)	150/1.001	150
	Drum diameter (mm)	21.7	
	Writing speed (m/s)	10.202/1.001	10.202
	Tape dimensions		
Tp	Track pitch (μm)	10.00	
Ts	Tape speed (mm/s)	18.831/1.001	18.831
qr	Track angle (deg)	9.1668	
Lr	Effective track length (mm)	32.890	
qe	Effective wrap angle (deg.)	174	
Wt	Tape width (mm)	6.350	
He	Effective area lower edge (mm)	0.560	

16 kbit capacity and stores record dates and photo time/date (stills) along with DV format data like ID, mode, size and tape type. More details of mini-DV tape, cassettes and capabilities will be found in the next chapter.

Digital-8 format

The Video-8 and Hi-8 analogue formats are much better suited to portable (camcorder) operation than VHS ones; the smaller cassette makes for a smaller and lighter machine, even than those VHS ones using the C (compact) cassette. The Video-8 formats were developed specifically for portability, while VHS was adapted from the basic large cassette/homedeck concept developed by JVC in the 1970s. Sales of Video-8 and Hi-8 camcorders flourished, and greatly outsold VHS-C and S-VHS-C in the market, to be themselves overtaken by DVC format in the fullness of time. In an attempt to maintain their market lead, and to provide a 'bridge' between analogue and digital systems for existing users, the inventor of Video-8, Sony, introduced a new 'dual' format called Digital 8. D-8 decks (virtually all of them in camcorder form) can replay analogue Video-8 and Hi-8 recordings as well as their own digital recordings, using the same head pair, drum and tape wrap, and the same tape types. During playback the type of signal on tape is automatically detected and the replay circuits switched accordingly. If required an analogue signal off tape can be output in digital form, using an A-D converter inside the camcorder.

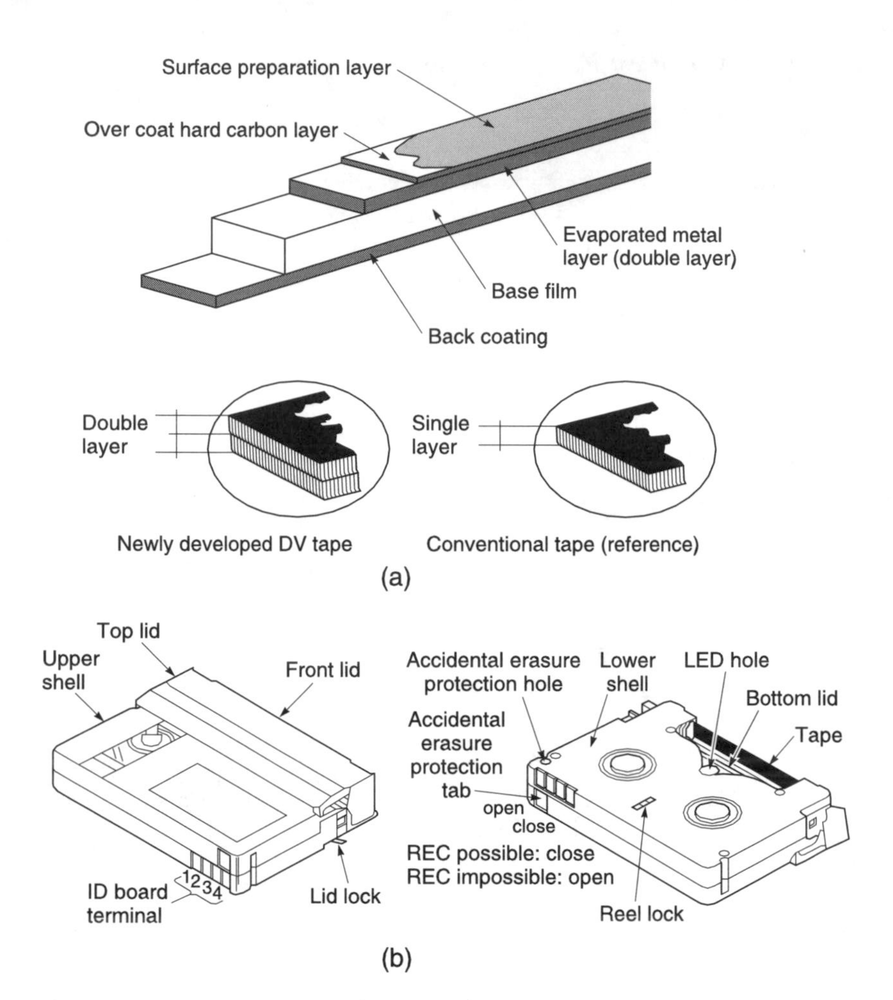

Figure 24.5 *DVC tape (a) and cassette (b)*

D-8 signal processing

Figure 24.6 shows the stages in the Digital-8 recording system. The CCD imager's output is cleaned up, gain-controlled, then A-D converted within IC502 into a 10-bit datastream, sampled at 13.5 MHz. IC251 separates Y and RGB data to derive Y, Cb and Cr values, all at 8-bit depth and the latter two at 3.375 MHz sampling frequency. Also within this chip is carried out the 'steady-shot' process in which the image is stabilised by a fast pan/tilt reading operation from a picture memory bank with a larger effective area than the useful picture size. The next chip in line, IC351, is similar to one used in DVC-format camcorders: it performs blocking, data shuffling, digital picture effects and D-A conversion for use in the viewfinder and to produce an E-E (electronics-to-electronics) output monitoring signal.

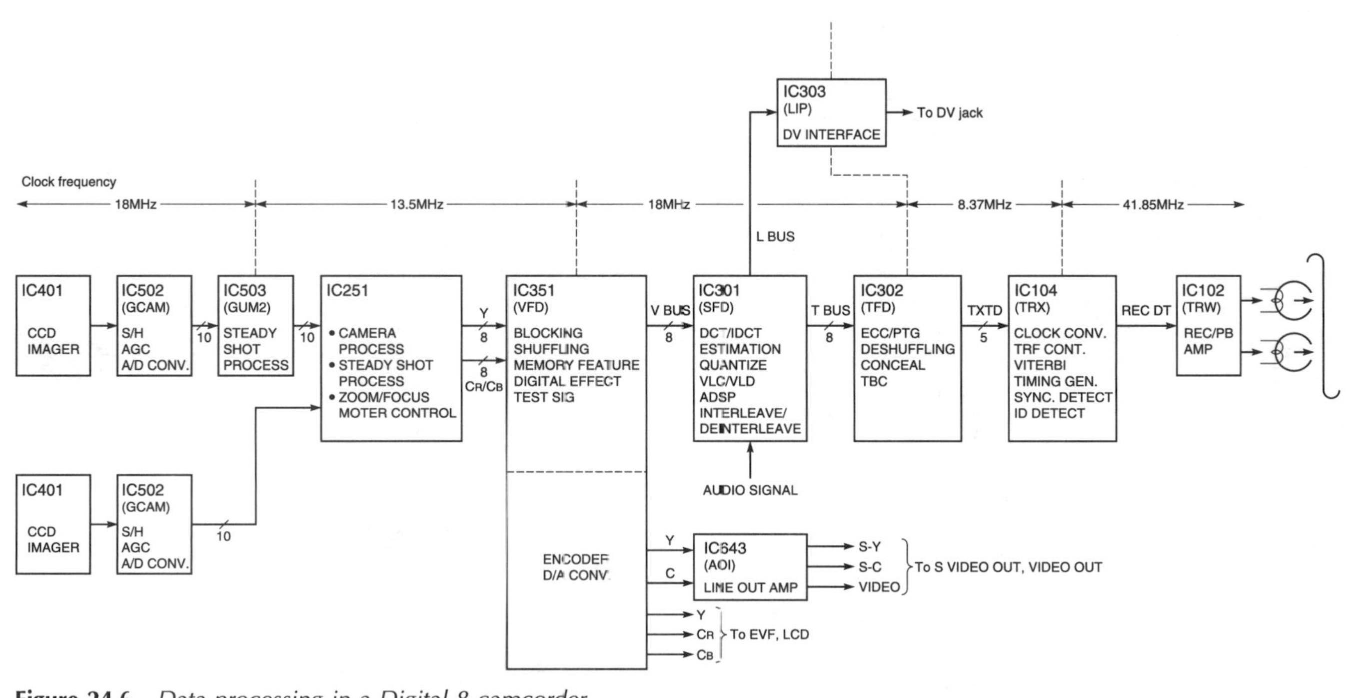

Figure 24.6 *Data processing in a Digital-8 camcorder*

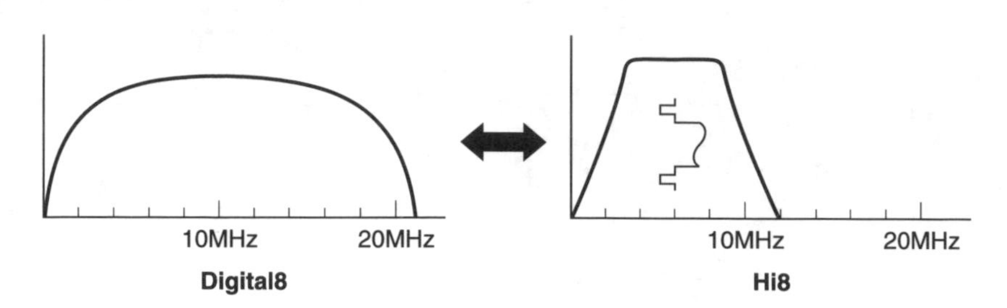

Figure 24.7 *Comparison of recording spectra of Digital-8 and Hi-8*

IC301 is concerned with the bit-reduction process. Here the data undergoes DCT, quantisation, VLC and framing (Chapter 12 refers) to achieve data compression to about one-fifth of the original content. The 'condensed' video data is joined by the audio bitstream inside IC301 for passage into IC302, where takes place error-correction coding, deshuffling and encoding into a form suitable for recording on tape. Further processing takes place inside IC104, primarily to convert the data from 5-bit parallel form to a serial (one line) datastream for application via recording amplifier IC102 to the video heads. For this application an FE (flying erase) head is also present on the drum to wipe off previously recorded data. The spectrum of the D-8 RF signal on tape is shown in Figure 24.7.

D-8 playback

During replay of digital video tapes, many of the same ICs are used, with their roles reversed to perform the inverse functions to those carried out during record; the same memory chips etc. also serve. Thus in Figure 24.6 IC102 now operates as head preamplifier; IC104 carries out clock conversion, synchronous detection etc; IC302 shuffles the data and provides error-correction and concealment; IC301 takes care of inverse DCT, de-interleaving and data expansion; and IC351 de-shuffles the data, D-A converts it and encodes the Y and C components into PAL (or other analogue) form to produce signals for the viewfinder and the AV-out ports. The digital in/out port interface is IC303, shown at the top of Figure 24.6.

Analogue replay

When the playback circuit detects the presence of an analogue (Video-8 or Hi-8) off-tape signal at the head preamplifier a different replay circuit is switched into line. Here the signal is A-D converted so that the Y-FM and chroma colour-under signals can be separated and processed in the digital realm, including chroma-crosstalk removal, Y-signal conditioning and encoding to PAL (or NTSC) standard. The process is shown in Figure 24.8, where it can be seen that IC251 is the main operator, receiving and passing on analogue signals, but having a digital 'core'. Analogue processor IC201 processes AFM sound carriers for passage to the stereo

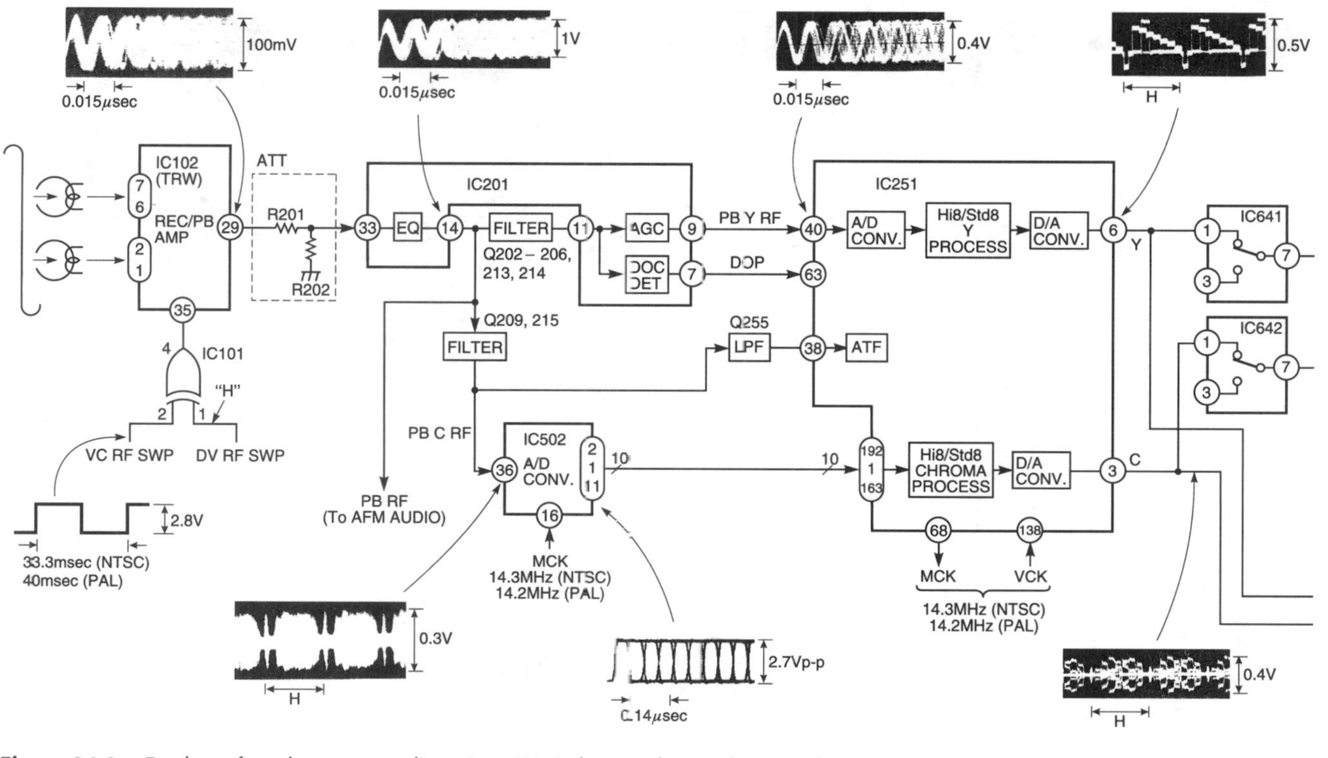

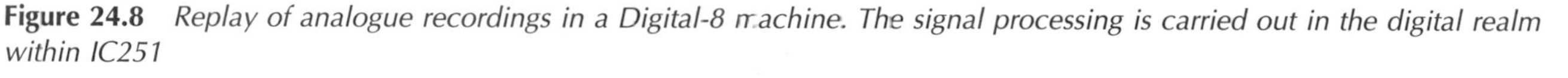

Figure 24.8 *Replay of analogue recordings in a Digital-8 machine. The signal processing is carried out in the digital realm within IC251*

audio section, as well as ATF tones to operate the tracking servo in analogue replay mode.

D-8 recording system

For compatibility the Digital-8 head drum must necessarily conform to the 40 mm diameter of the Video-8 format. Figure 24.9 compares D-8 and DVC-format tape-track configurations. While the PAL-DVC system records each TV frame in 12 tracks, D-8 lays down a frame in six tracks, each divided into two consecutive sub-tracks. To achieve this the D-8 drum speed is 4500 rpm. Figure 24.10 shows at the top a comparison between the NTSC and PAL variants of analogue Video-8 format, with one track per frame in each case: for NTSC the tape speed is slower and the tracks narrower than for PAL. The lower section of the diagram illustrates that tape linear speed is virtually the same in Digital 8 for both TV systems, with five (NTSC) or six (PAL) tracks per frame, all tracks being 16.34 microns wide. Both trace 150 tracks/sec, 30 five-track frames for NTSC and 25 six-track frames for PAL. Digital-8 video heads have a width of 19 microns, a little wider than the D-8 track pitch (each new head sweep erases the excess of its predecessor during record), and somewhat narrower than the Video 8/Hi-8 tracks of any analogue tape which may be presented. In the latter case the heads sweep along the centre of the pre-recorded tracks, the signal they produce displaying slightly higher S/N ratio than would come from a

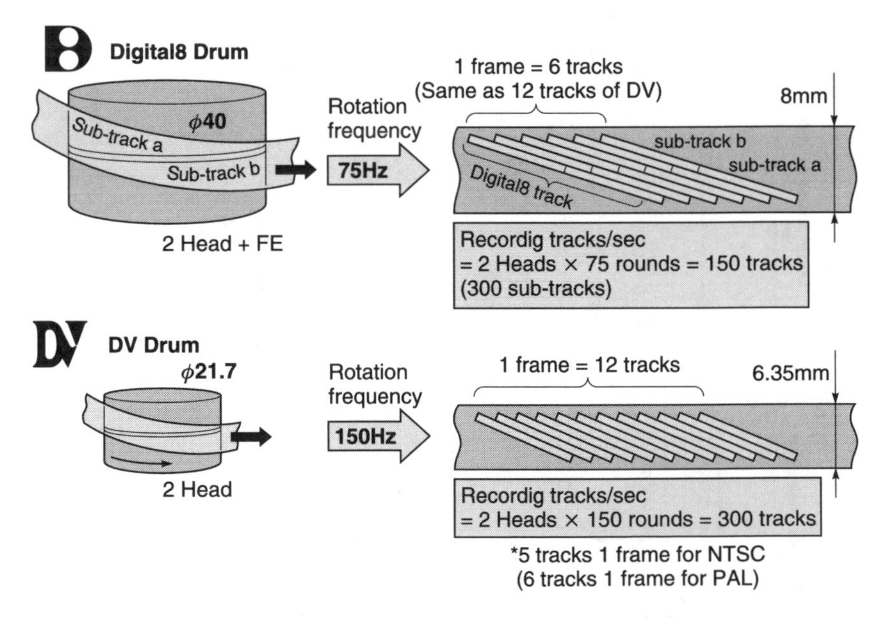

Figure 24.9 *Digital-8 and DVC head drums and tape track formations compared*

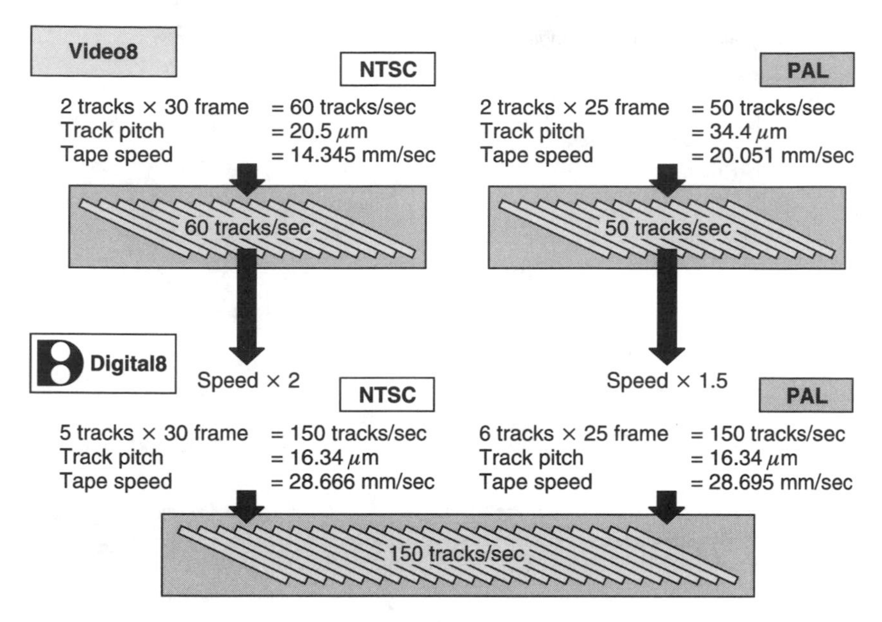

Figure 24.10 *Analogue and digital 8mm formats: NYSC and PAL variants*

Table 24.3 *Track-scanning in three formats using 8 mm tape*

Format	*Track Pitch*	*Rotation speed*
Digital8 (NTSC/PAL)	16.34 μm	4500 rpm
Hi8/Standard8 (NTSC SP)	20.5 μm	1800 rpm
Hi/Standard8 (PAL SP)	34.4 μm	1500 rpm

correct-width head. Table 24.3 shows the track pitch for D-8 and the variants of Video-8: the drum is designed to rotate at 1500 and 4500 rpm in European D-8 decks. Figure 24.11 shows the disposition of the three heads on the 40 mm-diameter drum, and their azimuth settings, ±10° for compatibility with Video-8 and Hi-8 recordings. The track patterns on tape for D-8 and Video-8 are shown in Figure 24.12. The track-following servo system for D-8 is the same as that already described for DVC: 465 and 697kHz tones are recorded in alternation with no-tone head-sweeps in a four-track sequence, ATF error being detected only in playback of no-tone tracks. When replaying an analogue recording the servo system switches to Video-8 ATF operation as described in Chapter 17 of this book.

Because of the two sub-tracks per head sweep of the D-8 tape pattern, and the fact that drum rotational speed is half that of a DVC drum, the off-tape RF signals during digital replay are identical to those of DVC as Figure 24.13 shows.

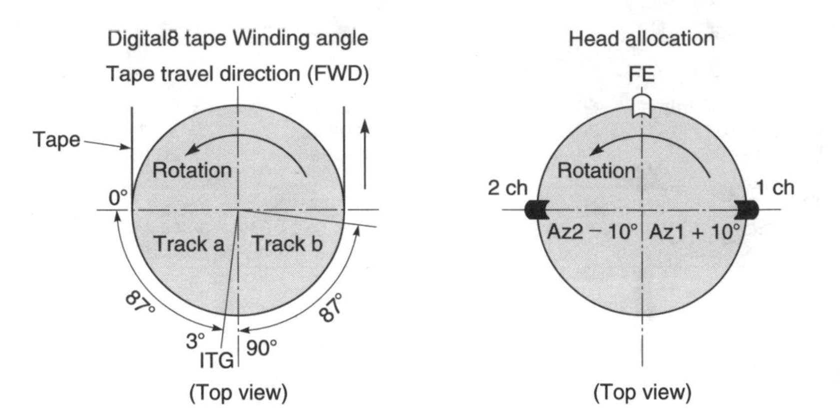

Figure 24.11 *Head disposition and azimuth cuts on Digital-8 head drum*

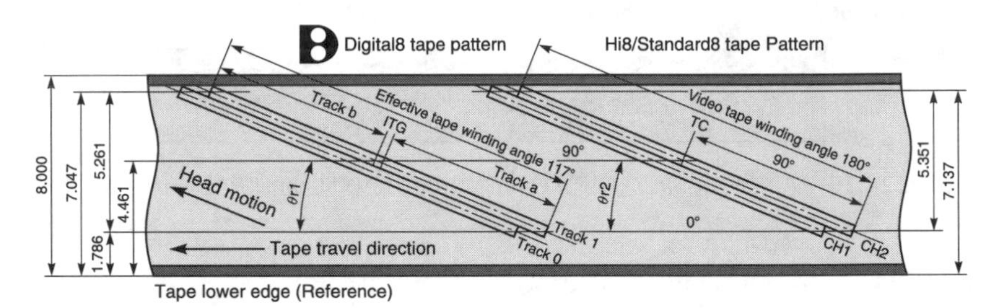

Figure 24.12 *Recording patterns on tape for digital and analogue recordings on 8 mm tape*

Deck mechanisms and control

Digital video decks use very similar mechanisms to those described earlier for other 'small-tape' formats; indeed the first D-8 camcorders utilised an identical deck to those fitted to contemporary Video-8/Hi-8 machines, differing only in the head drum assembly. System control in digital decks and camcorders follows the same principles as those used in analogue recorders, and described in Chapter 21. Likewise the servo systems, whose requirements and functions are the same, save only for the higher drum speed and the much closer tracking tolerance due to the very fine track pitch in digital tape-scanning systems.

Formats compared

A comparison between the two DV formats described so far and the established high-band analogue systems is given in Table 24.4, where it can be seen that the digital systems score in terms of picture and colour definition, audio performance and convenience of the cassette.

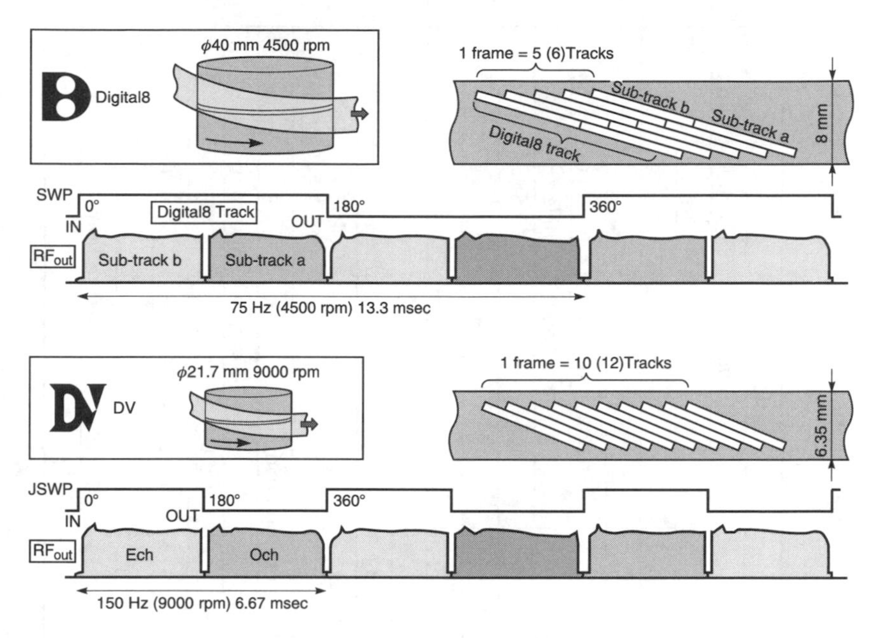

Figure 24.13 *Comparison of Digital-8 and DVC drum speed and track readout. Figures in brackets are for PAL/625*

D-VHS

Whereas DVC and D-8 formats are primarily designed for camcorder applications (though lending themselves well to home use) D-VHS, introduced in 1999, has a different purpose. It maintains compatibility with existing VHS and S-VHS tapes, while providing what might be called a digital storehouse for *data*: indeed D-VHS stands for *Data* VHS, and the storage capacity of a D-VHS cassette (DF-420 type) is very high at 44 Gbytes, corresponding to 70 CD-Roms or nine DVDs. This can be used in various ways, depending on the tape speed selected: modes range from $3\frac{1}{2}$ hours recording of high-definition pictures to 49 hours of relatively low definition ones, corresponding to tape speeds of 33.34 and 2.38 mm/sec respectively.

Cassettes and compatibility

D-VHS cassettes are physically identical to VHS ones, save only for a special identification hole in their bases, used for auto-switching between formats. D-specification tape is similar to ferric-oxide high-grade S-VHS type; indeed D-recordings can be made on S-VHS tapes, and both of them are less expensive than the metal-evaporated types used with other digital tape formats. The mechanical deck, too, is the same as an ordinary VHS one, with 62.5 mm drum diameter and a stationary audio/control/erase head stack. For use with analogue (VHS, S-VHS) recordings the drum is fitted with two heads mounted at 180° and having the standard ±6° azimuth cuts; two additional 'analogue' heads may be carried,

Table 24.4 *Digital and high-band analogue tape format characteristics. Key: *1, DVC cassette; *2, Mini-DVC cassette; *3 VHS-C cassette; <PAL-625 system>*

	DV	Digital8	Hi8	S-VHS
Luminance signal recording system	Digital		Analog (FM modulation)	
Horizontal resolution	About 500 lines		About 400 lines	
Sampling frequency	13.5 MHz		–	
Quantized bit number	8 bits		–	
Chroma signal recording system	Digital (color difference)		Analog (low-frequency band conversion)	
PB chroma signal band	About 1.5 MHz		About 0.5 MHz	
Sampling frequency	3.375 MHz		–	
Quantized bit number	8 bits, each		–	
Transfer rate (video)	25 Mbps		–	
Audio recording system	Digital (16 bits/12 bits PCM)		Analog (FM recording)	
Tape width (mm)	6.35	8		12.65
Cassette size (mm)	125 × 78 × 14.6 (*1) 66 × 48 × 12.2 (*2)	95 × 62.5 × 15		188 × 104 × 25 92 × 59 × 23 (*3)
Signal recording system	2 rotation heads			
Drum rotation speed	9000 rpm	4500 rpm	1800 rpm <1500 rpm>	
Tape speed (mm/sec)	18.812	28.666	14.345 <20.051>	33.35 <23.39>
Track pitch (SP)	10 μm	16.34 μm	20.5 μm <34.4 μm>	58 μm <49 μm>
Drum diameter (mm)	21.7	40		62
After-recording	○ 12 bits	×	PCM (option)	
Cassette memory	○	×		

as they are on high-end VHS and S-VHS decks. By these means, both record and playback can be performed in either VHS or S-VHS modes, from and to analogue or digital devices. Operation in digital mode uses a separate 180° head-pair on the drum; these have ±30° azimuth settings, the same as VHS Hi-Fi heads and thus offering shared use between these two modes – in D-mode the VHS Hi-Fi function is not used, and sound data goes onto the tape by the same route as the vision information.

Track configuration

In D-VHS mode the tape-track width is 29 microns, roughly comparable with VHS-LP mode. The tape speed varies according to the operating mode, ranging from 33.34 mm/sec to 2.38 mm/sec. In STD (standard) speed the tape moves at 16.67 mm/sec, for which all D-VHS decks are equipped; the provision of other speeds depends on the purpose and design of the machine. Auto-selection of tape speed (and thus picture quality) becomes possible when DTV signal sources incorporate a 'flag' to indicate their bit-rate. The drum speed remains constant in all tape speeds/modes at 1800 rpm, and the recording rate is always 14.1 Mb/sec per track. In the lower speed modes, then, the rotating heads are switched off during one or more half-revolutions of the drum; recording of a track is only carried out when the tape has moved sufficiently for the new track to be drawn alongside its predecessor at 29 microns pitch. This intermittent recording system requires the use of a short-term data store in the record circuit: Figure 24.14 shows how a relatively low data rate is time-compressed and released to the head in 'bursts' to coincide with recording duty cycles. Illustrated here is a compression ratio of 2:1 to store on tape a 7 Mb/sec datastream in LS2 (half-speed) mode, with the same recording rate of 14.1 Mb/sec. In this example nothing is lost because all the available data goes onto tape. If a recording speed is chosen which is too slow to accommodate the input data rate, the recording system automatically compresses the data to match the 14.1 Mb/sec writing rate, and replay picture quality falls off correspondingly. LS2 mode requires the presence of three digital heads on the drum to maintain correct azimuth in adjacent tracks.

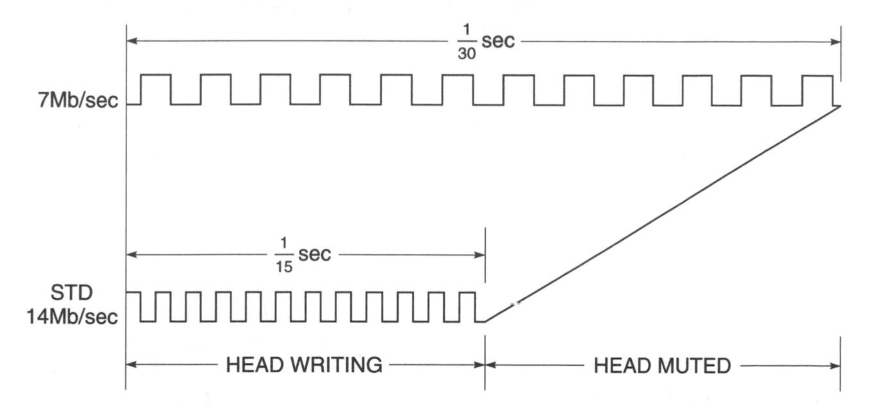

Figure 24.14 *Time compression for long-play modes in D-VHS format*

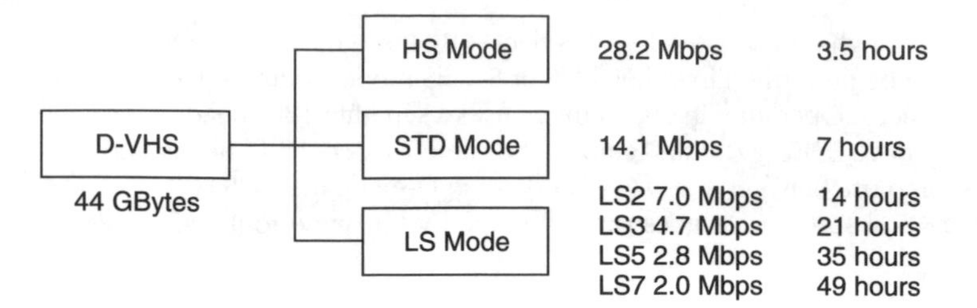

Figure 24.15 *D-VHS playing times and modes: the trade-off between tape consumption and picture quality*

An advanced version of D-VHS (HS mode) employs *four* digital heads on the drum. Here the tape speed is increased to 33.34 mm/sec, halving the cassette's running time to $3\frac{1}{2}$ hours, but now permitting *two* standard tracks to be recorded in unit time, with the drum speed still at 1800 rpm. The data-rate onto tape becomes 28.2 (2 of 14.1) Mb/sec, sufficient for HD-TV or high-throughput data to be recorded. The capacity of the tape remains at 44 Gbytes, as indeed it does in the progressively slower tape speeds of LS3, LS5 and LS7, the latter involving a tape speed of 2.38 mm/sec, a tape-running time of 49 hours and a maximum data rate of 2 Mbit/sec. D-VHS decks equipped with two digital heads are capable of running in STD and LS3 modes, see Figure 24.15. On replay the contents of the tape are sampled, initially to ascertain whether an analogue or digital recording is present. In the former case differentiation is made between VHS and S-VHS, SP and LP, and the servos, heads and replay electronics switch accordingly; in the latter case the D-heads are switched into circuit and 'strobed' according to the tape speed, now being auto-selected to match that used during record. In replay of 'slow' modes the data-stores are brought into play, too, to turn the intermittent bursts of data from the replay heads into a continual stream. Table 24.5 shows the basic specifications of D-VHS recording modes, while Table 24.6 indicates typical usage of the various recording speeds available to this flexible format.

The D-VHS track configuration is shown in Figure 24.16, where it can be seen that all data, vision and sound, is recorded in a single track-sweep, so that video insert or audio dubbing is not possible. Each track begins with a short burst of sub-data (housekeeping, search and address data) and finishes with parity codes used to check and repair data corruption during replay. The CTL track guides the servo and auto-tracking sections of the control system, and carries sophisticated index marks, vital to navigation of this type of tape system, with its long (compared to disc) access time.

Bitstream recording

D-VHS utilises bitstream recording, a method of capturing compressed or processed (e.g. encrypted) data like that of a digital broadcast on the tape directly as digital data, and replaying it in the same state. This bitstream recording does not involve functions like A-D and D-A conversion, digital compression/decompression or descrambling, the circuitry for which is incorporated in (e.g.) DTV receivers. D-VHS bitstream recording is compatible with MPEG2, the standard

Table 24.5 *Recording specifications for six modes in D-VHS*

		STD Mode	HS Mode	LS Mode			
				LS2	LS3	LS5	LS7
Head configuration		Rotating 2-head	Rotating 4-head	STD mode +1 head	Same as STD mode		
Recording time							
Using DF-420		7 hours	3.5 hours	14 hours	21 hours	35 hours	49 hours
Using DF-300		5 hours	2.5 hours	10 hours	15 hours	25 hours	35 hours
Track compo-sition	Tape speed (mm/sec.)	16.67	33.35	8.33	5.55	3.33	2.38
	Head azimuth	±30°					
	Drum rotation speed	1800 rpm					
	Tracking system	CTL track system					
Recording specifi-cation	Main data input rate	14.1 Mbps	28.2 Mbps	7.0 Mbps	4.7 Mbps	2.8 Mbps	2.0 Mbps
	Sub-data input rate	0.146 Mbps	0.292 Mbps	73.0 Kbps	48.7 Kbps	29.2 Kbps	20.9 Kbps
	Recording rate	19.14 Mbps	38.28 Mbps	19.14 Mbps			
	Track structure	1 sector					
	Sync block length	112 Bytes					
	Inner correction code	RS code					
	Outer correction code	RS code					
	Code word shuffling	6 track					
	Modulation system	SI-NRZI					

processing and delivery system for DTV, but can only be used with DTV receivers when they are fitted with digital in/out ports of a suitable type. This type of interface would open the way for relatively inexpensive D-VHS decks, dedicated to DTV programme storage. As Table 24.6 showed, whole DTV multiplexes can be recorded as a single bitstream if required, or single programmes selected (perhaps

Table 24.6 *D-VHS mode application chart*

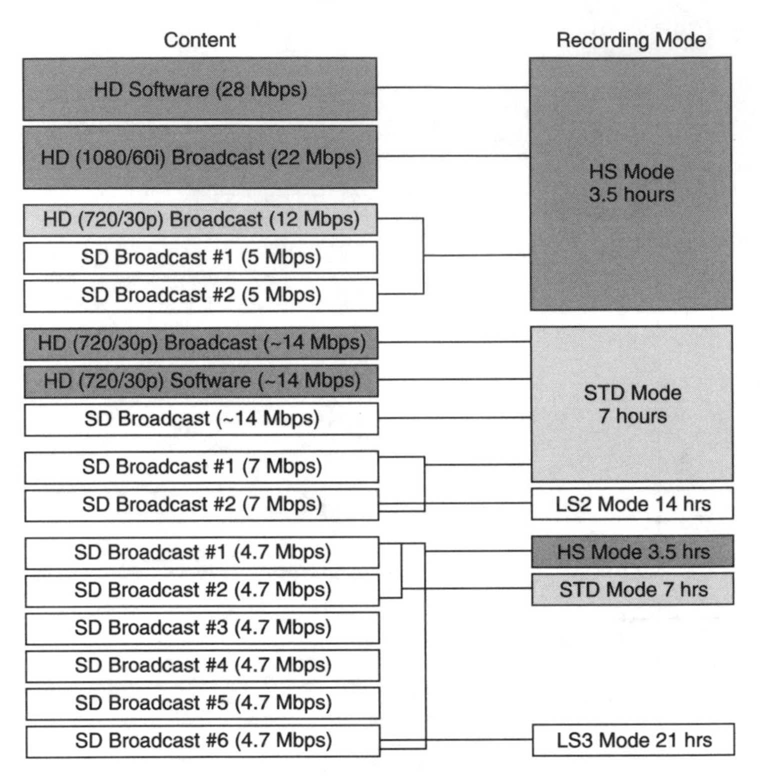

automatically, timed, or 'themed') with the help of a transport demux chip like that described on page 189. The 14.1 Mb/sec capability of STD mode in D-VHS is a reasonable match to the 15 Mb/sec norm for MPEG2's maximum transfer rate in MP @ ML, main profile at main level.

The heart of a full-spec D-VHS machine is a MPEG2 codec (coder-decoder) as shown in Figure 24.17. The codec interfaces several possible types of input source and signal with the D-VHS format while maintaining compatibility. The DV interface (Firewire, i-Link, IEE1394) is bi-directional and suited to camcorders and computers: we shall meet it again below, and in Chapter 28.

S-VHS processing

The presence of digital stores and processing in a D-VHS machine permits sophisticated processing of analogue replay signals. The technology used by JVC is called *Digipure*, consisting of a wide timebase corrector (TBC) to remove signal jitter in playback; 3-D colour, for good colour separation and image-edge enhancement; digital picture noise reduction to improve Y and C S/N ratio by 3 dB; and '3R picture system', enhancing luminance detail by the application of edge-correction. Also available to analogue playback mode are variable-speed

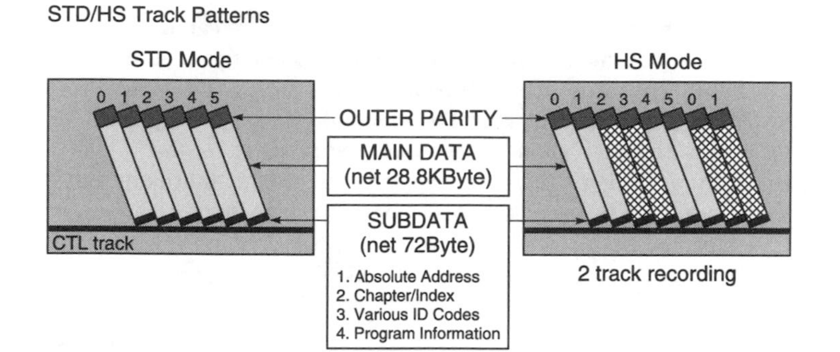

Figure 24.16 *Tape-track configurations for D-VHS*

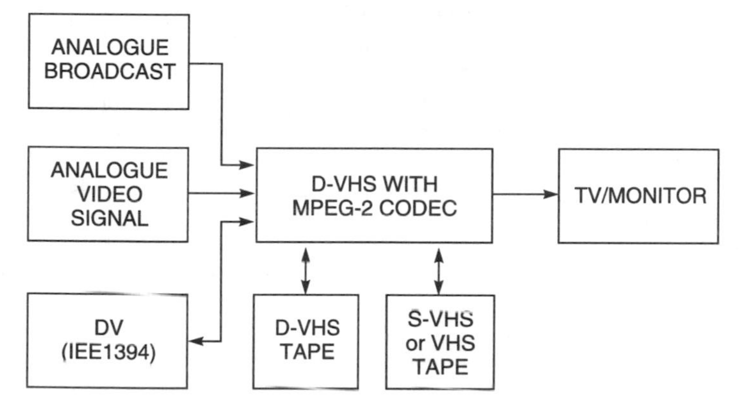

Figure 24.17 *Illustrating the versatility of D-VHS format*

picture search, slow motion and frame advance – only still picture and picture search are available in digital mode.

Digital home editing

Until the advent of MPEG technology and large data stores in computers and on digital tape decks, all home video editing was an 'on-line' process; selectively copying from one tape to another in real time, and losing sound and vision quality in the process. A second-, and particularly third-generation copy showed considerable impairment. At first the process was entirely manual, whereby the copy machine was paused while each required section of the source tape was found and set to replay. We saw in the last chapter how the on-line editing process was automated, first with stand-alone editors and then with home computers, all based on the concept of the user compiling an EDL (edit-decision list) whose instructions are carried out by the editing machine. It manipulates the transport controls of both source and recording decks, cued by a timecode or a picture-frame count from the source deck.

As computer operating (clock) speeds increased, and more capacious hard-disk stores were fitted, it became possible to edit 'off-line'. Here the picture and sound information from the source tape is assembled in compressed-digital form on a computer's hard disk, complete with the (human) editor's instructions. All the camcorder's shots can be copied to the computer's data store, then edited with single-frame accuracy using easily-operated software: sound, transitions, titles and special effects are added, then the 'master movie' can be output in a variety of formats to a tape, TV set or other storage/display/transmission device. The first stage in the process is a 'capture card', which plugs into a spare slot in the computer.

Capture card

The capture card processing – for an analogue input – starts with a fast A-D converter, providing YUV 4:2:2 data. Its sampling rate (hence final picture quality) can usually be selected in software, primarily to suit the resolution available from the source; and the computer/HDD capabilities. Typical resolution settings are 384 × 288 pixels (low band/VHS/Video-8 standard) and 768 × 576 pixels, corresponding to TV broadcast standard. Each pixel data-bunch contains data on its Y, U and V content. Following sampling and quantisation, data compression takes place to reduce the bit-rate without badly impairing the image quality: this reduction is essential to accommodate a useful programme period on the hard disk. Most analogue capture cards work to Motion-JPEG (M-JPEG) standard, with which the data rate can be reduced by a factor of up to 100, though of course there's a trade-off between compression rate and moving-image quality. At compression ratios up to 5:1 the results are very good; between 5:1 and 15:1 some deterioration in quality is perceptible, depending on motion rates, picture 'busyness' and original image quality. Compression ratios higher than 20:1 give progressively poorer results. The data-reduction system is again like that described in Chapter 12, with the pixels being processed in 8 × 8 blocks and applied to a DCT conversion stage.

The video data is now applied to the computer's data bus, along with audio data, ideally produced and processed on the same card to avoid lip-synchronisation problems. The interleaved audio and video (AVI) data forms a file which can be written to the hard disk.

Editing

The editing process is conducted by a special software program with which the user's requirements are specified in terms of trimming and combining clips, adding transitions, effects, titles and so on. Sophisticated editing programs offer many effects and facilities: batch capture, animation, 'morphing', 'paints', chroma-key, filters, image re-sizing, etc. The final picture/sound programme schedule is built up on a *timeline*, a series of horizontal on-screen rows, each representing a video, audio or effects track, and progressing in time from left to right. This timeline can be scaled as required, ranging from the entire required 'movie' to just a few frames, seconds or minutes. All these instructions are stored with frame/timecode markings, but not yet executed.

Rendering

The rendering process is a long one, during which all the instructions are carried out on a frame-by-frame basis, pulling sound and vision data off the hard disc, processing it as required, and then progressively reassembling it back onto the HDD, which ideally is a separate one from that fitted as a part of the computer's basic system. HDDs are available in multi-disk form containing several tens of Gbyte capacity and with fast data-transfer rates. The programme-time capacity of a disk depends on the capture resolution and data-compression rate chosen by the user in software: typical figures are 12 min per Gbyte for low-band VHS/Video-8 quality (15:1 compression ratio) and six minutes per Gbyte for high-band S-VHS/Hi-8 quality at a compression ratio of 8:1. During replay the edited material comes off the hard disk as an AVI file for de-interleaving, reverse DCT processing and data-expansion: these take place on the capture card (*codec*) where the data is finally D-A converted and output as composite video or S-signals and baseband audio, a process called printing to video.

Integrated off-line editor

The digital editing system described above can be carried out in a dedicated unit which contains all the elements of a computer, but without facilities like monitor, keyboard and printer ports, and loaded only with the required editing software, codecs and interfaces – plus a very large capacity HDD block. It is simpler to install and operate than a system using a general-purpose PC. The first unit of this type on the market was the *Casablanca* by Macro Systems.

DV Editing

The A-D and data-processing involved in off-line computer editing of analogue video necessarily introduces degradation of picture and sound. Where the footage is captured in a camcorder to DVC tape the data compression ratio is fixed by the system at 5:1, with an excellent, tailor-made data-reduction algorithm. So long as this data is fully preserved during the editing and storage phases it becomes possible to carry out 'transparent' editing, entirely in the digital realm, and with no degradation at all of picture and sound quality, no matter how many generations of dubbing takes place. This is achieved with a DV capture card having a Firewire/i-link/IEE1394 input/output port. It contains no A-D/D-A converters or data-compression systems, acting merely as a buffer/interface between the serial input line and the computer's data buses. Of course it carries out many other functions in the realm of software/instruction implementation.

To complete the 'transparent' editing system, the camcorder or video deck involved needs to have a Firewire input facility so that the edited material can be exported and stored on tape in digital form. Where the video camcorder or deck does not have from-the-factory provision for Firewire input (a 'political' issue concerned with import-duty tariffs) it can often be enabled in software by the use of a computer or other 'widget' device.

25 Tape formats – systems and facilities compared

In preceding chapters we have examined VCR principles and circuits in general terms, taking examples from all home formats to illustrate the techniques used. Where differences between formats have arisen in the text they have been surprisingly few. In this last chapter we shall expand on the format differences and look into their significance, both from the point of view of the consumer and those concerned with the technicalities of the machines.

There are four major analogue home videotape formats in the consumer field: VHS, S-VHS, Video-8 and Hi-8. VHS owes its origin to JVC, while the Video-8 formats are later developments, using more advanced features – we have met ATF, RCTC, AFM and PCM sound in previous chapters. Judged subjectively on normal programme and home-movie material, the performance of the low- and high-band variants of VHS and Video-8 respectively is quite similar in terms of picture resolution and S/N ratio.

Over the years the competing formats have shaken down into well-defined categories. Video-8 is favoured for camcorder use, while VHS is the system for general-purpose homedeck use, and virtually the only one in which pre-recorded cassettes are produced for sale and rental. The high-band variants Hi-8 and S-VHS find a relatively small market, even though they are capable of producing better pictures than the others. This suggests that the public are more concerned about price than image quality! A tribute to Hi-8 is its gradual acceptance by TV broadcasters for certain types of programme, typically ENG (electronic news-gathering) and low-budget 'community' contributions.

The existence of four incompatible formats – and two cassette-size variations for VHS – is inconvenient for the user and sometimes exasperating for the trade, but is indicative of the atmosphere of free enterprise and unrestricted marketing which we enjoy in the free world. The spur of competition, as in other spheres, confers a seldom-realised advantage for the consumer in that he/she is able to purchase and enjoy a very advanced and sophisticated VCR, bristling with features, at an artificially low price. The situation has arisen because manufacturers and retailers have played leapfrog amongst themselves, in their respective spheres of features and profit margins, in an effort to promote themselves in the marketplace. Manufacturers have the long-term objective of building their reputations, pet formats and market-share; while retailers, locked in the stranglehold of High Street competition, have the simple objective of staying in business by maintaining a high volume of low-profit turnover.

Let us now examine the main features of each analogue format, whose tape cassettes are illustrated in Figures 25.1 and 25.2.

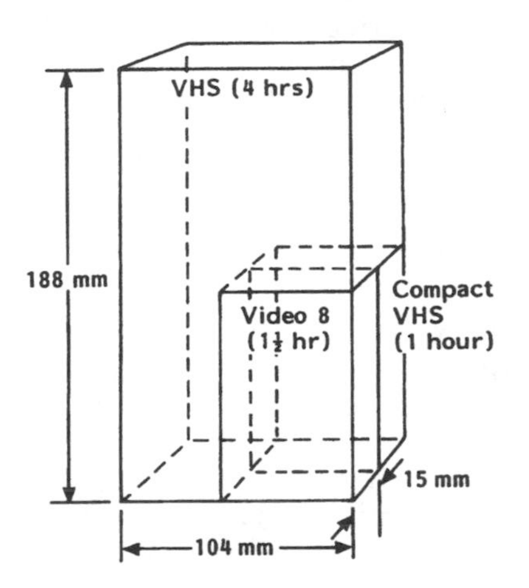

Figure 25.1 *Cassette packages and playing time compared. The running times given are for SP operation, and are doubled in LP mode*

Figure 25.2 *Cassette sizes compared. Clockwise from bottom right are Video-8, VHS, VHS-C and an audio Compact Cassette, shown here for size comparison*

VHS

VHS stands for Video Home System: this format was designed by the JVC company in Japan. It was released in Europe in 1978. VHS enjoyed the backing of one of the leading UK rental companies of those days, Thorn–EMI–Ferguson, and this largely accounts for its fast and deep initial penetration into the UK market, aided by the British habit of renting rather than buying television equipment. VHS is the most widely-used tape system, with about 750 million units in use worldwide, and the format has evolved to take in LP, Hi-Fi sound, S-VHS, VHS-C and D-VHS. It was in VHS form that the first 'mobile' VCRs appeared.

VHS-C

A clever adaptation of the standard tape and cassette package is VHS-C (Compact) which permits a small and light VHS comcorder. The cassette is a small (92 mm × 59 mm × 23 mm) housing containing up to 1 hour's worth of standard 12.7 mm VHS tape. It fits a small camcorder weighing less than 1 kg, and incorporating a small head drum, thin direct-drive motors and a solid-state image sensor. Back at home the small cassette is loaded, piggy-back style, into a normal-size adaptor shell for replay or editing in the standard VHS machine.

Long-play

Both camcorders and homebase machines, VHS and Video-8, are available with a dual-speed option. On record they are switchable betweeen standard and slow speed, and during replay can automatically recognise which recording speed was used in any cassette offered to them.

In LP mode the capstan speed is halved – to double the playing time of a standard cassette to a maximum (e.g. VHS E240 tape) of eight hours. Early dual-speed machines had separate head pairs for SP and LP modes; later designs incorporate both SP and LP heads in a single pair of ferrite chips; current practice is the use of the *same* pair of heads in both modes. Here the head width is a compromise; in SP, narrower-than-standard tracks are written, leaving a gap between them. In LP, each recorded track is wider than required, and its excess width is cut down by the erasing action of the next head sweep – see Figure 17.22 and associated text. VHS-LP video tracks are 24.5 microns wide, and the sound and control tracks are recorded in the normal way at the slower speed. The track patterns, then, represent a 'telescoped-up' version of Figure 17.5 with video tracks and control pulses now occupying half the linear tape space of those of a standard VHS recording.

Performance-wise, LP mode is surprisingly good. Video S/N ratio is slightly impaired due to the narrow video tracks; but sound HF response is dramatically cut, where longitudinal tracks are used. Many non-critical viewers cannot tell the difference between SP and LP on replay, so long as it comes from a good tape in good condition. The maximum eight-hour capability of VHS-LP comes into its own when used in conjunction with multi-event timers, and allows, for instance, eight separate one-hour recordings to be made without human intervention.

Even though LP performance is good, its use in camcorders and homebase machines is not recommended unless the programme will be viewed once (non-critically!) before being discarded.

In VCRs and camcorders which use a single pair of compromise-width heads, (most of them, in practice), the still-frame picture performance, especially in SP mode, is very poor due to the impossibility of keeping a narrow head sweep aligned with a narrow tape track when the scanning angle is changed by the cessation of tape movement: this was explained in Chapter 17 and Figures 17.18 and 17.20.

S-VHS

Super-VHS is an advanced 'high-band' variant of the established format. It uses a high FM carrier, with deviation (Figure 25.3) from 5.4 to 7 MHz and more 'meat' in the video sidebands. To permit this an advanced videotape formula is used in a cassette of conventional size, shape and running time. A very small video head-gap, and 'fine grain' magnetic tape permits a baseband video frequency response approaching 5 MHz, and on-screen resolution better than 400 lines. Noise ratio is also better than that of the low-band formats.

One of the problems of PAL and NTSC colour encoding is the presence of cross-colour effects, discussed in some detail at the end of Chapter 4. Where the programme is originated in these forms little can be done to prevent cross-colour – it stems from the decoding process at the receiving end. For other applications (home videography, MAC and digital TV transmissions, pre-recorded movies on tape etc.) the S-VHS system has provision for complete separation of Y and C (luminance and chrominance) signals throughout the record and playback

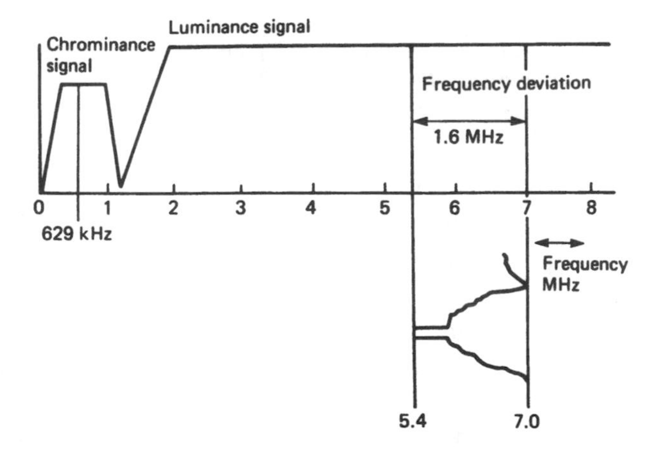

Figure 25.3 *Recording signal spectrum for S-VHS format*

processes. Where the TV receiver or monitor is equipped with a suitable input socket (S-terminal) and a large screen, S-VHS format is capable of a picture performance superior to that available from a terrestrial broadcast system.

Video-8

The Video-8 format uses 8 mm-wide tape, from which it gets its name. Unlike the others this system was developed by a consortium of companies and accepted for use by over 127 of the world's major audio/video manufacturers. It was the first domestic video system to use digital audio recording, and the first to be designed from the outset for an alternative mode of operation – as a high-quality digital sound-only recording system. Its other advantages are a small cassette (Figure 25.2), facilitating miniaturisation of the equipment, be it portable or homebase type; a flying erase head for good edits, also featured in some VHS equipment; and the exploitation of new tape and head materials and techniques for better performance.

Although offered in homebase form for tabletop use, Video-8 is seen mainly as a camcorder format, whose primary advantages are excellent sound, light weight, and high performance – the Video-8 manufacturing companies are particular experts with lenses and TV image sensors.

The Video-8 tape wrap was shown in Figure 19.9. Although the format specifications do not in any way dictate the mechanical layout of the machine, some early Video-8 deck designs used a combination of M-wrap and U-wrap techniques as shown in Figure 25.4. At *a* is shown the situation where the casssette has just been lowered onto the deck: the pinch-roller and several tape guides have penetrated the cassette behind the front tape loop. In diagram *b* the first (M-loading) phase is complete: two guides have drawn a loop of tape away from the cassette, VHS-fashion. At this point the loading ring starts to rotate anticlockwise, further wrapping the tape around the head drum, borne on a guide/post ahead of the pinch roller, and prevented from folding back on itself by further ring-mounted guides, diagram *c*.

Current machines use an M-wrap system, like that of Figure 17.6, the first phase of which involves the entry of the *cassette* into the machine.

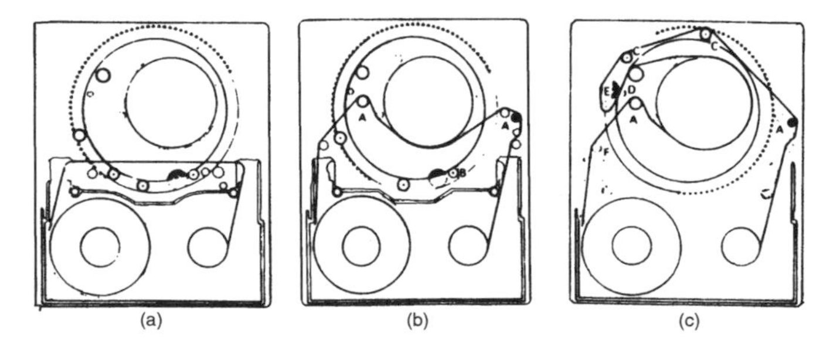

Figure 25.4 *Video-8 tape threading: (a) cassette in; (b) first stage; (c) wrap completed. Compare with Figure 17.6*

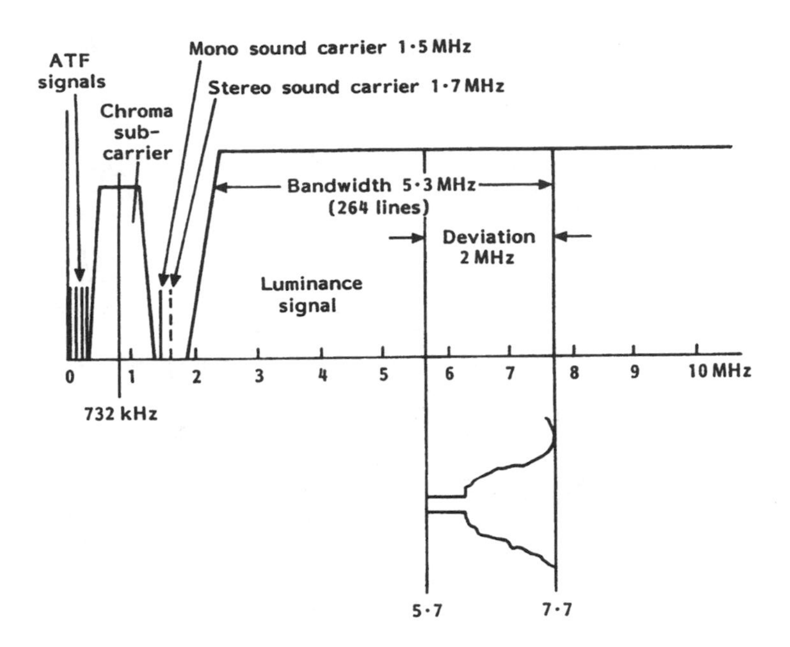

Figure 25.5 *Recording signal spectrum for Hi-8 format*

Hi-8

As with VHS, development of the basic Video-8 format led to the production of a high-band variant, with picture resolution exceeding 400 lines under favourable circumstances. This, the Hi-8 variant, has higher FM record frequencies and a greater sideband spread than Video-8, as shown in figure 25.5. It's used with ME (metal evaporated) and high-grade MP (metal powder) tapes. In camcorders, which is the form in which nearly all Hi-8 hardware is produced, the lens and CCD image sensor are upgraded to accommodate the higher resolution capability, and the bandwidth of the luminance path widened to suit. As with S-VHS, provision is made for separate handling of Y and C video components throughout the chain, and for their passage through S-links to suitably-equipped TVs, monitors and copy-VCRs.

In other respects, Hi-8 is similar to Video-8 format, with the same cassette dimensions, sound systems, tracking and transport arrangements.

Format comparison

The salient features of the competing formats are shown in Table 25.1 which includes details of some obsolete formats for comparison purposes. The VCR, VCR-LP, Betamax and V2000 formats are really only of academic interest now, but are included to show the progress made over the years.

Table 25.1 *Formats compared: the main parameters of past and present home formats. All but Video-8 use 12.7 mm ($\frac{1}{2}$ in) tape. The high-band variants S-VHS and Hi-8 have the same physical characteristics as the standard VHS and Video-8 bases shown here. It is interesting to see (right-hand column) how the information density on tape has steadily increased over the years*

	Linear tape speed, cm/sec	Video writing speed, m/sec	Video track width, microns	Azmuth offset of video heads	Head drum diameter, cm	Video track angle to tape	Sound track width, mm (mono or L & R)	Max playing time, hours	Information density on tape (Hrs per sq.metre)
VCR	14.29	8.10	130	±15°	10.5	3.69°	0.7	1	0.16
VCR-LP	6.56	8.10	85	±15°	10.5	3.71°	0.7	2	0.33
VHS	2.34	4.85	49	± 6°	6.2	5.33°	1.0	4‡	0.93
BETA	1.87	5.83	33	± 7°	7.45	5.97°	1.05	$3\frac{1}{2}$	1.16
V2000	2.44	5.08	23	±15°	6.5	2.65°	0.65	8*	1.79
Video 8	2.01	3.12	34	±10°	4.0	4.92°	N/A	$1\frac{1}{2}$	1.70
VHS-LP	1.17	4.85	24.5	± 6°	6.2	5.35°	1.0	8‡	1.86
V8-LP	1.01	3.12	17	±10°	4.0	4.95°	N/A	3	3.4

***Flip-over cassette: 2 × 4 hours. ‡ 5-hour tapes are available**

Regarding performance, too much emphasis should not be put on the details shown in Table 25.1, since it is very difficult, even for those experienced in domestic VCR use, to tell the formats apart by watching replayed pictures from any of them, so long as the comparison is like for like in high-band or low-band. In average domestic conditions, any slight performance differences between the various systems are sometimes 'swamped' by other shortcomings; a chain is only as good as its weakest link, and many factors beyond the control of the VCR manufacturer can mar performance. A weak aerial signal impairs S/N ratio; worn tape or dirty heads cause excessive drop-out; maladjustment of the tracking control, or a deck in need of servicing leads to mistracking. The TV set in use may not be in perfect condition, and maladjustment of the focus control, a worn tube or incorrect tuning can easily halve the definition available from the VCR!

Special features

The provision of extra features has a value which varies greatly from user to user, and in some cases the advantages of these are more apparent than real. All VCRs are capable of recording and replaying TV programmes well, and for the reasons just given, and the fact that signal-processing circuit design does not vary much between machines (leaving out the special case of high-band formats), a £250 model is unlikely to give markedly better performance on normal programme replay than one costing £150. Most domestic VCR use is for record and replay of films, TV serials and sports events, and replay of commercially-recorded cassettes which are purchased or rented, and for these a basic machine is quite adequate.

Most of the extra cost of full-feature VCRs, then, goes into the provision of editing, still-frame, hi-fi stereo sound, Nicam decoders and trick-picture effects using either special head arrangements or field-store memories, with the basic electronics and mechanics largely unchanged. In many households the use made of some of these features and facilities does not justify their cost. Certainly good freeze-frame, frame-advance and slow-motion facilities are very useful in educational spheres and for sports enthusiasts. Such things as high-band formats and Nicam hi-fi sound are completely lost if the machine is used with a small-screen mono-sound TV set, however, so careful thought at the time of purchase can save unnecessary expense. The same is true of camcorders. High-band models give a very worthwhile increase in definition, again of little use with a small-screen TV, but coming into its own with large playback screens and where editing or copying is envisaged. For the latter, too, it's essential to choose a camcorder with an edit-control port if it may ever be used in conjunction with any form of auto-editing system. The value of in-camera 'effects' is somewhat in doubt once the novelty has worn off, but the value of a long zoom range lens and good handling features is never so. Also worth bearing in mind is the audio performance of the two competing camcorder formats, and particularly the limited bandwidth and S/N ratio of VHS/VHS-C camcorders which use longitudinal audio tracks, especially in LP mode – this was illustrated in Figure 19.10. VHS hi-fi and all 8 mm systems have better sound, virtually independent of tape speed. Sound apart, there is little *intrinsic* difference in performance between the two formats in their high- and low-band variants, though the 8 mm systems currently offer a wider range of features and types due to the larger number of manufacturers now supporting them.

Digital tape formats

As in other spheres of home entertainment equipment, digital tape recording gives better pictures and sound than analogue. DVC format offers 500 line picture resolution, compared to the 400 and 240 of high- and low-band analogue formats. It eliminates the colour-smear and -bleed effects inherent in conventionally recorded pictures, along with cross-colour, noise, mistracking, jitter and other defects. It facilitates 'lossless' copying and editing with an off-line computer system as described in the previous chapter. And it has the smallest cassette of all the videotape formats, making for very light and compact camcorders. Since DVC camcorder prices are comparable with those of high-end analogue ones, it is difficult now to find any argument in favour of buying the latter.

Comparisons between DVC and D-VHS formats are not really valid because they serve very different purposes as we saw in the last chapter. A more valid comparison is between D-VHS and disk-based data storage systems, where the 44.4 Gbyte capacity of D-VHS is now challenged by computer-type HDD units, which (with optical discs like DVD) have the tremendous advantage of virtually instant access; a tape system will always have a long access time, measured in several minutes in worst case, but still acceptable where the order of replay is usually the same as on record – like TV programmes! The sophisticated 'tape navigation' system used as in D-VHS format helps to mitigate the slow access of this medium.

To redress the access-time drawback, D-VHS has the advantage of being able to save, swap and archive programmes on cassette as well as merely 'timeshift' them, something not possible with a magnetic HDD. S-VHS/D-VHS tapes are relatively cheap, and D-VHS machines likewise enjoy the economy which comes from the use of existing decks and mechanisms. The other virtue of D-VHS, of course, is its backward-compatibility with the hardware and software of the previous quarter-century.

Table 24.4 compares the DV formats with high-band analogue systems. Some comparisons between tape and optical-disc systems are given towards the end of the next chapter.

26 *DVD players*

Video discs have been around for a long time. The first optical system (*Laservision*) was introduced by Philips in 1972, using an analogue recording system of variable pit length and spacing on an aluminised plastic disc of 300 mm diameter, rotating at 570–1500 rpm. It was scanned by a 1 mW 'red' laser to give a maximum playing time of one hour per side, with much better picture quality than was available from other video-storage media at that time. Even so, it never sold in large quantities: because of its restricted playing time; its inability to offer a home-recording facility; and the limited availability of pre-recorded software, especially compared to VHS. It finally became a 'niche' product for home-cinema enthusiasts before it was superceded by the DVD system which will be the subject of this chapter. During the transitory period some dual-system players were made and marketed, notably by Pioneer.

DVD disc

The abbreviation DVD stands for Digital Versatile (as opposed to Video) Disc, indicating that the system can be used for storage and replay of data other than that relating to video and audio. Even so, it is not the same as that used for computer data storage (CD-ROM and DVD-RAM) which does not have so much capacity as a DVD disc. Many modern computers, however, are capable of playing DVD discs in the same tray as is used for CD-ROM types. The DVD disc is the same in size and appearance as an audio CD, at 12 cm diameter and 1.2 cm thick. For both, the programme data is 'printed' into the disc in the form of a spiral of tiny pits in its optical surface, and is read off by reflecting laser light from it. The DVD is capable of carrying a great deal more data than a conventional audio CD or analogue Laserdisc, however, by virtue of its smaller and tighter-packed pit-track: DVD pit length is 0.4 micron at a track pitch of 0.74 micron, see Figure 26.1 for a comparison with CD and LD topography. The storage capacity of DVD is 4.7 Gbytes per side, about seven times that of a CD, and this – with the use of MPEG-2 data-compression technology – can offer over two hours of high quality (500+ lines resolution) pictures and multi-channel digital surround sound. In addition to these it is possible to provide eight different language soundtracks and up to 32 subtitling tracks, all selectable by the

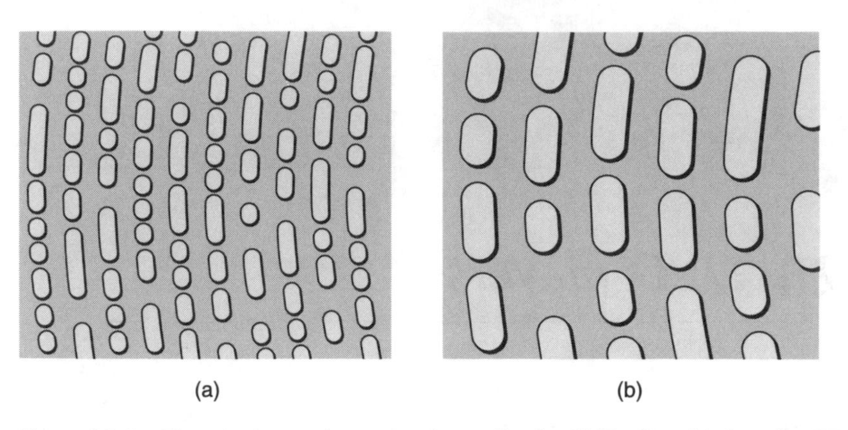

Figure 26.1 *The pit-size and -spacing is smaller for DVD discs (a) than for CD and LD (b), permitting longer playing time*

user. The very small pit structure is addressed by a sharply-focused light beam from a short-wavelength (640 nm) red semiconductor laser.

A standard DVD disc, like an audio CD, LDV or CD-ROM, has only one playing surface. Because it is made from two 0.6 mm discs bonded back-to-back, however, it is possible to make a double-sided DVD, and to automatically play both tracks sequentially in a machine with two laser assemblies or a flip-over-disc mechanism. The format also allows for two pitted surface layers per side, the upper of which is translucent; the required layer is selected by mechanically adjusting the focal point of the laser's objective lens, as shown in Figure 26.2. The maximum data-storage capacity, then, is 17 Gbyte in a double-sided double-layer disc, providing 8 hours playback capacity.

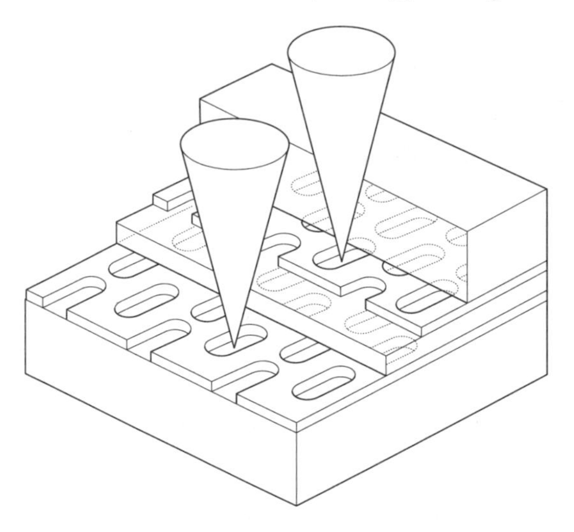

Figure 26.2 *In double-layer DVDs the laser beam is focused on the required surface, the upper of which is translucent*

Light path

Figure 26.3 gives an idea of the optical system of a disc player. This OPU (optical pick-up unit) sits below the disc on a swinging (radial) arm, and its bottom-most component is the laser itself, whose light output is regulated in intensity by an inbuilt photodiode in a feedback loop. The beam passes through a special prism on its way up to the collimator lens whose job it is to gather the diverging rays and direct them via the quarter-wave plate (acts as polarising filter) to the objective

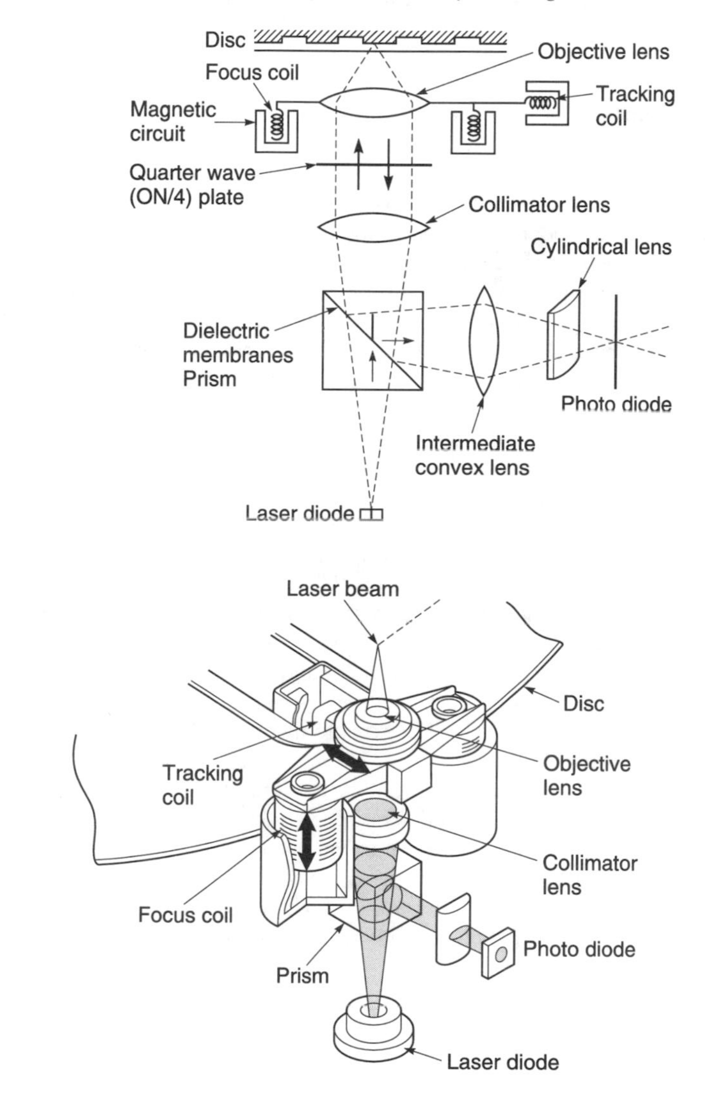

Figure 26.3 *(a) Light path for optical disc reading; (b) The main components of the optical block in a disc player*

lens. The focal point of this lens is the disc surface, wherefrom light is reflected back in the absence of a pit, but scattered in the presence of one.

On its return path, then, the light is modulated. It passes back through the objective lens, quarter-wave plate and collimator to the prism, whose 45° membrane surface reflects it – with some loss – out sideways to a vertically-mounted convex lens. This focuses the return beam onto the surface of a photodiode matrix which picks up the modulated light beam via a cylindrical lens for optical (coma) correction. The photodiode array is called an OEIC (Optical Electronics Integrated Circuit). We shall see in a moment the functions of this and the focus and tracking coils in the diagram.

Servo and control

The mechanical aspects of a DVD player are similar in essence to those used in audio CD players – indeed both types of disc can be used in the DVD machine. Thus there is a spindle motor servo, working with a motor FG (see Chapter 20) and the data coming off disc to maintain a constant linear velocity (scanning speed) of about 4 metres/sec. The tracking servo's function is to keep the light beam centred on the pit track being read, achieved roughly by the radial drive motor and very finely by a moving-coil assembly attached to the objective lens and capable of moving it laterally to 'steer' the beam by a small amount – tracking accuracy needs to be much greater for DVD reading than for CDs. The lens is also under the (axial) control of a focus coil, similar in operation to that fitted to a loudspeaker, which forms the mechanical element of a third servo system, this one dedicated to keeping the light beam focused on the reflective surface of the disc. These are all pulse-counting systems of the type described on page 316, and the mechanical operating range of the tracking and focus servo systems are measured in microns, calling for a high degree of mechanical precision, stability and cleanliness. The signal from the disc has two states, light reflected and light diffused, governed by the pit pattern. It is intercepted by a 4-segment photodiode array, whose combined output renders signal data, and whose differential outputs provide positioning and focusing data for their respective servos. More information on the principles and mechanics of these opto-mechanical systems is available in other books published by Newnes.

System control for DVD is carried out by a microprocessor system of the type described in Chapters 8 and 21 of this book, with the same type of EEPROM data memory and serial clock/data internal communication and control lines. A typical DVD player has two CPUs, a main one to govern the deck functions and replay processors, and a 'slave' one to cater for such housekeeping functions as remote- and power-control, front panel indicators, control-key codec and output mode and switching. See Figure 26.5.

Signal processing

Figure 26.4 shows the main stages in DVD record and playback. Incoming video and audio signals are A-D converted, compressed in the MPEG-2 system

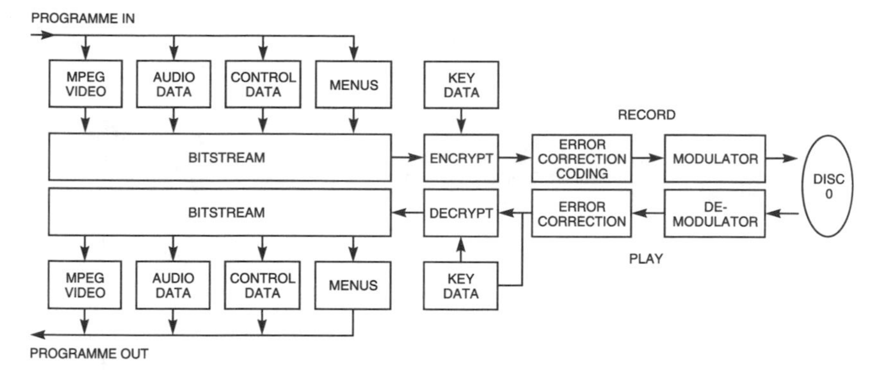

Figure 26.4 *DVD data processing in record and replay*

and coded for error protection in the way we have seen for (e.g.) digital tape recording in earlier chapters. The compression system used here is a variable-redundancy one, depending on picture content and the level of picture quality required; it varies from 1 to 10 Mbit/sec, averaging about 4.5 Mbit/sec. Menus, subtitles and control/housekeeping data are added to the sound and vision data to make a composite bitstream which is then encrypted to provide *regional coding* which we shall examine later. The error-correction system is based on a form of Reed-Solomon coding, and having had this added the datastream is modulated in 8-16 form to provide a single serial data stream which governs the length and spacing of the pits burned into the disc surface during record, or into the master disc from which the stamping die will be made in mass production.

In playback the data pulses coming from the photodiode matrix in the OPU are preamplified in a low-noise stage, demodulated and the tracking and focus feedback signals led off. The signal datastream undergoes processes which mirror those carried out during the conversion, compression and modulation stages of recording, to emerge as analogue signals from the D-A converter chips.

Player diagram

Figure 26.5 reproduces an overall block diagram of a Thomson CD player. The 'bitstream' block of Figure 26.4 is represented by the heavy lines in the centre of the diagram, and consists of a wide and fast data/control bus. Servo and mechanics are outlined on the left, controlled by ICs SIC 1 and MIC 1, the latter drawing on a 2K EPROM for operating instructions and user-programmed data. The off-disc signal is amplified and demodulated within RIC1, and its signal output passes to IC DIC 1, DVD/CD processor, working in conjunction with memory IC DIC 2.

The MPEG-2 decoding/expansion magic for both sound and vision is performed within decoder chip VIC 1; the expanded and corrected datastreams are built up inside the very large memory chip VIC 4. The video data passes to VIC 50,

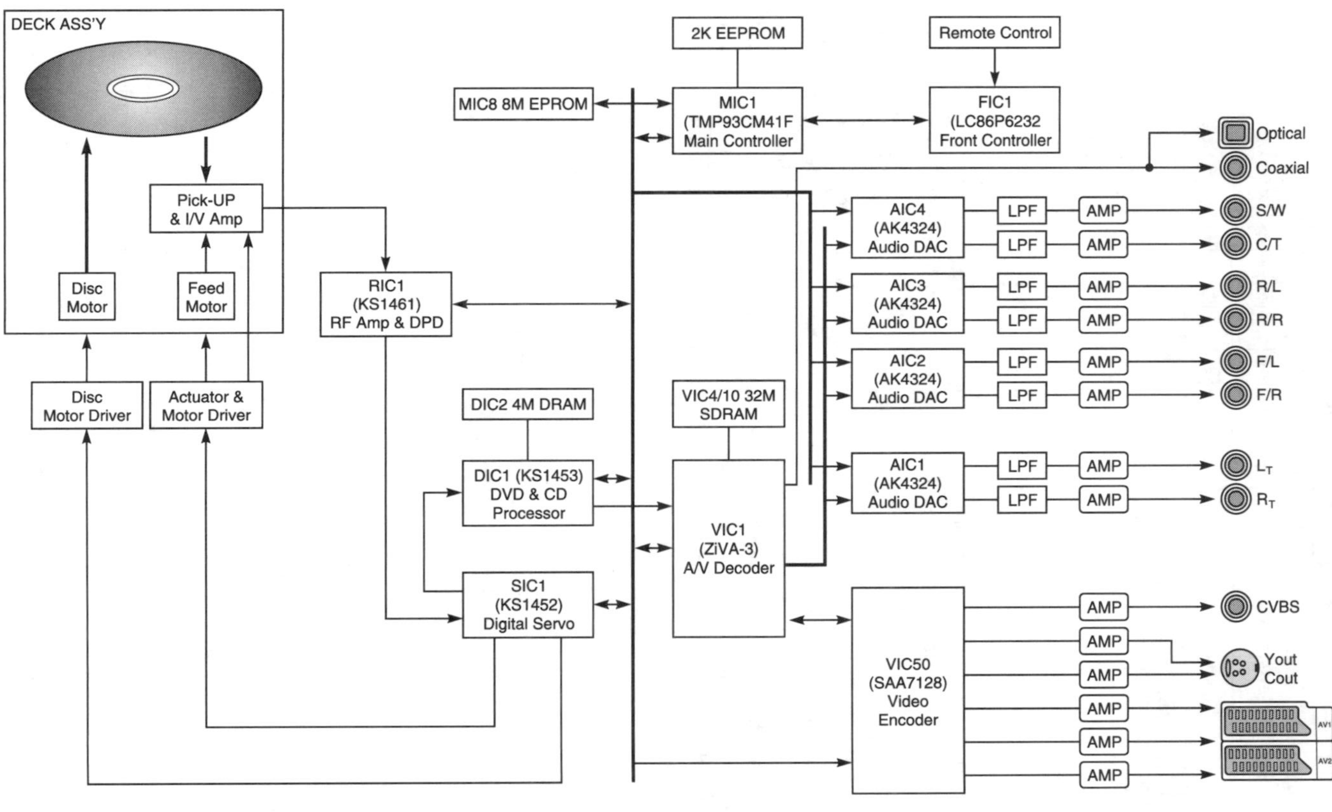

Figure 26.5 *Block diagram, much simplified, of commercial DVD player (Thomson)*

video encoder, from which Y, C, RGB and composite signals are routed to four output sockets. A second data bus from VIC 1 feeds four D-A converters AIC 1-4 with surround signals – a Dolby Digital 5.1 (see Chapter 11) decoder is built into VIC 1. D-A converters 2–4 each handle two channels, giving the 5.1 outputs (.1 is a subwoofer feed), while AIC 1 provides 'straight' stereo outputs Lt and Rt, from which a separate Dolby Pro-Logic decoder can extract Dolby Surround signals if required. As with many such models, this one can replay a wide variety of disc types: all surface variations of PAL and NTSC DVD; audio CD; Video CD; and S-video CD.

Regional coding

A 'flag' encoded onto the disc works in conjunction with a 40-bit 'random-key' data code encryption system to prevent disc replay where the player and disc are bought in different world regions: see Table 26.1. This facilitates the phased-release which the copyright-owners require. Copying from disc to another disc or tape is prevented by a Macrovision anti-copy system. Unfortunately it has been found easy by 'hackers' to overcome the encoding system – and the Macrovision protection – by modifying the software of the player, either by replacing the memory chip in which it's carried, or by re-programming the IC, usually to obtain 'Region 0' compatibility, which means that discs from any region can be played. In the UK this modification is popular amongst home-cinema enthusiasts who import discs from the USA, where the choice of films is wider and the certification rules different.

DVD features

DVD pictures are generally in widescreen (16:9) format; on some discs the flip-side contains the same film converted to 4:3 aspect ratio by a post-production 'pan and scan' operation. For some purposes a very long playing DVD can be made, with reduced picture quality and definition. The Dolby Digital sound system gives excellent quality audio reproduction, with no risk of inter-channel crosstalk because each of the six channels has its own separate 'slot' in the data-stream. Up to eight different languages can be selected from one disc, and up to 32 subtitle/caption tracks in different languages, independent of the audio selection.

Table 26.1 *Regional coding for DVD discs and players*

Region 1:	The USA and Canada
Region 2:	Europe, Japan and South Africa
Region 3:	Asia
Region 4:	Australia, New Zealand, Mexico and South America
Region 5:	Africa
Region 6:	China

Home-recordable DVD

Recordable DVD has several advantages over the long-established VHS tape system for home use. The picture from disc is better in terms of noise level and definition; the sound, too, is better, with multi-track/surround available as standard; there is virtually instant access to any part of the disc/programme; a better user interface; a wider choice of trade-offs between picture quality and playing time; and a more compact and easily filed and stored medium, similar to that of an audio CD.

Blank DVDs are pre-formatted with spiral grooves and tracks on which a red laser (similar in principle to those we have seen above, but more powerful) can write to the disc. The disc is made of a polycarbonate compound, coated with an alloy of 'phase change' material which can take on two distinct states, amorphous and crystalline. In its blank form the surface of the disc is in a crystalline state, shiny and reflective. During the writing phase a relatively powerful laser beam is pulse-modulated to raise the disc's surface temperature at the focal point sufficiently to melt the material, after which it quickly cools and solidifies, now in amorphous state with virtually no reflectivity to light. Each of these 'burn marks' has the same effect during replay as a pit on a conventional DVD.

Re-writable DVDs can be erased by passing them over a lower-powered laser beam whose focused spot-temperature is just sufficient to bring the disc surface layer back to its crystalline state and thus effectively erase the recording. The surface takes on its original mirror-like reflectivity, ready for a new recording to be 'burnt' into it.

DVD recorders of course use the same 8-16 bit modulation code and bitstream protocol as conventional discs to maintain compatibility with the standard format. They contain a complete MPEG2 encoder/data compressor, and – as with the DVC camcorders we examined in Chapter 24 – the processing ICs and their memory chips play a dual role, reversing during playback the functions they carried out on record.

27 *Care, operation and maintenance of VCRs*

A videocassette recorder is a complex ensemble of mechanical, electrical and electronic parts working to very high precision. As with any such equipment, careful handling and regular routine maintenance are necessary to maintain performance throughout the life of the machine.

Operation of the domestic VCR is relatively simple. Much thought has been given by designers to ease of operation, and the cassette system and syscon between them make the machine virtually foolproof. The VCR should be installed on a strong, level surface in a position where the ambient temperature is reasonably constant and the atmosphere clean – the 'chimney corner' is definitely out on both these counts! It is recommended to buy a dust cover – regular use of this will prevent ingress of atmospheric dust and soot particles, so prolonging the machine's life. In high ambient light, and particularly direct sunlight, the usable range of any infra-red remote control system will be greatly reduced, and this should be borne in mind during installation.

If the VCR is in close proximity to the TV receiver or monitor, mutual interference effects can take place due to direct radiation of signals between them. This leads to spurious effects ranging from patterning and striations to complete or partial loss of colour. While the specific cause of the effect may be obscure, physical separation of the TV and VCR will prove the point, and these symptoms can often be cured by installing a metal or foil screen between the two equipments, and earthing it if necessary.

Tape care

Considering its nature, video tape is surprisingly uncritical in storage and handling, largely due to the protective cassette housing. Cassettes should not be placed near magnetic field sources such as TV sets, large loudspeakers or transformers; they should be stored standing upright to prevent sag or cinching, in an even temperature away from dust.

When a tape is used for 'time-shift' or similar purposes, the temptation is to rewind it to the beginning before each recording so as to have maximum recording time available. This practice leads to disproportional wear on the first part of the tape, resulting eventually in noisy recordings with lots of drop-out. Very often a three-hour tape is used for general purposes and the first 30 minutes-worth can become worn before the tail-end of the tape has seen any action at all! It is far better to work through all the tape, using the counter memory, before starting again at the beginning.

Head-cleaning tapes

The use of proprietory cleaning tape cassettes for domestic VCR head-cleaning is something of an emotive issue, with contradictory claims being made for their effectiveness, and much more important, the possibility of their causing damage to the video heads, by their makers on the one hand, and VCR manufacturers and the service trade on the other. If a head-cleaning tape must be used, the soundest advice is to use a type marketed or recommended by the VCR manufacturer. Leaving the head-damage claims aside, however, there is no doubt that hand-cleaning of the heads (and the rest of the deck) is far more effective than any cleaning tape, although the former course is not open to the average VCR user, except by arranging it with a specialist service organisation.

Deck servicing

It must be emphasised here that any attempt to clean tape heads or adjust deck components by a person without the necessary knowledge and equipment is almost bound to end in disaster! One half-turn on a tape-guide adjustment screw can destroy the machine's compatibility, and the lightest of cleaning strokes in the wrong direction across a video head will destroy it; the repair bill could run to three figures. VCRs are not like the car or the central heating system!

Deck servicing resolves into three main areas, those of cleaning heads, tape path and deck surfaces; adjustment of tape guides and tensions to ensure compatibility; and attention to secondary mechanical parts such as brakes, clutches, drive wheels and operating levers and sliders. We'll look at each of these aspects in turn.

Cleaning

In Chapter 17 we studied the requirements of compatibility and saw that this calls for great accuracy in the laying-down and reading-off of video tape tracks. This is governed entirely by tape guides and head-drum ruler edge, and the angles and dimensions of these components are critical. After a period of use, dirt and grease deposits build up on their surfaces, and that of the head drum. These can obstruct or divert the tape path, and the result is mistracking. The first step in deck servicing, then, is to thoroughly clean the parts illustrated in the typical deck layout in Figure 27.1. They are the tape guides, rollers, ruler edge, head drum

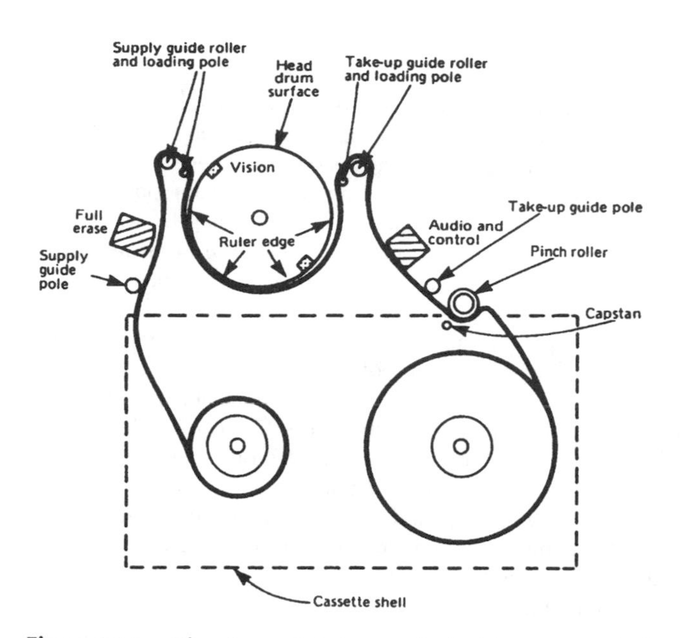

Figure 27.1 *Cleaning points in a typical videocassette deck. The shaded parts are the tape heads, and cleaning of these should be left till last*

surface, capstan and pinch roller. If necessary the latter should be replaced rather than refurbished. At the same time, residual dirt, tape particles and grease should be removed from all deck surfaces to prevent its later migration on to critical components. A cotton bud and methylated spirit are the best tools here, the moistened bud forming an effective wiper and collector of debris. When all surfaces are clean and guides, rollers and capstan polished, head cleaning can begin.

For the static heads (full-width erase, sound and sync) a perfectly clean cotton or buckskin cloth should be used, moistened with alcohol; a cotton bud is useful where head surfaces are difficult of access, as in some U-loading machines. Firm rubbing across the face of the heads is required, followed by final cleaning and polishing with a clean cloth or bud. The video heads are very fragile indeed, and cleaning should be undertaken with very great care. Manufacturers recommend the use of a lint-free (i.e. chamois or buckskin) cloth dipped in methanol, though many engineers prefer pure surgical spirit as it leaves less deposit. The cleaning motion should not be prolonged, and must always be made across the heads (i.e in the direction of the tape path), *never* vertically. This is illustrated in Figure 27.2. Suitable cleaning materials and solvents are available from VCR manufacturers. For inspection of heads and guides during cleaning a dental mirror is very useful, especially an illuminated type. After cleaning, it is important to allow all fluids to dry completely before tape loading, or the moist surfaces will 'pick up' the tape with dire results. A hairdrier can be used to speed up the drying process if necessary.

Tape guide alignment

There are several tape guides encountered by the tape on its path through the deck, as shown in Figure 17.9, and the most critical of these are the entry and exit guides at each end of the tape's wrap around the head drum. These determine the tape angle relative to the fixed video head sweep, and are made adjustable so that exact compliance with the format specification can be achieved. For correct setting-up, a standard is required, and this comes in the form of an *alignment* or *interchange* tape, pre-recorded at the factory on a precision standard machine under closely controlled conditions.

The entry guide determines head tracking angle at the point where the heads start their sweep of the tape, and examination of the RF *FM envelope* signal from the replay heads will indicate any fall-off in amplitude here due to guide misalignment. A good envelope signal is shown in Figure 27.3*a*, while *b* indicates low FM output at the LHS (start) of each envelope pattern due to incorrect entry-guide height. The exit guide is adjusted for optimum envelope shape at the end (RHS of oscillogram) of each head sweep, and *c* shows the effect of maladjustment here. Because of the FM limiting circuits employed on playback, a worst-case FM envelope shortfall of 40 per cent is tolerable before deterioration of the picture is noticed; this is illustrated in Figure 27.4.

The other tape guides are concerned mainly with defining the tape path across the static heads, and ensuring a parallel passage through the capstan/pressure roller combination. The adjustment of the sound/sync head is critical for height,

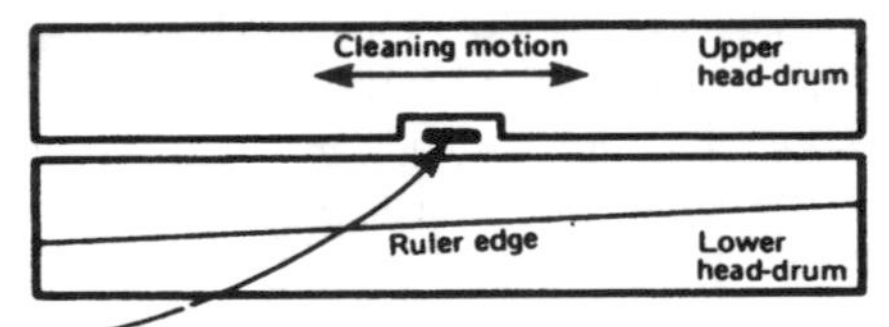

Figure 27.2 *The cleaning motion for video heads must be side-to-side. The video head is very delicate indeed; its width may be less than that of a human hair*

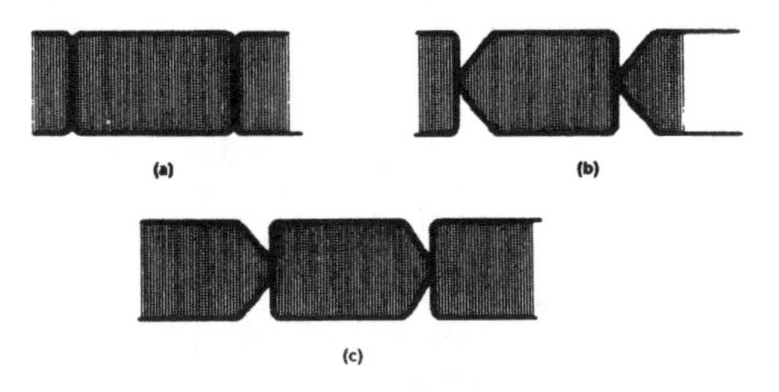

Figure 27.3 *RF envelope patterns: (a) shows a good envelope, (b) one resulting from a dirty or incorrectly adjusted entry guide, and (c) the effect of dirt or maladjustment in the exit guide*

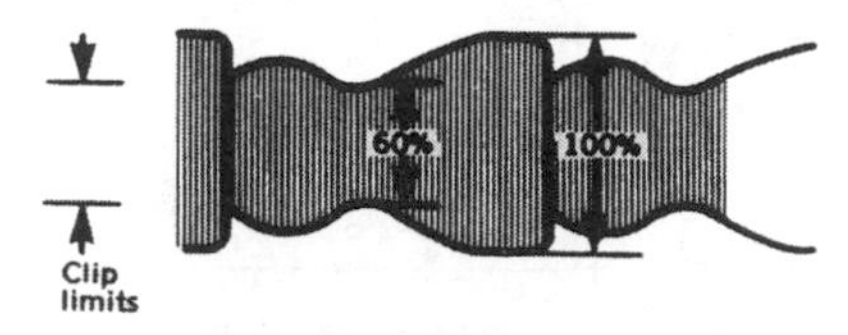

Figure 27.4 *The limiter will clip the FM luminance signal on replay, so that a shortfall of up to 40 per cent will pass unnoticed*

azimuth and vertical tilt, and these are set up (on the audio signal from the alignment tape) by adjustment of the head's base mounting screws, in similar (but more critical) fashion to audio-recorder heads.

Tape tension is set by adjustment of the tension servo, in whatever form it takes (see Chapter 20), for a specific back-tension, monitored by a suitable gauge supplied by the manufacturer. Take-up tension is determined by the friction in the take-up clutch. This is either fixed during manufacture or is adjustable in three or four pre-set steps. VCRs with separate reel motors may have an electrical pre-set TU tension adjuster.

Reel drive, alignment and braking

The final aspect of deck servicing is concerned with the reel transport mechanics and operating levers. Their intricacy will depend on the vintage of the machine; early types require lubrication and occasional physical alignment of spindles, levers and slider bars; and roughening and degreasing of rubber friction wheels and drive surfaces. Spirits can attack rubber compounds, and a mild detergent solution is recommended for cleaning and degreasing drive belts and rubber wheels. Sometimes this is easier said than done, and replacement rather than refurbishing of interwheels, brake pads and drive belts is recommended. Later machines with direct-drive systems are easier to maintain in this respect.

There are several other critical adjustment points in the video tape deck, such as reel disc height, tension pole positioning and video head eccentricity, and the setting of these is comprehensively covered in the relevant service manual for each model. Jigs and tools are available from VCR manufacturers to assist with precision adjustment, and a set of these for one particular machine is pictured in Figure 27.5.

Electrical adjustment

Many of the cautions mentioned above are also relevant to the electrical pre-set controls within the machine. Although electrical setting-up is a part of routine maintenance, electronic devices are not subject to wear; and little drift is encountered in modern circuits. Electrical adjustments tend to be specific to each model and format, so that little general guidance can be given here, except to say that service manuals give precise setting-up instructions to be used with the necessary test gear.

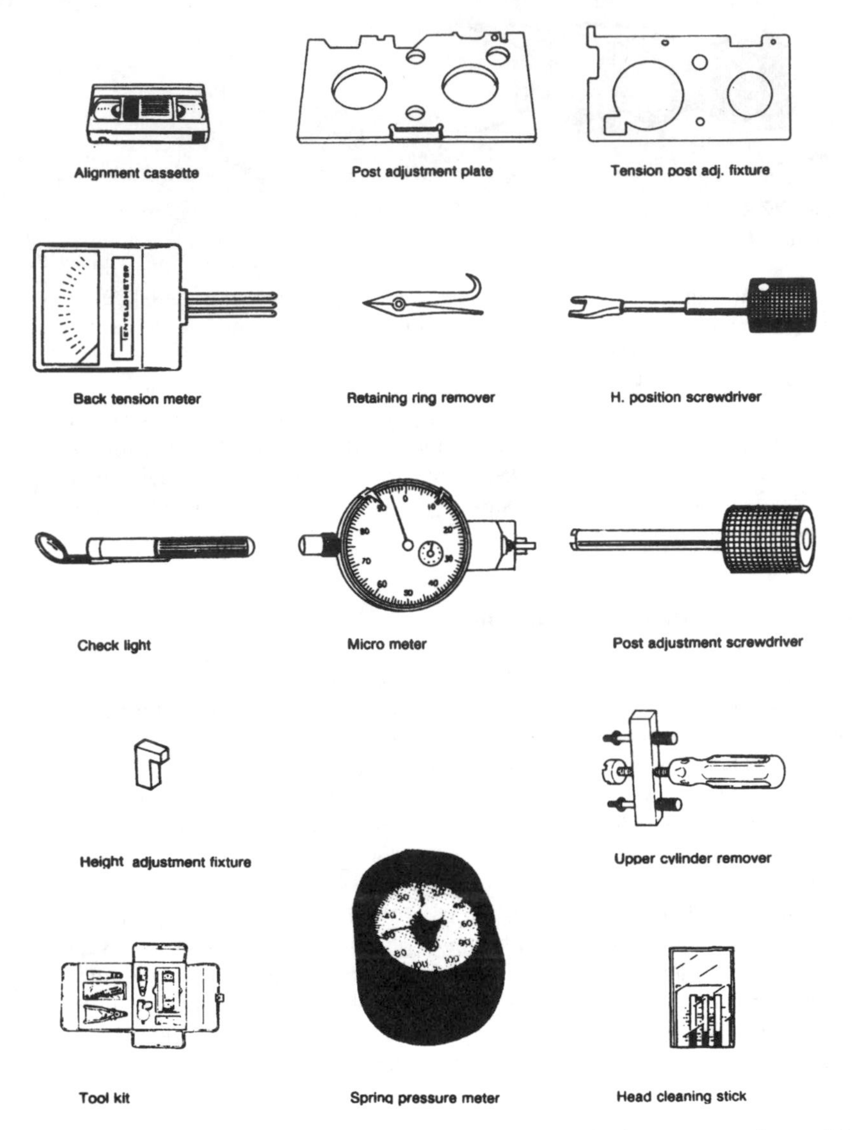

Figure 27.5 *Tools and jigs required for use in VCR deck servicing. The tool kit incorporates a tension/torque-measuring cassette*

Test equipment

For VCR circuit investigation and servicing, it is important that the test equipment used be adequate, accurate and dependable. An oscilloscope of minimum specification would be 10MHz bandwidth, 2mV sensitivity and dual-trace capability. High-impedance multimeters of both digital and analogue type are required, the former with a DC voltage accuracy of 0.5 per cent or better. For investigation of colour-under and servo electronics, a frequency counter is useful, preferably a type with $6\frac{1}{2}$ or more digits and an accurate timebase.

Necessary signal sources are a bench-type colour bar signal generator with split-field display and outputs at video (baseband) and UHF; an accurate signal generator with sine- and square-wave outputs; and a source of noise-free off-air transmission signals at UHF. For monitoring purposes a modern high-performance TV receiver is ideal, preferably with a video input in monitor fashion. A mains variac is very useful for investigating power-supply and excessive-loading faults; a mains isolating transformer is also important for safety, especially when dealing with the sort of switch-mode mains power supply described in Chapter 22.

In addition to these, alignment tapes and adjustment jigs are required as described earlier in this chapter, along with a full service manual for the machine under investigation. The final factor is a good working knowledge of the principles and practice of the VCR!

Output channel adjustment

The best method of connection between a VCR and TV is an AV cable as described in the next chapter. When this cannot be arranged for any reason and where the VCR modulator cannot be turned off it may be necessary to adjust the VCR's output frequency to avoid clashes with other RF signals. The latter may come from terrestrial analogue or digital broadcasts picked up by the aerial or from other 'local' equipment which also has an RF output in the same band. The effects of an RF-channel clash is patterning (dots, wavy or herringbone lines) on either the VCR picture or that from the adjacent carrier. Where a terrestrial digital transmission is involved, the 'random' effect of its modulation, falling on or near the VCR's carrier, can impart to the latter's picture an effect like RF noise, easily mistaken for poor signal strength. The effect of a VCR's AM carrier signal clashing with a terrestrial digital transmission ranges from 'blocking' and freezing to a complete loss of digital reception, due to data corruption.

To avoid these effects it is necessary to move the VCR's output carrier frequency to a 'quiet' part of the UHF band, best found by first tuning the TV to a point where no vestige of signal is present, and then adjusting the VCR (or satellite box, for that matter) output channel to match. Early-model VCRs have an adjustment trimmer – for use with a fine-bladed miniature screwdriver – adjacent to the aerial sockets on the rear panel. Later models use a frequency-synthesised carrier oscillator whose output channel can be pre-set anywhere in the UHF band in user-software. Setting instructions are given in the user's handbook.

28 *Interconnection and compatibility*

As the count of electronic 'black boxes' in the home increases, so do the number and complexity of the connections between them. The evolution of technology has brought with it progressively better and lower-loss signal linkages; over the years we have moved from RF to composite video and then to S-video coupling for the picture, and now use RGB links for some equipment. The advent of digital transmission and recording equipment, and the convergence of TV and computer technology and systems has spawned the domestic use of Firewire/IEE1394 data coupling. And increasingly we exchange control commands and data between boxes for the purposes of editing, automated recording, function switching and the like. In this final chapter we shall examine the pros and cons of the various coupling and connection systems used in the domestic environment, and also see why some equipment fails to work even when it *is* hooked up correctly – or appears to be.

Mains power

None of the equipment described in this book draws much power from the mains. Even so, it is important that good reliable contact is maintained, especially since most 'boxes' are microprocessor-controlled, and these chips are particularly vulnerable to the mains-borne spikes, 'glitches' and mutual interference which can come from the use of cube-shaped multi-way adapters and the like. A good mains power distributor is a high-quality multiple socket-strip, best incorporating a spike-suppressor or 'power cleaner' device. The most vulnerable boxes, computers, should ideally be given a wall socket of their own.

RF connections

For full versatility in recording and viewing of broadcast transmissions, the UHF aerial feed must be looped through the VCR (and terrestrial DTV box if present) on its way to the TV set. Some equipment (particularly satellite boxes) can generate interference in the UHF band, in which case the use of double screened cable like the CT100 type used for satellite downleads can be helpful when used for inter-equipment links working at UHF. Even though VCRs and terrestrial DTV receivers must have a UHF input, it is far better that their modulators are switched off, and that their signal outputs go via AV (usually SCART) cables to avoid carrier clashes and consequent interference in the UHF band; these boxes are often designed to act as 'switching centres', automatically routing the AV signals as required. When RF output carriers must be used, the channel-setting advice given on page 405 should be followed.

Where a VCR and/or DTV box is required to feed several screens simultaneously, perhaps at widely different points, RF distribution can be used; if more than two TV sets are involved, a multi-output UHF amplifier will be necessary, and by this means many receivers can be fed, at great distances if required. This system does not permit stereo sound to be conveyed, and may involve difficulty in setting carrier frequencies to avoid interference, but has the advantage of simplicity and convenience. Figure 28.1 shows a local UHF network.

Very often a household distribution system involves the main TV in the lounge and a second set in bedroom or kitchen, from where it's required to 'drive' the equipment in the living room. It can be achieved by the use of a *remote control extender* which has an infra-red receiving eye near the second TV to relay command codes – either via the RF cable or by a wireless link – back to a mains-powered repeater in sight of the equipment to be controlled. Sky digital receivers have a built-in facility to power and operate a 'Remote Eye' which is simple to connect using just a co-ax cable, but whose operation is limited to the sat-box itself.

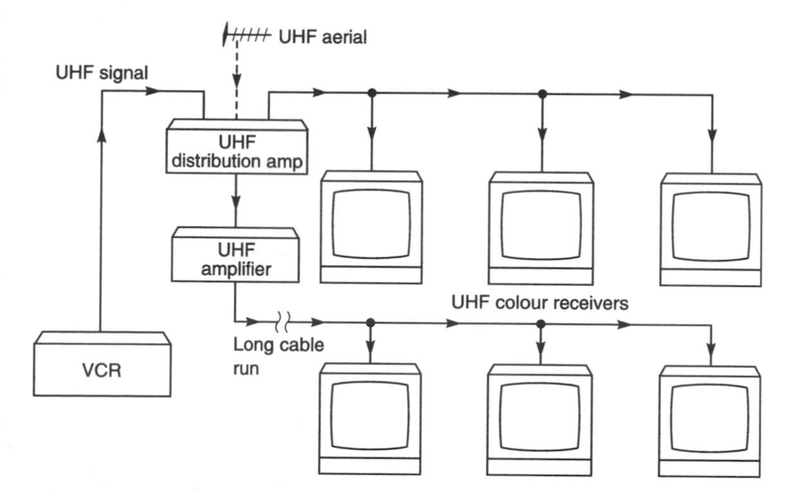

Figure 28.1 *RF signal distribution system for sound and vision*

Baseband distribution

Better picture and sound quality in a 'local network' comes from using a baseband distribution system as shown in Figure 28.2. This also gives the opportunity to distribute stereo and Dolby Surround sound where it's present, while avoiding the risk of interference and patterning. Some applications outside entertainment TV may not require sound, dispensing with the dotted components in the diagram.

SCART connections

The 21-pin SCART coupling system is a versatile one. Its plug connections are shown in Figure 28.3 together with pin assignments, showing that it can convey composite, S-, and RGB video (we'll examine these shortly) as well as stereo sound. Pin 8 carries out some control functions: when it goes high (9–12 V) it switches the TV to AV input, or a standing-by satellite receiver to loop-through mode. At a level of about 6 V it signals the presence of a wide-screen picture from VCR or sat-box to a suitably equipped TV, which switches its scan generators accordingly. In S-mode (see below) the luminance and sync signals are passed out on pin 19 and received on pin 20, while pin 15 conveys the C signal at a burst level of 300 mV. Composite/S (Y-C) switching is here carried out at pin 16. TVs, VCRs and satellite receivers can be programmed in user-software for the various signal modes, and it is important that the boxes at both ends of the cable are in agreement as to the mode in use; and that the best possible mode is selected, depending on the signal and source in use and the capabilities of the equipment involved. Thus RGB is appropriate to DVD players and large-screen TVs, while composite video is used by low-band (VHS, Video-8) video machines.

Inexpensive SCART coupling leads have neither screened signal conductors nor separate ground paths for them, giving rise to crosstalk. This is manifest as floating bars, outlines and/or colours on vision, and buzz/crosstalk on sound. The picture problems can often be solved by cutting off the connection to pin 19 in the plug at the TV end, so long as no video-out signal is required from the TV tuner.

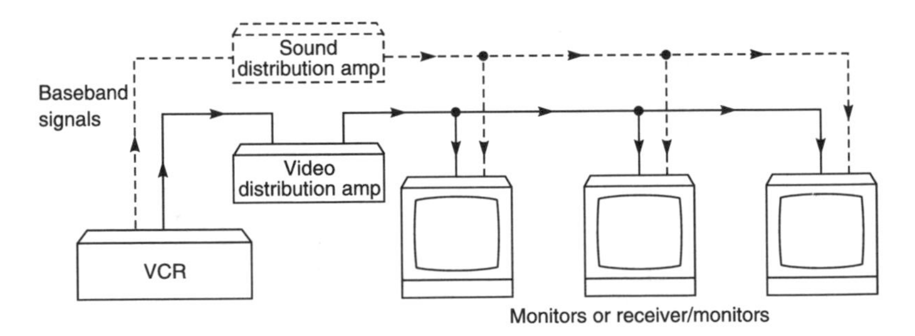

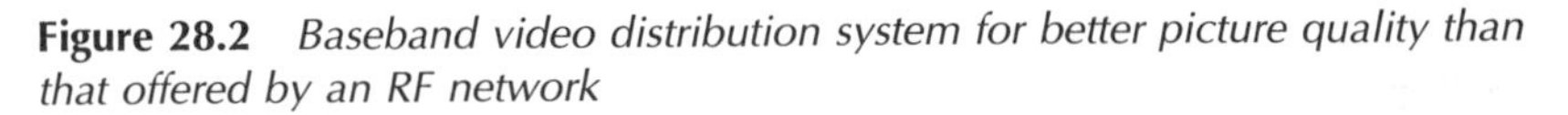

Figure 28.2 *Baseband video distribution system for better picture quality than that offered by an RF network*

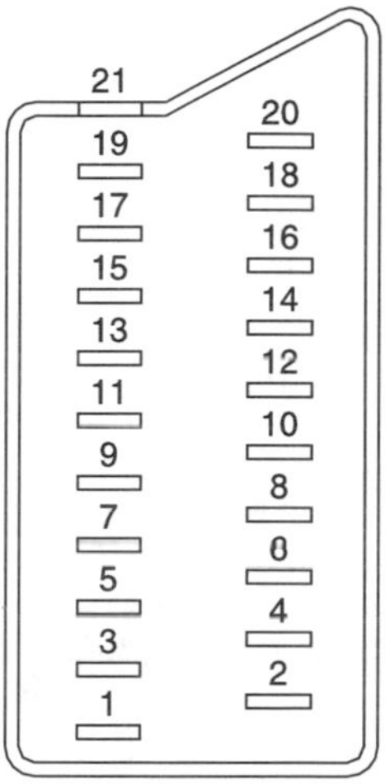

Pin No	Signal	Signal level
1	Audio output B (right)	Standard level: 0.5V rms Output impedence: Less than 1kohm*
2	Audio input B (right)	Standard level: 0.5V rms Output impedence: More than 10kohm*
3	Audio output A (left)	Standard level: 0.5V rms Output impedence: Less than 1kohm*
4	Ground (audio)	
5	Ground (blue)	
6	Audio input A (left)	Standard level: 0.5V rms Output impedence: More than 10kohm*
7	Blue input	0.7 +/– 3dB, 75 ohms positive
8	Function select (AV control)	High state (9.5-12V): AV mode Low state (0-2V): TV mode Input impedence: More than 10K ohms Input capacitance: Less than 2nF
9	Ground (green)	
10	Open	
11	Green	Green signal: 0.7 +/– 3dB, 75 ohms, positive
12	Open	
13	Ground (red)	
14	Ground (blanking)	
15	Red input	0.7 +/– 3dB, 75 ohms, positive
	(S signal Chroma input)	0.3 +/– 3dB, 75 ohms, positive
16	(Blanking input Ys signal)	High state (1-3V) Low state (0-0.4V) Input impedence: 75 ohms
17	Ground (video output)	
18	Ground (video input)	
19	Video output	1V +/– 3dB, 75 ohms, positive sync 0.3V (–3 + 10dB)
20	Video input	1V +/– 3dB, 75 ohms, positive sync 0.3V (–3 + 10dB)
	Video input Y (S signal)	1V +/– 3dB, 75 ohms, positive sync 0.3V (–3 + 10dB)
21	Common ground (plug, shield)	

Figure 28.3 *SCART connector: layout, pinning and signal levels. Pins 10 and 12 are used by manufacturers for various purposes, mainly system control*

The appropriation of SCART pins 10 and 12 by manufacturers for various purposes can lead to trouble, so its best to either cut them or use leads in which they are not connected – unless the connection specifically requires their use. Indeed a fully-wired SCART lead, heavy, thick and relatively expensive, is best avoided unless the application calls for it, see Figure 28.3. Low-level applications involving just composite video and mono or stereo sound are best served by a lead with screened conductors at pins 1/2/3/6/19/20, (basically three crossed-over signal feeds) plus a link between pins 8. The control function of SCART pin 8, pulling the TV into AV mode, can be a nuisance where for example it's required to watch a TV programme while recording another from satellite. Often there's a TV/SAT key on the sat-box's remote control to overcome this, but if necessary the lead can be cut from SCART pin 8, when signal input selection is under the sole control of the TV set's remote handset.

Video coupling

There are three main modes of conveying an analogue video signal. We shall look at their characteristics in turn.

Composite video

This is the most common form of picture signal, representing the PAL (or other) video signal as broadcast by analogue terrestrial transmitters. It carries luminance, chroma and sync, corresponding to our Figure 3.16. While it can convey signals between any two pieces of video equipment it is mainly applicable to low-band VCRs and signals which come from analogue transmitters, satellite and ground-based. The picture from a composite signal is limited in bandwidth/definition, and can display cross-colour and other spurious effects.

S-video

Similar to a composite one, the S (separate Y-C) coupling has separate leads and connections for luminance/sync and chroma components. It has the advantage – where the Y and C signals have not come via a broadcast transmission – of avoiding crosstalk effects like cross-colour, and it affords a little more picture definition by virtue of the absence of Y/C separating filters. It is applicable to high-band (S-VHS, Hi-8) camcorders and VCRs, also satellite receivers and DVD players where an S-output port is provided. As well as SCART links, S-video can be carried in a specially-designed plug/socket connector as shown in Figure 28.4. This cannot convey audio as well, so S-connection requires a separate audio link, typically in two (L, R) phono leads.

RGB

RGB offers the best-possible video coupling mode for domestic applications, but is only applicable where the signal is generated on the spot, as it were: DVD players, DTV receivers and (where a suitable scan-converter is present, see below) computers. Here there are separate paths for each primary colour red, green and

Pin No.	Signal	Level
1	Ground Y	
2	Ground C	
3	Y (S signal) input	1V ± 3dB 75 ohm, positive Sync. 0.3V −3 + 10dB
4	C (S signal) input	0.3V ± 3dB 75 ohm, positive Sync.

Figure 28.4 *Pinning and signal levels in the S-Video (Y-C) connector*

blue, with virtually no bandwidth restriction inherent in the linking system or mode. In home-video applications it is usually carried in SCART links as shown in Figure 28.3, with sync at 300mV peak on pin 20 (in) and 19 (out). VCRs do not deal in RGB signals, though some models have a loop-through-SCART facility for them.

Audio coupling

Sound signals are coupled between boxes at a standard level of 500 mV rms, most commonly in the SCART connector, but also in phono links, using sockets for both input and output with a colour code of red for right channel and white for left channel or mono: a *yellow* phono socket generally carries video signals.

Dolby Pro-Logic and other audio decoders and processors commonly form part of a home-cinema ensemble. Although they are concerned, for operational purposes, solely with the sound signal, it's necessary to include them in the video path on its way to the TV or monitor so that their captions and on-screen displays can be shown on screen. This enables the user to set up the system, check its mode and status, and get an indication of signal source. These processors/amplifiers are well equipped with input and output sockets for this purpose, and can form a convenient remote-controlled 'switching centre' for the various components of a complete AV system.

High-level audio links

Some audio processor boxes are equipped to exchange 'composite' (multi-channel) sound signals in digital datastream form: the DVD player shown in Figure 26.5 is one such with optical (passes over a fibre lightguide) and RF/coaxial outputs on the right of the diagram. Here the complete audio bitstream is modulated onto an optical or RF carrier for lossless passage to a suitably equipped surround sound decoder.

Speaker wiring

Apart from ensuring that each speaker is hooked to the correct socket – corresponding to its position in the room – the most important aspect of loudspeaker wiring is correct *phasing*, in which the + terminal of each speaker is connected to the + terminal on the box. Incorrect speaker phasing gives rise to strange and unnatural effects in the reproduced stereo or surround sound field. Loudspeaker power and impedance matching is also important: too 'small' a loudspeaker may get damaged at high volume levels, while too low a loudspeaker impedance – for instance by connecting two 4 Ω units in parallel – risks damage, at high volume level, to the output section of the amplifier in the box.

Cordless loudspeakers and headphones save the need to run audio cables in the viewing/listening room. They depend on infra-red or RF transmitters at the box, whose emissions are picked up by mains and battery-operated receiver/demodulator/amplifier systems respectively. The carrier frequency for IR types is 2.3/2.8 MHz, and for RF systems 863-4 MHz, with the latter generally giving the best results.

Firewire

Other names for the Firewire link system are IEE1394 and i-Link: they all amount to the same thing in practice, but different equipment manufacturers prefer different names! Firewire is a very fast (up to 400 Mbit/sec) serial data link originated by Apple Computers as a LAN (Local Area Network) protocol for computers, and now adopted as a standard interface for domestic digital exchange between boxes: it can carry entertainment, communication and computing data. In current domestic technology it is applicable to DVC and D-VHS video machines, primarily where 'lossless' editing is required, when the whole process can be carried out in the digital domain, see page 381.

The Firewire data is carried by a cable having two individually-screened twisted-pair cables, one for NRZ (non-return-to-zero) data and one for strobe pulses as shown in Figure 28.5. Exclusive-OR gating of the two pulse trains provides the system clocking pulses. The pulse amplitude of each train is 220 mV, centred on a bias voltage of 1.86 V. The datastream has two components: a relatively slow asynchronous unidirectional pulse train for control purposes; and, time-interleaved with it, a very fast isochronous one which carries 'payload' data in the form of variable-length packets, with header, ident, address, data and error-check components. A negotiation and arbitration process is used for access control in a Firewire network, not dissimilar in its operation and effect to that employed in the I^2C internal communication system we met in Chapter 8.

The Firewire link system is bidirectional, with a capability of 63 devices on a single bus, and a vast number of them on a bridge-bus system. Firewire ports are provided on some computers, on DV-interface computer cards, on DVC and D-VHS cassette decks, printers and (especially for D-VHS recorders) next-generation

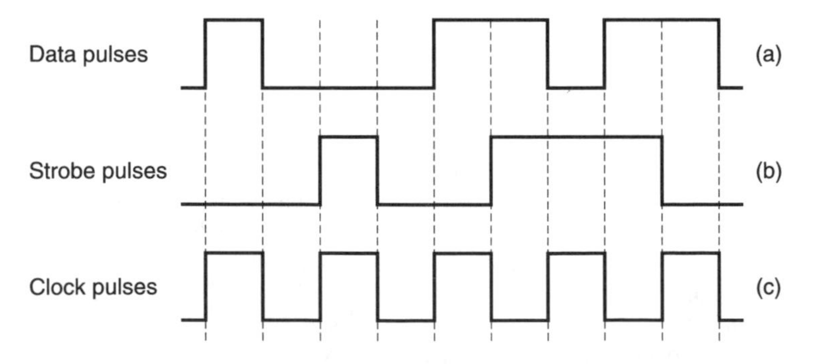

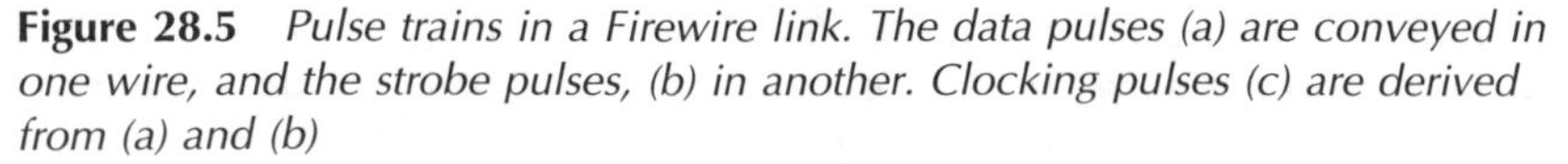

Figure 28.5 *Pulse trains in a Firewire link. The data pulses (a) are conveyed in one wire, and the strobe pulses, (b) in another. Clocking pulses (c) are derived from (a) and (b)*

digital TVs and set-top boxes. As we saw at the end of Chapter 24, many camcorders sold in the UK require the DV-in facility to be 'enabled', usually in software.

Edit control links

For the purpose of auto-editing, the video source and destination boxes (usually a camcorder and VHS deck or computer respectively) need control ports via which the editing machine can govern the decks' transport control system. The first established of these was Sony's LANC (Local Area Network Control) protocol, implemented by a 5-pin or 3-pole minijack (*Control L*) connection system. It carries serial control data, similar to (but not directly compatible with!) the I^2C internal control bus used in TVs, VCRs etc. The data is arranged in a series of byte (8 bit)-long words which convey sender and recipient ident, mode, data, status and command pulse trains, at a peak level of about 3.5 V peak.

Panasonic's system, clumsily known as a 'Panasonic 5-pin', even though it is sometimes used in a Panasonic 11-pin connector (!) is exclusive to this company, and also uses a serial data format for editing-control commands. The 'freelance' market has produced a translator to interface Panasonic and LANC systems. Finally, JVC came up with their JLIP (Joint Level Interface Protocol) concept for editing, computer linking and remote control applications. It uses a simple serial data format with eight data, start, stop and parity bits, coupled by a 3.5 mm 4-pole (tip plus three rings) mini jack plug/socket connector. It is not compatible with the Sony or Panasonic systems.

In addition to these, many video decks and camcorders, especially older models, are fitted with simple pause control jacks in 2.5 or 3.5 mm form, which can be used for *synchro-edit*, whereby the transport controls of one deck control the mechanical functions of another, one of them usually being a camcorder. A more sophisticated development of this, using the same ports and connection, is RAE (random assemble edit) in which a sequence of clips can be programmed to be assembled on the tape of one machine by controlling the deck of the other. Synchro-edit and RAE are generally only possible between machines of the same (or 'cloned') make.

Satellite wiring

The use of dual satellite LNBs, DiSEqC switching systems and small-scale satellite-signal distribution networks was described in Chapter 13, page 201.

Compatibility

There is some confusion over the question of interchangeability of VCR machines and tapes between different countries of the world, probably arising from the fact that the major formats are used world-wide, and it's reasonable to suppose that, for instance, a VHS recording made in the USA will replay on a UK-market machine. In fact all the basic formats are manufactured in export versions for different world markets, but within each format some differences are present in the electronics to cater for variations of colour encoding (PAL, SECAM,

Table 28.1 *World TV broadcast standards*

PAL B/G is mainly used in Western Europe but is common throughout Afghanistan, Albania, Algeria, Australia, Austria, Azores, Bahrain, Bangladesh, Belgium, Canary Islands, Denmark, Ethiopia, Finland, Germany, Ghana, Iceland, India, Indonesia, Israel, Italy, Jordan, Kenya, Kuwait, Libya, Malaysia, Malta, Mozambique, Netherlands, New Zealand, Nigeria, Norway, Oman, Pakistan, Portugal, Qatar, Sierra Leone, Singapore, Spain, Sri Lanka, Sudan, Swaziland, Sweden, Switzerland, Tanzania, Thailand, Turkey, Uganda, United Arab Emirates, Yemen Arab Republic, Yugoslavia, Zambia, Zimbabwe

PAL I is used in Great Britain and in some African countries, Angola, Botswana, Gambia, Great Britain, Hong Kong, Ireland, South Africa, Zanzibar

The following can playback NTSC 4.43 programmes via a video recorder but cannot receive off air broadcasts

PAL M
Brazil

PAL N
Argentine, Paraguay, Uruguay

PAL D/K
China, North Korea, Romania

NTSC 3.58 is mainly used in America and The Caribbean Bahamas, Barbados, Bermuda, Bolivia, Burma, Canada, Chile, Columbia, Costa Rica, Cuba, Dominican Republic Ecuador, Guatemala, Haiti, Honduras, Jamaica, Japan, Mexico, Nicaragua, Panama, Peru, Philippines, Puerto Rica, South Korea, Sunman, Taiwan, Trinidad and Tobago, USA, Venezuela

SECAM B/G
Cyprus, Egypt, Greece, Iran, Iraq, Lebanon, Libya, Mauritius, Morocco, Saudi Arabia, Syria, Tunisia

SECAM L is used in France and in other countries to receive French broadcasts France, Luxembourg, Monaca, Russia

SECAM D/K
Berlin, Bulgaria, Cameron, Central African Republic, Chad, Congo, CSSA, Djibouti, Equatorial Guinea, French Guyana, Gabon, Guinea, Hungary, Ivory Coast, Mali, Malagasy, Martinique, Mauntania, Mongolia, Niger, Poland, Romania, Senegal, Togo, Upper Volta, Russia, Vietnam, Zaire

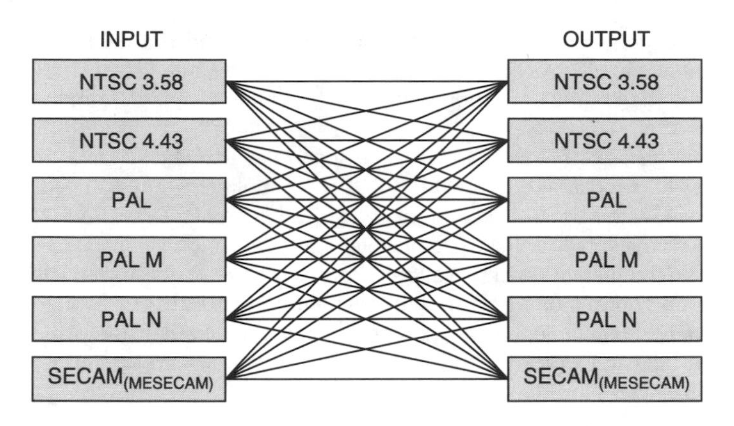

Figure 28.6 *VCRs with internal field-store digital transcoders can convert between any of the world's video standards*

NTSC), line frequency (525 or 625 lines) and field rate (50 or 60 Hz), and these determine the circuit design of servos, colour-under systems and still-frame facilities.

The different *transmission* standards between countries add further complication, this time in the peripheral circuits of the VCR such as tuners, sound demodulators and RF modulators. Vision and sound modulation systems differ, and VHF bands are used for broadcasting in many parts of the world. As a result UK-type machines will only work in Eire, Hong Kong and South Africa and a few other places, though the tapes are interchangeable with all countries using PAL/625/50 standards. Table 28.1 gives an idea of world standards. Some manufacturers market multi-standard machines which go some way towards solving the compatibility problems arising out of differing TV systems, but unless they contain complete digital field-store memories (Figure 28.6) to act as standards-converters they can still only replay a signal in the form in which it was recorded.

Many VHS VCRs on the UK market can replay NTSC tapes by auto-switching the tape- and head-drum speed and changing the operating mode of the colour-under replay system, but they do not alter the field rate, so the TV or monitor in use must be able to cope with 60 Hz scanning. Most large-screen types do, thanks to inbuilt multi-standard ICs in timebase and colour-decoder sections, and this also fits them for display of Region1/NTSC pictures from DVD players, American camcorders etc. It is not possible to record (e.g. from a disc or another VCR) NTSC video on these machines as they stand.

Standards conversion

To convert NTSC signals into PAL form or vice-versa, stand-alone transcoders are available from accessory suppliers: the picture quality depends largely on how many picture fields they can store in digital form. Also on the market are

PAL-SECAM transcoders, much cheaper because they do not need a field-store system, merely an analogue process to change the colour encoding system. They are mainly used in the UK in connection with French satellite reception from the Telecom craft at 5°W.

Computer monitors have higher field and line scan rates than TV systems, so they are not directly compatible, even with RGB video coupling. Again, 'gadgets' come to the rescue, with a PC to video interface, typically compatible with VGA, SVGA, and XGA formats, converting to composite or S-video, and retailing at less than £200, though the picture quality available from a TV screen does not come up to that of the computer monitor itself. Similarly, and still under £200, a video to VGA converter is available, taking composite video input, decoding and re-formatting the signal to pass out via a standard 15-pin D connector for PC and monitor.

Index